LED PACKAGING FOR LIGHTING APPLICATIONS

LED PACKAGING FOR LIGHTING APPLICATIONS

DESIGN, MANUFACTURING AND TESTING

Sheng Liu
School of Mechanical Science and Engineering
and Wuhan National Laboratory for Optoelectronics
Huazhong University of Science and Technology
Wuhan, Hubei, China

Xiaobing Luo
School of Energy and Power Engineering
and Wuhan National Laboratory for Optoelectronics
Huazhong University of Science and Technology
Wuhan, Hubei, China

Chemical Industry Press

John Wiley & Sons (Asia) Pte Ltd

This edition first published 2011
© 2011 Chemical Industry Press. All rights reserved.

Published by John Wiley & Sons (Asia) Pte Ltd, 1 Fusionopolis Walk, #07-01 Solaris South Tower, Singapore 138628, under exclusive license by Chemical Industry Press in all media and all languages throughout the world excluding Mainland China and excluding Simplified and Traditional Chinese languages.

For details of our global editorial offices, for customer services and for information about how to apply for permission to reuse the copyright material in this book please see our website at www.wiley.com.

All Rights Reserved. No part of this publication may be reproduced, stored in a retrieval system or transmitted, in any form or by any means, electronic, mechanical, photocopying, recording, scanning, or otherwise, except as expressly permitted by law, without either the prior written permission of the Publisher, or authorization through payment of the appropriate photocopy fee to the Copyright Clearance Center. Requests for permission should be addressed to the Publisher, John Wiley & Sons (Asia) Pte Ltd, 1 Fusionopolis Walk, #07-01 Solaris South Tower, Singapore 138628, tel: 65-66438000, fax: 65-66438008, email: enquiry@wiley.com.

Wiley also publishes its books in a variety of electronic formats. Some content that appears in print may not be available in electronic books.

Designations used by companies to distinguish their products are often claimed as trademarks. All brand names and product names used in this book are trade names, service marks, trademarks or registered trademarks of their respective owners. The Publisher is not associated with any product or vendor mentioned in this book. This publication is designed to provide accurate and authoritative information in regard to the subject matter covered. It is sold on the understanding that the Publisher is not engaged in rendering professional services. If professional advice or other expert assistance is required, the services of a competent professional should be sought.

Library of Congress Cataloging-in-Publication Data

Liu, S. (Sheng), 1963-
LED packaging for lighting applications : design, manufacturing and testing / Sheng Liu, Xiaobing Luo.
p. cm.
Includes bibliographical references and index.
ISBN 978-0-470-82783-3 (hardback)
1. Light emitting diodes–Design and construction. 2. Light emitting diodes–Computer simulation. 3. Electronic packaging. 4. Electric lighting–Equipment and supplies. I. Luo, Xiaobing, 1974- II. Title.
TK7871.89.L53L58 2011
621.3815'22–dc23

2011015480

Print ISBN: 978-0-470-82783-3
ePDF ISBN: 978-0-470-82784-0
oBook ISBN: 978-0-470-82785-7
ePub ISBN: 978-0-470-82840-3
Mobi ISBN: 978-1-118-08295-9

Contents

Foreword

By Magnus George Craford

LEDs have been commercially available for nearly 50 years but for most of that time they were low power and relatively inefficient devices that were primarily used for indicator applications. Initially LEDs were red only, but these were soon followed by green and yellow devices and finally about 15 years ago by blue and, with phosphors, white devices. During these years the efficiency increased by a factor of 10 every 10 years. Ten years ago high power (one watt) white LEDs were introduced with efficiencies high enough that there began to be serious discussion about using LEDs for solid state lighting (SSL) applications. Over the last decade performance has continued to increase from about 20 lumens/watt to over 100 lumens/watt today. The power handling capacity of packages has increased, with packages which can handle over 10 watts now available. There is no longer any question about whether LEDs will be important for SSL. I believe the only question is when LEDs will dominate all lighting applications. China has been one of the leaders in pushing for the rapid adoption of LEDs in order to save energy and to provide improved illumination for its vast population. Over the next decade we are sure to see an explosion of new applications, and new package types to enable those applications. A critical issue is to develop packages and systems that enable efficient and reliable solutions to lighting problems. If systems are unreliable it will slow down the adoption of LEDs to the detriment of everyone. There are many books about LEDs but most of them focus on the chip and epitaxial materials technology. Books of this type focusing on packaging and applications are badly needed to help engineers and scientists use LEDs in the most effective manner possible, and to ensure the rapid adoption of efficient LED technology around the world.

The authors Professors Sheng Liu and Xiaobing Luo have done a thorough job of discussing the optical design, thermal management, and reliability of high power LEDs and systems. System reliability is only as good as the weakest link and it is critical for system designers to understand all aspects of the system. The authors have also adopted experience gained in silicon technology to the field of high power LEDs. I am happy to see this book completed and feel that it will be an important addition to our field.

Dr. Magnus George Craford
Recipient of the 2002 USA National Medal of Technology
Member of Academy of Engineering of the USA
IEEE Life Fellow
Former Chief Technology Officer
Solid State Lighting Fellow
Philips Lumileds Lighting
Palo Alto, California, USA

Foreword

By C. P. Wong

Design is a multi-disciplinary activity that relies on the expertise of the engineering profession and is supported by the methodology and innovations developed within the fields of science. The integration of science, engineering, and end applications has produced remarkable changes in the end users. This system integration can be demonstrated by the evolution of understanding of solid state physics and compound semiconductors, the development of epitaxial layers, the design of LED devices, and the applications to packaged modules and light fixtures.

The most popular methodology of design is named Design for X (DFX, here X refers to manufacturing, assembly, testing, reliability, maintenance, environment, and even cost), which has been widely adopted by those multinational and many small high tech start-up companies. The design methodology is being adjusted to meet the requirements of a full life cycle, so called "concept/cradle-to-grave" product responsibilities, coined by Dr. Walter L. Winterbottom of Ford Science Lab.

An LED packaging module and the related application systems, like any other electronic systems, involve a lot of manufacturing processes from epitaxial growth to chip manufacturing to packaging and to final fixture assembly, and extensive reliability testing for extended life goals of many critical products such as those used for road lighting, automotive lighting, and so on. Defects in terms of dislocations, voids, cracks, delaminations, and microstructure changes can be induced in any step and may interact and grow in subsequent steps, imposing extreme demands on the fundamental understanding of stressing and physics of failures. Currently, the testing programs have been extensive to assure reliability during the product development. An iterative, build-test-fix-later process has long been used in new product development; significant concerns are being raised as cost effective and fast time-to-market needs may not be achievable with such an approach. In terms of high reliability, system hardware design, manufacturing and testing are costly and time consuming, severely limiting the number of design choices within the short time frame, and not providing enough time to explore the optimal design. With the current situation of three to six months for each generation of LED devices, it is challenging to achieve truly optimal and innovative products with so many constraints in design. Design procedure must be modified and DFX must be used so as to achieve integrated consideration of manufacturing processes, testing, and operation.

Professors Sheng Liu and Xiaobing Luo have been promoting the new design method in the past many years to help assist in material selection, manufacturing yield enhancement, and appropriate rapid reliability assessment when the packaging module and system are subjected to uncertainties of material selection, process windows, and various service loadings. All these issues must be addressed prior to hardware buildup and test. The authors have demonstrated excellent examples for optical, thermal, and reliability aspects. Application of specific LED packaging (ASLP) is indeed an example of a careful design and consideration of packaging integration. Its three-in-one, four-in-one and five-in-one modules conceived by the authors' group are very likely to be widely used by this fast growing industry. They will be popular choices in terms of performance and cost for those traditional light fixture companies, as they represent the true nature of integration in microelectronics, MEMS, and optoelectronics/LED, to name a few fields. Detailed modeling of manufacturing processes such as wire bonding for LED has been shown to be important and the co-design of the LED chip and packaging indeed show the importance of concurrent consideration of traditionally divided product chains and provides a new direction for further improvement of optical performance.

This book focuses on LED packaging for lighting applications and illustrates the importance of packaging and the power of integration in the packaging modules and lighting application systems by the authors' pioneering efforts. Packaging has been ignored from the whole system development in the past and the authors explore four functions of packaging in this book: powering, signal distribution in terms of both optical and electrical signals and quality, thermal management, and mechanical protections. The authors describe their contributions in detail and provide guidance to those in the field and present a design approach that must ultimately replace the build-test-fix-later process if the efficiencies and potential cost benefits of high power LED based systems are to be fully realized.

C.P. Wong
IEEE Fellow
Member of Academy of Engineering of the USA
Former Bell Labs Fellow
Regents' Professor, Georgia Institute of Technology, Atlanta, GA 30332

Foreword

By B. J. Lee

Since the invention of the light-emitting diode (LED) by Holonyak and Bevacqua in 1962, the LED has experienced great breakthroughs particularly with the invention of nitride blue LED in the early 1990s by Nakamura from Nichia Corporation of Japan. Ever since then, the average LED usage in each household has grown more than tenfold.

There was no LED industry until mid-1960s. Nevertheless, the worldwide LED device production values have reached $7 billion in year 2008. It is estimated that the compound annual growth rate will be more than 20% for LED applications in the next 10 years. This LED revolution has been affecting the lives of many people around the world. In 2005, global lighting consumed 8.9% of total electric power in a whole year and contributed 0.63% of the GDP, according to Dr. Jeff Tsao from Sandia National Laboratories of the USA. In 2050, the contribution to GDP can reach as high as 1.65%.

The LED chip cannot operate by itself without connection to outside driver circuits. Light, as an optical power coming out of the LED chip, must be extracted efficiently. In order to maintain a good efficacy through a whole temperature range, the generated heat from LED chips must be dissipated as efficiently and as quickly as possible. Finally, mechanical protection must be adopted to prevent the chip from being damaged or being gradually degraded in subsequent testing or operations in harsh environments. While chip manufacturers are making efforts to improve the optical, electrical, and thermal performances of LED chips by various approaches, including better light extraction, improved crystal quality, uniform current spreading, and using substrate with good thermal conductivity, the corresponding packaging technologies must be developed to make use of these chip improvements. Collaborative designs between chip manufacturers and packaging vendors are thus required to take advantage of each others' progress, which has been mostly ignored among LED communities.

In addition, optical design has been playing an important role in improving the LED efficacy. How light can be re-directed efficiently out of the active layers becomes crucial. It involves light extraction, phosphor coating, and secondary optics. More and more emerging applications require certain light emission patterns, which is posing the demand for a matched secondary optical lens along with the original first level optics. Due to the point source characteristics of the LED chip, the glare issue has prompted the LED community to develop highly efficient secondary lenses to reduce the glare effect. In the meanwhile, the cost pressure is always there when compared to the traditional lighting sources, so the functional integration (either monolithically or in a hybrid way) seems to be essential.

Reliability has been the focus of LED industries. LED has been regarded as a more reliable light source compared with traditional light sources. To make LED worthy of the fame, systematic reliability monitoring will be needed before the products are shipped out. Usually, reliability tests are very costly and time-consuming. Therefore, rapid reliability evaluation will be an alternative. It will be difficult to develop a more efficient reliability evaluation method if there still exists poor failure analysis methods, lack of the appropriate test methodology and standards, and shortage of an effective approach to evaluate the safety of the lighting fixtures.

Based on all these issues, there was an urgent need, both for industry and for academy, for a comprehensive book covering the current state-of-the-art technologies in the design of LED packaging for solid state lighting applications. This book has been written in such a way that readers can quickly learn about the fundamental theories and problem-solving techniques, as well as understand the design trade-offs, and finally make accurate system-level decisions.

Dr. Sheng Liu and Dr. Xiaobing Luo have done a significant amount of work and brought together all the useful information from the latest technical publications related to LED packaging. They have together written this technical book entitled *LED Packaging for Lighting Applications*, an informative book for both industrial and academic users. It is appropriate to be used either as an introductory book for those who are just entering this field or as an up-to-date reference for those who have been engaged in LED packaging and lighting module/system development for some time.

This book covers the subject of LED packaging and related lighting applications on several key topics – high power packaging development trends, optical design of high power LED packaging modules with the focus on the integration of secondary optics into the device packaging, thermal management, reliability engineering including the analysis of failure mechanisms and method of quick evaluation, advanced design of LED packaging applications, and, in the final chapter, an introduction to standards and measurement methods including LED street lights. I am also delighted to find that both authors have made significant efforts to discuss the connection between chip level design (such as surface roughening and color uniformity) and packaging efficiency. I would like to join the authors in hoping that this book will attract the attentions of engineers and applied scientists working in this field, as well as faculty and students, to become aware of the design challenges that must be overcome in order to provide the best products to the market. Let us work together to achieve a greener Earth by lighting the Earth by LED.

B. J. Lee
Chairman, Epistar Corporation

Preface

Eight years ago, when the first author was a newcomer in LED packaging, he thought that LED packaging must be easier than integrated circuit (IC) packaging, as there were only two input/output (IO) terminals for most LED packaging, while for IC packaging, IOs tended to be in the hundreds and thousands and there were already many different packaging types such as plastic packaging, ball grid array, flip-chip, wafer level packaging, chip-scale packaging, and so on. After visiting many leading packaging houses and going through details of packaging in our laboratory and at those collaborating companies, he found that the packaging of LED was not that easy. According to the classical definition of conventional IC packaging, there are four major functions of a packaging: powering, signal distribution, thermal management, and mechanical protection. Powering has actually become a bottleneck for the claimed long life of the LED. What is also unique about the packaging is that light and color associated with light are new. Thermal management is also challenging due to the requirement for a lower junction temperature, which is related to both the chip design and packaging design. Mechanical protection is also important due to the natural use of those materials with poor adhesion and possible poor material handling in the early stage of process development. We began to be very interested in high power LED packaging in 2005 and have spent a lot of efforts since then. It has been the belief that the knowledge learned in the past 20 years in IC packaging can be applied to LED packaging and in particular the concept of system in packaging (SiP), which is still a hot research topic in IC packaging and a useful industrial practice as well, can be further developed in LED packaging. This book intends to assemble what we have learned in the past few years into a useful reference book for both the LED and IC packaging communities, with the hope that the results to be presented are going to benefit engineers, researchers, and young students.

Therefore, this book focuses on solid state lighting by light emitting diode (LED) and it is intended for design engineers, processing engineers, application engineers, and graduate students. It is also helpful for art designers for buildings, roadways, and cities. This book provides quantitative methods for optical, thermal, reliability modeling and simulation so that predictive quantitative modeling can be achieved. It proposes Application Specific LED Packaging (ASLP) to integrate the secondary optics into the first level device and modules. This book also further develops System in Packaging (SiP) for LED modules and applications and provides a co-design approach for the rapid design of module and lighting systems so as to minimize the time to market for LED products. Fundamental research is also presented to satisfy the interests of the active researchers.

Since the first light-emitting diode (LED) was invented by Holonyak and Bevacqua in 1962, the field has experienced great breakthroughs particularly in the early 1990s by Nakamura from

Nichia Corporation of Japan. Nakamura successfully prepared high-brightness blue and green LED in GaN-based materials. LEDs have made remarkable progress in the past four decades with the rapid development of epitaxy growth, chip design and manufacture, packaging structure, processes, and packaging materials. White LEDs have superior characteristics such as high efficiency, small size, long life, dependable, low power consumption, high reliability, to name a few. The market for white LED is growing rapidly in various applications such as backlighting, roadway lighting, vehicle forward lamp, museum illumination, and residential illumination. It has been widely accepted that solid state lighting, in terms of white LEDs, will be the fourth illumination source to substitute the incandescent lamp, fluorescent lamp, and high pressure sodium lamp. In the next five to eight years, with the development of the LED chip and packaging technologies, the efficiency of high power white LED will reach as high as 160 lm/W to 200 lm/W, which will broaden the application markets of LEDs furthermore and will also change the lighting concepts of our life.

There are already five books on this topic available to readers. They are *Introduction to Light Emitting Diode Technology and Applications* by Gilbert Held in 2008, *Light-Emitting Diodes* by E. Fred Schubert in 2006, *Introduction to Solid-State Lighting* by Arturas Zukauskas, Michael S. Shur, and Remis Gaska in 2002, *Introduction to Nitride Semiconductor Blue Lasers and Light Emitting Diodes* by Shuji Nakamura, and Shigefusa F. Chichibu in 2000, and *Power Supplies for LED Driving* by Newnes in 2008. However, all of them allocated a very small section to LED packaging and there is no book focusing on high power LED packaging for applications. The authors thought that this might be due to the highly proprietary nature of high power LEDs. In addition, there are no books dedicated to reliability engineering and standards. In recent years, China has been pushing hard for many demonstration projects in LED. Many lessons have been learned and there is an urgent need for both reliability and standards for both modules and light fixtures. Both authors feel obligated to explore these subjects and contribute to this community by sharing their recent findings so as to promote the healthy development of high power LED packaging and their applications. Chapter 1 provides an introduction of LED. Chapter 2 provides the fundamentals and development trends of high power LED packaging, demonstrating that LED development follows a similar trend to IC packaging. Optical design of high power LED packaging module is discussed in Chapter 3, with the focus on the importance of integration of secondary optics into the device packaging level and more integration of other functions to form more advanced modules. Chapter 4 is devoted to the basic concepts in thermal management. Chapter 5 is devoted to the reliability engineering of high power packaging with the preference of physics of failure based modeling and sensors based prognostics health management for LED systems and more robust models with more physical variables and integration of processing, testing, and field operation. Chapter 6 is devoted to the design of LED packaging applications to sufficient details. Chapter 7 provides an introduction to standards and measurement methods for some applications.

We hope this book will be a valuable source of reference to all those who have been facing the challenging problems created in the ever-expanding application of high power LEDs. We also sincerely hope it will aid in stimulating further research and development on new packaging materials, analytical methods, testing and measurement methods, and even newer standards, with the objective of achieving a green environment and eco-friendly energy saving industry.

The organizations that know how to learn about the design and manufacturing capabilities of high power LED packaging with high reliability have the potential to make major advances in developing their own intellectual properties (IP) in packaging and applications, to achieve benefits in performance, cost, quality, and size/weight. It is our hope that the information presented in this book may assist in removing some of the barriers, avoid unnecessary false starts, and accelerate the applications of these techniques. We believe that the design of high power LED packaging for applications is limited only by the ingenuity and imagination of engineers, managers, and researchers.

Sheng Liu, PhD, ASME Fellow
ChangJiang Scholar Professor
School of Mechanical Science and Engineering
and Wuhan National Laboratory for Optoelectronics
Huazhong University of Science and Technology
Wuhan, Hubei, China

XiaoBing Luo, PhD, Professor
School of Energy and Power Engineering
and Wuhan National Laboratory for Optoelectronics
Huazhong University of Science and Technology
Wuhan, Hubei, China

About the Authors

Sheng Liu is a Changjiang scholar professor of Mechanical Engineering at Huazhong University of Science and Technology and he has a dual appointment at Wuhan National Laboratory for Optoelectronics. He was once a tenured faculty at Wayne State University. He has over 19 years of experience in LED/MEMS/IC packaging. He has extensive experience in consulting with many leading multinational and Chinese companies. He won the prestigious White House/NSF Presidential Faculty Fellow Award in 1995, ASME Young Engineer Award in 1996, NSFC Overseas Young Scientist Award in 1999 in China, IEEE CPMT Exceptional Technical Achievement Award in 2009, and Chinese Electronic Manufacturing and Packaging Technology Society Special Achievement Award in 2009. He has been an associate editor of *IEEE Transaction on Electronic Packaging Manufacturing* since 1999 and an associate editor of *Frontiers of Optoelectronics in China* since 2007. From 2006 to 2010, he was one of the 11 National Committee Members in LED at the Ministry of Science and Technology of China. He obtained his PhD from Stanford University in 1992, his MS and BS degrees from Nanjing University of Aeronautics and Astronautics in 1986 and 1983 respectively. He was an aircraft designer at Chengdu Aircraft Company for two years. He is currently also an ASME Fellow. He has filed and owned more than 100 patents in China and in the USA, and has published more than 400 technical articles, given more than 100 keynotes and invited talks, edited more than nine proceedings in English for the ASME and the IEEE.

Xiaobing Luo is a professor in Huazhong University of Science and Technology (HUST), Wuhan, China. He works at the School of Energy and Power Engineering and Wuhan National Lab for Optoelectronics in HUST. He received his PhD in 2002 from Tsinghua University, China. From 2002–2005, he worked in the Samsung Advanced Institute of Technology (SAIT) in Korea as a senior engineer and obtained SAIT Best Researcher Award in 2003. In September 2005, he was back in China and became an associate professor. In November 2007, he became a full professor after exceptional promotion. His main research interests are LED and electronics packaging, heat and mass transfer, and MEMS. He has published more than 50 papers, applied and owned 40 patents in the USA, Japan, Korea, Europe, and China.

Acknowledgments

Development and preparation of *LED Packaging for Lighting Application* was facilitated by a number of dedicated people at John Wiley & Sons, Chemical Industry Press, and Huazhong University of Science and Technology. We would like to thank all of them, with a special mention for Gang Wu of Chemical Industry Press and James W. Murphy of John Wiley & Sons. Without them, our dream of this book could not have come true, as they have solved many problems during this book's preparation. It has been a great pleasure and fruitful experience to work with them in transferring our manuscript into a very attractive printed book.

The material in this book has clearly been derived from many sources including individuals, companies, and organizations, and we have attempted to acknowledge the help we have received. It would be quite impossible for us to express our appreciation to everyone concerned for their collaboration in producing this book, but we would like to extend our gratitude. In particular, we would like to thank several professional societies in which we have published some of the material in this book previously. They are the American Society of American Engineers (ASME) and the Institute of Electrical and Electronic Engineers (IEEE) for their conferences, proceedings, and journals, including *ASME Transactions on Journal of Electronic Packaging*, *IEEE Transactions on Advanced Packaging*, *IEEE Transactions on Components and Packaging Technology*, and *IEEE Transactions on Electronics Packaging Manufacturing*. Many important conferences such as the Electronic Components and Technology Conference (ECTC), and the International Conference on Electronic Packaging Technology & High Density Packaging (ICEPT–HDP) are also acknowledged for allowing the reproduction of some of their publication materials.

We would also like to acknowledge those colleagues who have helped review some chapters in the manuscript. They are Professor Ricky Lee of Hong Kong University of Science and Technology, Professor Dexiu Huang and Professor Liangshan Wang of Huazhong University of Science and Technology (HUST), Professor Jiangen Pan of Everfine Optoelectronics of China, and Dr. Shu Yuan of HK ASTRI. We would like to thank them for their many suggestions and comments which contributed tremendously to this book. Their depth of knowledge and their dedication have been demonstrated throughout the process of reviewing this book.

We would also like to thank Huazhong University of Science and Technology (HUST), Wuhan National Laboratory for Optoelectronics, the School of Mechanical Science and Engineering, and the School of Energy and Power Engineering for providing us with an excellent working environment to make this book possible. Without being able to recruit outstanding students with cross-disciplinary background, it would have been impossible to include the high quality of information regarding optics, thermal management, materials processes, reliability, and intelligent control. To the best knowledge of the authors, HUST is one of the very few schools which have implemented this new policy, which was initiated by President Peigen Li to allow one faculty to recruit students from different schools or departments. We would like to express our appreciation to those who have worked in LED for many years, such as Mr. Zhijiang Dong, and Dr. Caixia Jin of Wuhan AquaLite Company. We also appreciate the help from Guangdong Real Faith Opto Inc. and Guangdong Real Faith Lighting Fixtures Inc. for collaborating in our research and providing us with many packaging module samples and light fixture prototypes. In particular, our thanks go to Mr. Chunxiao Jin, and his engineers for many useful discussions in the industrial implementation of Application Specific LED Packaging (ASLP).

We would like to register our thanks to our outstanding students for their work in contributing material to this book. They include Kai Wang, Zhongyuan Liu, Fei Chen, Shengjun Zhou, Zhaohui Chen, Zhangming Mao, Han Yan, and Pei Wang.

We also appreciate the Chinese Electronics Society and its Electronic Manufacturing and Packaging Branch led by Professor Keyun Bi for providing us with many technical and academic exchange opportunities.

We would also like to acknowledge the support of many funding agencies in the past years such as the USA National Science Foundation, USA SRC (Semiconductor Research Corporation), National Natural Science Foundation of China, The Ministry of Science and Technology of China, Hubei Department of Science and Technology, Wuhan Science and Technology Bureau, Guangdong Department of Science and Technology, Foshan Bureau of Science and Technology, and Nanhai Bureau of Science and Technology. The authors also appreciate an excellent learning environment created by China Solid State Lighting Alliance in the past few years by organizing many activities and conferences. The authors are delighted to have worked with outstanding people such as Ms. Ling Wu, Dr. Jiming Li, Mr. Jun Yuan, Mr. Yubo Fan, Mr. Bo Geng, and many office assistants involved in conference and training under Ms. Ling Wu's leadership, and those other members of the National Expert Committee for SSL. Sheng Liu enjoyed working with them during the past years and learned a lot in different aspects of SSL by many stimulating discussions. Working and socializing with them has been a privilege and a pleasant experience.

Finally, Sheng Liu would like to thank his parents, Mr. Jixian Liu and Ms. Yanrong Shen, his wife, Bin Chen, and his daughter Amy Liu, his son Aaron Liu, and XiaoBing Luo would like to thank his parents, Mr. Junsheng Luo and Ms. Daxiang Shen, his wife, Ling Deng, and his daughter Yanran Luo for their love, consideration, and patience in allowing them to work on many weekends and late nights for this book. It is the authors' simple belief that the contribution of this book to the LED lighting and packaging industry is worthwhile, in this rapid development of solid state lighting, and will continue to be worthwhile to our civilization for so many years to come. The authors would like to dedicate this book to their families.

Sheng Liu, PhD, ASME Fellow
Chang Jiang Scholar Professor
School of Mechanical Science and Engineering
and Wuhan National Laboratory for Optoelectronics
Huazhong University of Science and Technology
Wuhan, Hubei, China

XiaoBing Luo, PhD, Professor
School of Energy and Power Engineering
and Wuhan National Laboratory for Optoelectronics
Huazhong University of Science and Technology
Wuhan, Hubei, China

1

Introduction

1.1 Historical Evolution of Lighting Technology

In the history of human development, lighting sources have experienced numerous changes initially from collecting natural fire sources to making fire by drilling wood. The development of lighting has witnessed the progress of human history. Fire plays an important role in human history in that it provides humans with food, warmth, and brightness. The use of fire follows the tremendous progress of human civilization. Prior to the eighteenth century, fire had always been a lighting tool for humans, the form of which developed from torch, animal oil lamp, and vegetable oil lamp to the candle, and later to the widely used kerosene lamp. Humans have never stopped exploring new lighting methods. During the use of oil lamps, the wick developed from grass to cotton to multi-strand cotton. Around the third century BC, people made candles with beeswax. In the eighteenth century, candles had been made with paraffin, and mass production of candles was enabled by using machines. In the ninteenth century, the British invented the gas lamp that was originally used as a street lamp. Because of its flickering flame, and the harmful gases that would be produced when it was extinguished, the gas lamp was not very safe and was very dangerous for indoor uses. However, through improvements, the gas lamp replaced the kerosene lamp in tens of thousands of households. These light sources all depended on the flames of the burning materials to provide light. In the eighteenth century, the invention of electricity greatly promoted the development of society, bringing new opportunities for the provision of lighting. In 1809, David Humphrey in Britain invented the arc light, using an electrical light source that was produced by the separation of two contacting carbon rod electrodes after electrifying the electrodes in the air [1]. It was used in public and was the first electric light source for practical lighting before the invention of the incandescent lamp. However, because burning produced a hissing sound and the light was too bright, it was not appropriate for indoor lighting. In 1877, a Russian invented the electric candle by modifying the structure of the arc light, but its performance was not improved. At that time, many scientists began to explore a new, safe, and warm light source.

After a long time trial, the US inventor Thomas Edison lit the world's first lamp that had practical value in October 21, 1879. During this process, Edison conscientiously summarized the failures of previous trials in the manufacturing of electrical light and developed a detailed

LED Packaging for Lighting Applications: Design, Manufacturing and Testing, First Edition. Sheng Liu and Xiaobing Luo.
© 2011 Chemical Industry Press. All rights reserved. Published 2011 by John Wiley & Sons (Asia) Pte Ltd.

experimental plan. Edison experimented with a variety of plants, and decided to use bamboo thread after it had been carbonized. The available lighting time after the production of the electric bulb increased to 1200 hours. The use of this bamboo thread light lasted for more than 20 years. In 1906, Edison used a tungsten filament to improve the quality of the electric bulb, and this is the incandescent lamp that has been used till now.

In 1959, the halogen tungsten circulation theory was discovered to help invent halogen tungsten lamp. Its luminous efficiency was better than the ordinary incandescent lamp.

The invention of the incandescent lamp illuminated the world, but from the perspective of energy utilization, there existed a serious drawback. Only 10–20% of the power had been converted into light, the remaining power being dissipated in the form of heat. Scientists began a new journey to explore new lighting lamps in order to make better use of energy. In 1902, Peter Cooper Hewitt invented the Mercury lamp, the photovoltaic efficiency of which was then greatly enhanced, but with obvious drawbacks. It radiated a large amount of ultra-violet rays, which was harmful to the human body and the light was too strong. Therefore it was not widely used.

In 1910, the Neon light was put into use. The light was emitted by the cold cathode glow discharge in the high-voltage field of a low pressure inert gas in the glass tube. The spectral properties of the inert gas determined the color of the neon.

Mercury lamps aroused many scientists' interests further. They found that as long as the inner wall of the mercury lamp tube was painted with a fluorescent material, then, when the ultraviolet rays of mercury were projected on it, the large amount of harmful ultraviolet rays would be excited into visible light. However, due to the poor start-up device of mercury, the scientists encountered serial failures during actual operation. In 1936, George E. Inman and other researchers produced fluorescent lamps which were different from mercury lamps by using a new start-up device. This fluorescent lamp was made by filling a glass tube with a certain amount of mercury steam, coating phosphor inside the tube wall, and installing one filament at each end of tube as an electrode. This light was brighter than the incandescent. It had a higher efficiency of power energy conversion, larger illumination area, and could be adjusted into different light colors, therefore it went into the homes of ordinary people right after it came out. Because the ingredients of fluorescent lamp were similar to those in daylight, it had been called the "daylight lamp".

Mercury in the fluorescent tube caused environmental pollution, therefore, scientists and manufacturers of lighting began to seek new lighting sources. In the late 1960s, a high pressure gas discharge lamp such as a high pressure sodium lamp and metal halide lamp emerged, which are shown in Figure 1.1.

1.2 Development of LEDs

As far back as 1907, Henry Joseph Round found that the SiC crystal emitted yellow light when studying the non-symmetrical current path on the contacting point of silicon carbide. The first diode should be called the Schottky diode instead of the *p–n* junction diode. The real application of the principles of semiconductor light-emitting into a light-emitting diode (LED) began in the early 1960s. The General Electric's Nick Holonyak Jr. used GaAs to develop the first commercial red GaAsP LED by using vapor-phase epitaxy. At that time, production yield was very low and the price was very high. In 1968, Monsanto in the USA became the first commercial entity to produce LEDs. It began to establish a factory to produce low-cost GaAsP

Figure 1.1 High-pressure sodium lamp.

LEDs, opening a new era of solid-state lighting. Between the years of 1968 and 1970, sales of LEDs doubled every few months. During this period, this company cooperated with Hewlett-Packard to reduce the LED production costs and improve performance. Commercialized GaAsP/GaAs LED devices produced by them became the leading products on the market. However, the luminous output of these red LEDs in that period was 0.1 lm/W, much lower than 15 lm/W of the average incandescent. Monsanto's technological backbone, M. George Craford, has made great contribution to the development of LEDs. He and his colleagues successfully developed yellow LED in 1972. The method they adopted was to grow a nitrogen-doped GaAsP excitation layer on the GaAs substrate. Almost at the same time, ZnO doped red GaP LED and the N-doped GaP green LED devices appeared with a liquid phase epitaxy (LPE) growth. Therefore, Monsanto's research team could produce red, orange, yellow, and green LED devices through doping nitrogen to GaAsP by adopting vapor phase epitaxy method.

In 1972, the Hamilton Company produced the first digital watch with an LED display. In the mid-1970s, the portable digital calculator was produced by the Texas Instrument Company, and Hewlett-Packard had a seven-segment digital display composed of red GaAsP LEDs. However, the power consumption of an LED display at that time was very large. Therefore, the demand for power consumption of the liquid crystal display screen (LCDs) that appeared in the late 1970s was very strong. In the early 1980s, LCDs soon replaced LEDs in calculators and watches display.

The company producing the first color televisions, Radio Corporation of America (RCA) in July 1972 adopted metal halide vapor phase epitaxy growth (MHVPE) and Mg-doped GaN film to obtain blue, violet light with the 430 nm emission wavelength. A major technical breakthrough in the early 1980s was the development of an AlGaAs LED that could emit light with a light-emitting efficiency of 10 lm/W. This technology progress enabled LED to be used in outdoor sports information displays, as well as installation of light equipment such as a center high mount stop lamp (CHSML) in cars.

From the late 1980s to 2000, as a result of new LED technologies such as AlGaInP material technology, multi-quantum wells excitation region, and GaP transparent substrate technology, the size and shape of the naked chip (that is, a chip not packaged with other materials) have been further developed. In the early 1990s Hewlett-Packard and Toshiba successfully developed GaAlP LED devices by using metal organic chemical vapor deposition (MOCVD) technology.

Because of its high light-emitting efficiency and wide color range, it gained extensive attention and was developed rapidly, especially after Craford *et al.* successfully developed transparent substrate technology. The light-emitting efficiency was improved to be 20 lm/W, exceeding the light efficiency of incandescent. Recently, the use of technologies such as the flip-chip structure has further improved luminous efficiency. In 1993, Shuji Nakamura *et al.* in Nichia Company of Japan used two-flow MOCVD technology to solve the annealing process of *p*-type InGaN material. After that, they successfully developed ultra-high brightness blue LED devices with sapphire as substrates. Soon the green and blue-green LEDs were launched in succession. At that time, the high brightness GaInN green LEDs were widely used in traffic lights, but the application of early N-doped GaP green LEDs were limited because of their low luminous efficiency. In 1996, Nichia introduced white light-emitting diodes which used the blue LED chip covered with phosphor composed mainly of yttrium aluminium garnet (YAG). Not long after, the Cree Company in the USA also used blue green LED devices of InGaN/SiC structure with SiC as substrate. Through continuous improvement, the performance of this device is the same as the device of the sapphire substrate. In recent years, research in the ultraviolet (UV) LED technology also has made significant progress, so as to lay the foundation for new-type white light devices.

With the current advances in LED technology, more applications of white LED have gradually become possible, including indicators, portable flashlights, LCD screen backlight panels, automotive instrumentations, medical devices, road lamps, indoor lights, and so on. The industry predicts that white LED will be widely used in general lighting areas within the next 10 years. Figures 1.2–1.5 gives some typical applications of LEDs.

Figure 1.2 LED tunnel lighting in Shanghai Changjiang River Tunnel, China. (Reproduced with permission from www.gd-realfaith.com, Guangdong Real Faith Enterprises Group Co., Ltd., accessed April 12, 2011.)

Figure 1.3 LED tunnel lighting in Guizhou Province, China. (Reproduced with permission from www.gd-realfaith.com, Guangdong Real Faith Enterprises Group Co., Ltd., accessed April 12, 2011.)

Figure 1.4 LED road lighting in Guangdong Province, China. (Reproduced with permission from www.gd-realfaith.com, Guangdong Real Faith Enterprises Group Co., Ltd., accessed April 12, 2011.)

Figure 1.5 LED road lighting in Guangdong Province, China. (Reproduced with permission from www.gd-realfaith.com, Guangdong Real Faith Enterprises Group Co., Ltd., accessed April 12, 2011.)

1.3 Basic Physics of LEDs

An LED can convert electrical energy into light energy. The main functional part of LED is a semiconductor chip. Like a normal diode, the current flows from the p-side to the n-side. When an electron meets a hole, it falls into a lower energy level, and releases energy in the form of a photon. This is the principle of LED light emitting. The wavelength of the light emitted depends on the materials forming the P–N junction.

1.3.1 Materials

Different materials for manufacturing LED will generate photons with different energy levels, thus the wavelength of light could be adjusted. In early days, the LED industry used a $GaAs_{1-x}P_x$ structure to produce an LED of any wavelength in the scope from infrared light to green light theoretically. The typical ones include red $GaAs_{0.6}P_{0.4}$ LED, orange $GaAs_{0.35}P_{0.65}$ LED, yellow $GaAs_{0.14}P_{0.86}$ LED, and so on. Because gallium, arsenic, and phosphorus are used, these LEDs are commonly known as ternary LEDs. The latest technology is to use quaternary compound materials of AlGaInN of aluminum (Al), gallium (Ga), indium (In), and nitrogen (N) to produce a quaternary LED, which can cover the spectral range of all the visible light and part of the ultraviolet.

Figure 1.6 illustrates the relationship between the band-gap of compound semiconductors and the chemical bond. The light emission wavelength of group III nitride covers a wide range from infrared to ultraviolet wavelength.

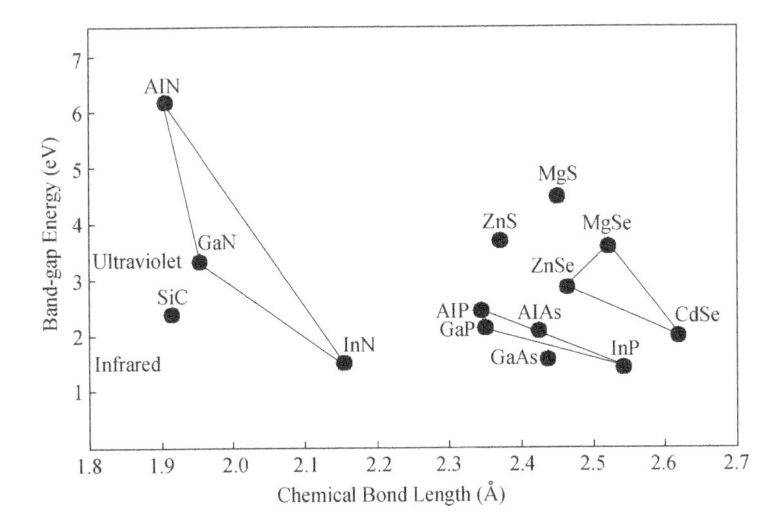

Figure 1.6 Relationship between the band-gap of compound semiconductor and the chemical bond length. (Reproduced with permission from W.S. Zhang and C.G. Liang, "Progress in visible LED," *Semiconductor Information*, **34**, 3, 1–9.)

In 1965 came the first commercial LED using GaAsP materials. The chip structure was similar to a common diode and its luminous efficiency was very low, with only a few lumens of red light emitted. With the development of semiconductor doping technology, Akira of Japan in 1985 prepared InGaAsP double heterostructure Lasers by the method of liquid phase epitaxy (LPE) [3]. The brightness of AlGaAs LEDs exceeded that of previous LEDs, and they were referred to as high-brightness LEDs subsequently [4]. Research in the 1980s was mainly concentrated on the P–N junction and the improvement of quantum efficiency. By the early 1990s, the research of alloy materials of quaternary III–V semiconductor InGaAlP had achieved success [5–7]. While people focused on internal quantum efficiency (IQE), some researchers turned their attention to the external quantum efficiency (EQE). Thus, the distributed Bragg reflector (DBR) and the micro optical resonator cavity (RC) were introduced in the substrate structure of semiconductor materials, which greatly improved the external quantum efficiency of LEDs [8–12]. Subsequently, the reflector was also improved to develop a wide or total reflection angle distributed Bragg reflector LED. Nakamura from Nichia Corporation of Japan completed another great breakthrough in the early 1990s. He successfully prepared high-brightness blue and green LEDs in GaN-based materials with luminous intensity exceeding 1 cd [13,14]. After that, a single quantum well structure was used to obtain blue, green, and yellow InGaN-based LEDs which exceeded 10 cd [15–17], extending light-emitting spectral region of LEDs from 650–560 nm to 650–470 nm. The LED enjoyed the fastest development in the 1990s in which the research was mainly focused on improving the external quantum efficiency and expanding the light-emitting spectral range to achieve tricolor (red, blue, and green) LEDs. The emission spectrum of LED has covered the entire visible spectrum area and the light emission efficiencies of green and pure blue LEDs are close to and catching up with that of the red LED.

Table 1.1 shows the materials used in the LED light-emitting layer, including the epitaxial substrate, production method, and the optical wavelength range. At present, the materials used

Table 1.1 Material used in LED light-emitting layer [2]

Light-emitting Layer	Substrate	Production Method	Light Color
GaP:Zn,O	GaP	LPE	Infrared, Red
AlGaAs	GaAs/AlGaAs	VPE (CVD)	
GaAsP			
GaP:N	GaP		
AlGaP	GaP		Amber, Orange, Yellow
GaInP	Sapphire		
InGaN	Sapphire/SiC	MOCVD(MOVPE)	Green, Emerald Green
ZnSe	ZnSe	MOCVD(MOVPE)	Bluish Green, Blue, Near Ultraviolet
SiC	SiC	MBE	
AlGaN	Sapphire/SiC	HVPE	
AlN	Sapphire		

in the LED mainly include AlGaAs material systems, AlGaInP material systems, AlInGaN material systems, as well as other material systems.

AlGaAs is suitable for manufacturing high-brightness red and infrared LEDs, mainly produced by the liquid phase epitaxy (LPE) method, using a double-heterojuction (DH) structure. AlAs, GaAs, and their alloys are mainly zincblende structure. Al_xGa_{1-x} As band-gap changes with x and light emission wavelength changes from 900 nm to 640 nm. The energy band structure of material changes from direct band-gap into indirect band-gap, when the x value exceeds 0.35. The epitaxial layer of the infrared region uses GaAs as a substrate. A high quality epitaxial layer is achieved due to the small lattice mismatch. When the light emission wavelength is 660 nm, light transmittance of GaAs substrate is very low, so that AlGaAs is used as a substrate. However, it is very difficult to produce AlGaAs substrates. Therefore, there have been few manufacturers to invest in its further development.

AlGaInP alloy is a new type of LED material used in recent years, which has a direct band-gap structure in a large range of group III element component. The lattice constant of $In_{0.5}(Ga_{1-x}Al_x)_{0.5}P$ alloy has a slight mismatch with that of GaAs substrates. As the x value adjusts from 0 to 0.6, the wavelength of light emission changes from 660 nm to 555 nm. In the red region to green region, high-efficient LED devices can be obtained. Because AlGaInP is suitable for high-brightness red, orange, yellow, and yellow green LEDs, and the life test results of red AlGaInP LED in high temperature and high humidity environment are better than red AlGaAs LED. There is a trend for AlGaInP materials to become the mainstream of red LED. The epitaxial growth of the AlGaInP system material in early days used liquid phase epitaxy and vapor phase epitaxy (VPE) technology, but it had been difficult to obtain high quality materials. Currently all the AlGaInP materials use MOCVD and molecular beam epitaxy (MBE) growth technology. AlGaInP quaternary alloy material contains both Al and In, therefore, growth temperature is important. The chemical bond of AlP is strong and chemical bond of InP is weak, so that the appropriate growth temperature range is narrow. As a reference, the growth temperature for MOCVD is about 700 °C, and the growth temperature for MBE is about 500 °C. High growth temperature is favorable for AlP, but it is easy for InP to desorb out from the surface at relatively high temperature, making it difficult for composition control.

Low growth temperature is suitable for the growth of InP, but the decrease of the diffusion length of Al atoms will lead to three-dimensional island growth.

By changing the x value, $GaAs_{1-x}P_x$ light emission wavelength changes from infrared to green. Metal organic chemical vapor deposition method is used to grow N-type GaAsP, then Zn is diffused to obtain P–N junction. When GaP is used as a substrate, 650 nm light emitted from the $GaAs_{0.45}P_{0.55}$ based active layer cannot be absorbed by the substrate. With the progress made in the technology of GaP single crystal substrate in recent years, and improvements in epitaxial growth light-emitting layer crystal performance and the doping technology, the brightness of $GaAs_{0.45}P_{0.55}$ based LED gradually increases, and the cost has lowered.

Gallium nitride and related group III nitride materials include the binary AlN, GaN, InN, the ternary InGaN, AlGaN, and quaternary InGaAlN. The energy band structure of group III nitride materials is direct band-gap. By adjusting the alloy compositions, continuously adjustable band-gap energy from 0.7–0.8 eV (InN), 3.4 eV (GaN), to 6.2 eV (AlN) can be obtained. Group III nitride therefore can cover such a wide spectrum from the ultraviolet to the visible light to the infrared. Because group III nitride material has characteristics such as high thermal conductivity, high electron saturation drift velocity, high critical breakdown voltage, and high fracture toughness to resist defects growth, it has therefore become a suitable material of choice for high-power LEDs. Due to the lack of suitable substrate materials, development of group III nitride-based materials has been limited over the years. Group III nitride materials are heteroepitaxially grown on a foreign substrate with a large lattice mismatch and thermal expansion coefficient mismatch between the films and substrates. Despite the large lattice mismatch and thermal expansion coefficient mismatch, GaN-based materials of device quality have been able to be grown on the sapphire and SiC substrates. At present, the commercially available gallium nitride devices are grown on sapphire and 6H-SiC substrates. Apart from these two substrates, people have also developed other substrate materials, such as ZnO, Si, MgO, and so on. In 1969, Maruska and Tietjen successfully prepared single crystal gallium nitride thin film [19]. However, the research development of gallium nitride materials was then very slow, as the result of some factors such as no suitable substrate materials, high N-type intrinsic carrier density, and the low P-type doping efficiency. After the 1990s, because of the utilization of a low temperature buffer layer and the breakthrough of P-type doping technology, the research on gallium nitride has been developed vigorously in a few leading countries, and brilliant achievements have been made. InGaN/GaN a multiple quantum well (MQW) is often used as an active layer in high-brightness green, blue, and violet LEDs, while GaN/AlGaN MQW in ultra-violet LEDs, and a high-temperature MOCVD method is used for mass production. The light-emitting efficiency is higher than that of AlGaAs, AlGaInP. The world's major companies have been actively involved in research and development of epitaxy technology of related materials and have made a few breakthroughs. InGaN based high-brightness blue and green LEDs have been commercialized in recent years.

The early research of blue light materials has been focused on group II–VI materials ZnSe and the group IV materials SiC [20–22]. Although the II–VI group material band-gap covers the entire visible spectrums and it is the direct band, since the existence of compensation effect, it is hard to achieve heavily doped. It is difficult to make a good P–N junction and the injection efficiency is not high. Blue LED of ZnSe material has the drawbacks of low reliability and short life. Currently LED of ZnO materials are also facing this problem. Although blue SiC LED is the earliest commercialized LED, its band-gap is indirect, and it is hard to obtain high-brightness devices.

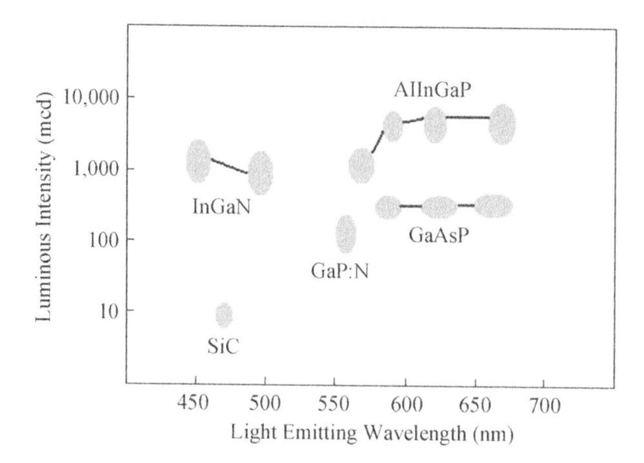

Figure 1.7 Relationship between light emitting wavelength and luminous intensity. (Reproduced with permission from W.S. Zhang and C.G. Liang, "Progress in visible LED," *Semiconductor Information*, **34**, 3, 1–9.)

At present, a high-brightness green LED is still facing the problem of low luminous efficiency, as shown in Figure 1.7. Light emitting efficiency of blue light InGaN material and red light AlInGaP material decreases when entering the green light emission region. Therefore, a high-brightness green LED has been a hot research topic.

1.3.2 Electrical and Optical Properties

(i) LED General Photoelectric Properties

The LED is one of the semiconductor P–N junction diodes. A P–N junction is formed when a P-type and an N-type semiconductor are in contact, as shown in Figure 1.8. If the N-type and P-type regions are made out of the same semiconductor material, the junction is called a homojunction. If the semiconductor materials are different, the junction is known as a heterojunction. When a P–N junction is first created, electrons from the N-type region diffuse in the P-type region. When a mobile electron meets a hole, both hole and electron recombine and vanish. For each electron hole that recombines, a positively charged dopant

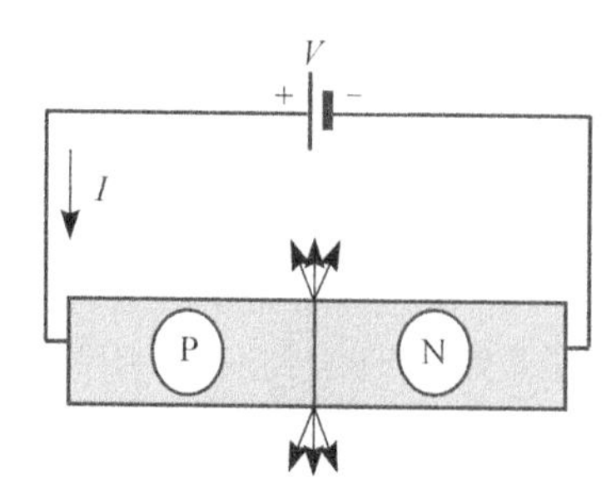

Figure 1.8 Light-emitting P–N junction.

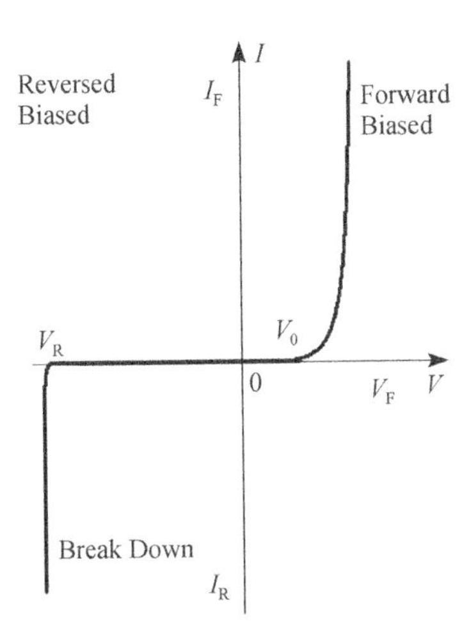

Figure 1.9 Current-voltage characteristics of a P–N junction.

ion is left behind in the N-type region, and a negatively charged dopant ion is left behind in the P-type region. As recombination proceeds and more ions are created, an increasing electric field develops through the depletion zone that acts to slow and then finally stops recombination.

In the forward voltage, the electron is injected from the N region to the P region, and the hole is injected from the P region to the N region. Electrons combine with holes to emit light.

Unlike a resistor, a P–N junction has a highly nonlinear current-voltage characteristic and is often used as a rectifier. LED has the I–V characteristics of a general P–N junction: forward conduction, reverse cut-off, and breakdown. In addition, under certain conditions, it also has light-emitting properties. The relationship between its voltage and current can be shown in Figure 1.9. The current-voltage characteristic of LED is related to the transport of carriers through the depletion region that exists at the P–N junction.

The current-voltage (I–V) characteristic of a P–N junction was first developed by Shockley and the equation describing the I–V curve of a P–N junction diode is therefore referred to as the Shockley equation. The Shockley equation for a diode with cross-sectional area A is given by [23]

$$\mathrm{I} = eA\left(\sqrt{\frac{D_\mathrm{p}}{\tau_\mathrm{p}}\frac{n_\mathrm{i}^2}{N_\mathrm{D}}} + \sqrt{\frac{D_\mathrm{n}}{\tau_\mathrm{n}}\frac{n_\mathrm{i}^2}{N_\mathrm{A}}}\right)\left(e^{eV/kT} - 1\right) \tag{1.1}$$

where e, V, k, T, n_i, N_D, N_A, D_n, D_p, τ_n, and τ_p are the elementary charge, applied voltage, Boltzmann constant, working temperature, intrinsic carrier density, doping carrier density, background carrier density, the electron and hole diffusion constants, and the electron and hole minority-carrier lifetimes, respectively. Under reverse-biased conditions, the diode current saturates and the saturation current is given by the factor preceding the exponential function in the Shockley equation. The diode I–V characteristic can be written as

$$\mathrm{I} = I_s\left(e^{eV/kT} - 1\right) \tag{1.2}$$

$$\mathrm{I_s} = eA\left(\sqrt{\frac{D_p}{\tau_p}\frac{n_i^2}{N_D}} + \sqrt{\frac{D_n}{\tau_n}\frac{n_i^2}{N_A}}\right) \tag{1.3}$$

Under typical forward-biased conditions, the diode voltage is $V \gg kT/e$, and thus $(\exp(eV/kT) - 1) \approx \exp(eV/kT)$. The Shockley equation can be rewritten, for forward-biased conditions, as [24]

$$\mathrm{I} = eA\left(\sqrt{\frac{D_p}{\tau_p}}N_A + \sqrt{\frac{D_n}{\tau_n}}N_D\right)\left(e^{e(V - V_D)/kT}\right) \tag{1.4}$$

The exponent of the exponential function in Equation 1.4 illustrates that the current strongly increases as the diode voltage approaches the diffusion voltage, that is $V \approx V_D$. The voltage at which the current strongly increases is called the threshold voltage and this voltage is given by $V_{th} \approx V_D$.

(ii) LED Electrical Parameters

Allowed power (P_m): the allowed maximum product of the forward DC voltage applied at both ends of the LED and the current flowing through it. If exceeding this value, LED chip cannot work normally.

Turn-on voltage (V_0): the voltage drop produced between two electrodes of LED chips when given a forward current. Generally, GaN LED turn-on voltage V_0 is between 3 V~4 V. When $V < V_0$, the applied electric field cannot overcome the barrier electric field due to the diffusion of the carrier, and the resistance of the LED is very high.

Maximum forward DC current (I_{Fm}): the allowed maximum forward DC current. If exceeding this value, the diode can be damaged.

Maximum reverse voltage (V_{Rm}): the maximum reverse voltage allowed for P–N junction applying reverse bias voltage. When the reverse bias voltage has increased so that $V = -V_R$, then I_R suddenly increases and the LED may be damaged by breakdown. Because different types of compound materials are used, various LED reverse breakdown voltages V_R are also different.

Light response time: the time delay of lighting and extinguishing the LED, as shown in Figure 1.10. Response time is closely related to carrier lifetime, junction capacitance of devices, and circuit impedance. LED lighting time is the time of rising t_r, which is the time taken from 10% of normal brightness when the power is on up to 90% of normal brightness reached. LED extinguishing time is the time of falling t_f, which is the time taken for the normal brightness reduced to 10% of the original one.

In the LED, the injected electron directly combines with the hole to emit light or is first captured by the radiative recombination center and then combines with the hole to emit light, as

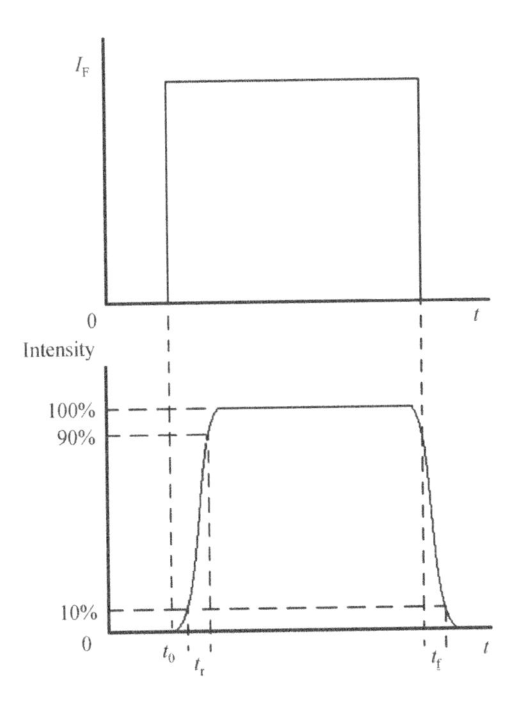

Figure 1.10 LED electro-optical response time.

shown in Figure 1.11. In addition to this recombination light emission, there are also some electrons that are captured by a non-radiative recombination center which is near the middle level between the conduction band and valence band, and then combined with the holes. The energy released is too small to form a visible light. The greater the proportion of radiative recombination electrons to non-radiative recombination electrons is, the higher efficiency the optical quantum has. Because recombination light emission only occurs in the diffusion region between the P-type region and the N-type region, that light is generated only within a few microns near the P–N junction. Theory and practice have proven that the peak wavelength λ of light is relevant to the band-gap E_g of semiconductor material in light emission regions, that is,

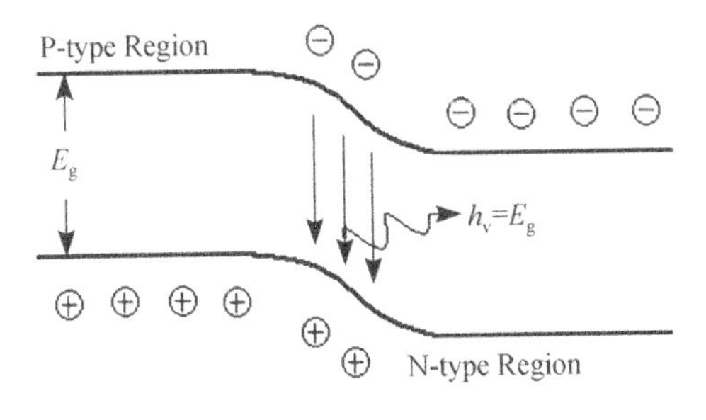

Figure 1.11 Band diagram of a light-emitting P–N junction.

$\lambda \approx 1240/E_g$ (nm). The unit of E_g is electron volt (eV). If visible light can be generated, E_g of semiconductor material should be between 3.26–1.63 eV.

(iii) LED Optical Parameters

Luminous flux (Φ_V): total radiation flux emitted by LED in the entire space per unit time. The luminous flux unit of visible light LED is lumen (lm). Considering that human eyes do not have the same feeling for visible light of different wavelengths, the International Commission on Illumination (CIE) summarized human eyes' sensitivity to monochromatic light of different wavelengths. The optical sensitivity experiment had been conducted by many observers and the resulting average was used. The CIE standard eye is an optical sensor with sensitivity corresponding to the function $V(\lambda)$: the maximal value occurs at a wavelength of $\lambda_m = 555$ nm. Luminous flux characterizes the radiation energy of the LED total light and symbolizes the strength and weaknesses of device's performance.

Spectral energy distribution: the spectral energy distribution is radiation energy distribution of spectrum of a certain range in the radiation wavelengths, because the light emitted by a light-emitting diode is not a single wavelength. At the same time, the luminous intensity or optical power output of the LED changes with the wavelength. A spectral distribution curve is shown in Figure 1.12. When the curve is established, the related chromaticity parameters such as the device's dominant wavelength, color purity, and so on are then determined. The LED spectral distribution is dependent on the type, the property and structure of P–N junction, such as the epitaxial layer thickness, doping impurities, and so on of the compound semiconductor used in preparation, but has nothing to do with the device's geometry and package.

Light intensity (I_V): usually referring to the light intensity in normal direction. It is an important property characterizing light emission strength of light emitting devices. It has strong directivity due to the role of the convex lens: the light intensity in the normal direction maximum with horizontal angle of 90°. Lambert is the most common light intensity distribution of the LED and can be expressed as: $I(\theta) = I_0\cos\theta$, where I_0 is the light intensity

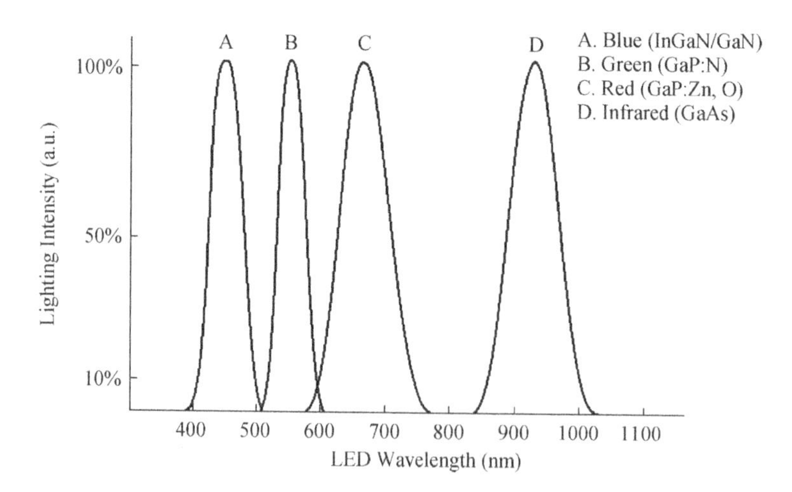

Figure 1.12 LED spectral distribution.

Table 1.2 Coordinates in 1931CIE-RGB system and 1931-XYZ system [25]

	RGB System Chromaticity Coordinates			XYZ System Chromaticity Coordinates		
	R	G	B	X	Y	Z
(R)	1	0	0	0.7347	0.2653	0.0000
(G)	0	1	0	0.2737	0.7174	0.0089
(B)	0	0	1	0.1665	0.0089	0.8246

in the normal direction. If radiation intensity in one direction is (1/683) W/sr, then light emitting is 1 candela (cd). Because the general low power LED has a small light intensity, the light intensity commonly uses millicandela (mcd) as unit.

Peak wavelength (λ_p): The wavelength corresponding to the maximum value of the spectral radiant power. Some LEDs may not emit only monochromatic light, that is, they may have more than one emitting peak. Dominant wavelength is introduced to describe the LED color characteristics. Dominant wavelength is the wavelength of main monochromatic light emitted by the LED.

Spectral half-wave width ($\Delta\lambda$): The interval between two wavelengths corresponding to half of the radiation power of peak emission wavelength, it represents the spectral purity of light-emitting diodes. On both sides of the peak of the LED spectrum line $\pm\Delta\lambda$, there are two points whose light intensity is equivalent to half of the peak (maximum light intensity). These two points correspond to $\lambda_p - \Delta\lambda$, $\lambda_p + \Delta\lambda$. The width between them is called the spectral width, also known as half-power width or half-height width.

Half-value angle ($\theta_{1/2}$) and visual angle: The angle between direction in which the luminous intensity value is half of the axial intensity and axial (normal) direction. Visual angle is twice of half-value angle.

The 1931CIE-XYZ system, in accordance with the wavelength of three primary colors red, green, and blue recommended by CIE have their wavelengths of 700 nm, 546.1 nm, and 435.8 nm, respectively, and their coordinates in the 1931CIE-RGB system and the 1931-XYZ system are shown in Table 1.2. and Figure 1.13.

Compared to other semiconductor materials, group III nitride materials have a strong polarization electric field and piezoelectric effect. The absolute value of piezoelectric constants of group III nitrides is more than ten times larger than the traditional arsenic compounds, which are known as the largest piezoelectric constants among semiconductors. For these reasons, spontaneous polarization of group III nitride and piezoelectric polarization have larger impact on the devices' optoelectronic performance than any other group III–V compounds. This will have a significant impact on physical properties and device performance of nitrides materials. Because the electric field can affect the shape of the band-edge and the distribution of carriers in nitride heterojunction, the direction of spontaneous polarization is determined by its polarity and is sensitive to structural parameters. The direction of the piezoelectric field is determined by whether the material is subjected to tensile stress or compressive stress. In the currently used MOCVD technology, high-quality nitride materials are grown along the (0001) axis.

Changes of band-edge caused by internal spontaneous polarization and piezoelectric field of gallium nitride have significant impact on its optical properties. As a result of the Stark and Franz–Keldysh effects, effective band-gap width of gallium nitride red shifts, the recombination rate of electrons and holes will drop due to the spatial separation of holes and electrons.

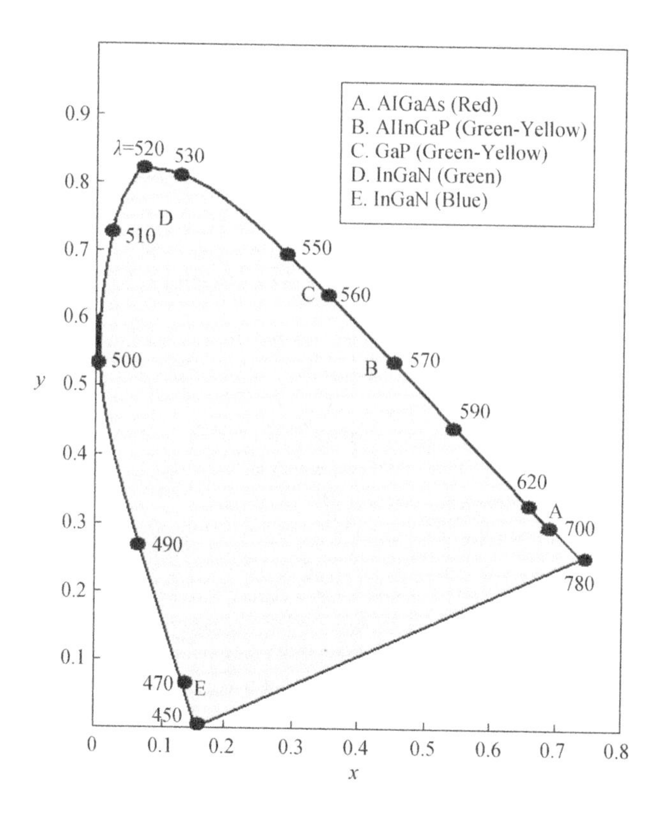

Figure 1.13 CIE1931 chromaticity diagram [25].

These physical effects will change the electroluminescent energy of GaN or InGaN quantum wells and the recombination rate of carriers of group III nitrides. A strong internal electric field would improve the accumulation of electrons and holes in the interface of AlGaN/GaN. This effect can be used by a heterostructure field-effect transistor.

The electrical properties of gallium nitride material are major factors influencing the device. Because quality of gallium nitride prepared by different research groups varies greatly, reports on its electrical properties are often different in the literature. With the continuous improvement of gallium nitride crystal quality, the understanding of gallium nitride has been better developed. Unintentionally doped gallium nitride shows N-type in a variety of situations. The intrinsic carrier concentration of the gallium nitride sample reported in recent years can be reduced to about $10^{16}\,cm^{-3}$ [26,27]. It is generally believed to be caused by nitrogen vacancies [28–30]. The P-type samples prepared under the general circumstances are highly compensated [31,32], and have been able to control the doping concentration in the range of $10^{18}\,cm^{-3}$ through the low-energy electron beam irradiation [33] or thermal annealing treatment [34].

Carrier mobility is a basic parameter very important for determining working characteristics of semiconductor devices. In theory, Littlejohn *et al.* calculated the gallium nitride low-field mobility at room temperature. When they calculated ionized impurity concentration of gallium nitride as $10^{20}\sim10^{17}$ cm^{-3}, its mobility will be 100–1300 cm^2/V × s [35]. In experiments, material quality depends closely on the substrate chosen. If the substrate varies greatly with lattice constants of an epitaxial layer, in order to accommodate the lattice mismatch, many extended defects will be formed and lead to decreasing of Hall mobility, such as gallium nitride directly grown on sapphire at high temperature whose Hall mobility is only 10~30 cm^2/V × s. Pre-deposition of the AlN buffer layer prior to the growth of gallium nitride can greatly raise the Hall mobility of gallium nitride, which is 350–400 cm^2/V × s at room temperature. Using gallium nitride as the buffer layer can further raise the gallium nitride mobility. Nakamura *et al.* used MOCVD to grow the buffer layer of gallium nitride at low temperature with 20 nanometers and grow high temperature GaN whose Hall mobility at room temperature reached 900 cm^2/V × s and mobility at 70 K was 3000 cm^2/V·s [36].

Many researchers studied optical properties of group-III nitride. Maruska and Tietjen first measured the direct band-gap of gallium nitride to be 3.39 eV [37]. Monemar reported that the band-gap of gallium nitride below 1.6 K to be 3.503 eV [38]. Many people have studied the relationship between band-gap and temperature. Different researchers assume that the relationship between gallium nitride band-gap energy and temperature is compliance with the following formula:

$$E_g = E_{g0} - \frac{\alpha_c T^2}{T - T_0} \tag{1.5}$$

where E_{g0} is the band-gap of GaN at absolute zero degree Kevin, α_c and T_0 are the correction factors. The values calculated by different researchers are different, as shown in Table 1.3.

The photoluminescence peak of gallium nitride at room temperature are usually the only band-edge and peak at around 3.4 eV, the yellow band peak at around 2.2 eV, and sometimes the blue band peaking at around 2.9 eV can be observed as well. The sample band-gap peak with poor crystal quality forms higher defect levels whose side band of photoluminescence peak at room temperature weakens and the yellow band strengthens. Thus, people use the ratio of gallium nitride band-edge peak to the yellow band intensity to measure the crystal quality. Yellow bands are likely to appear in single crystal gallium nitride and gallium nitride thin films grown in different ways. Studies have shown that pre-reaction of the growth process has a great influence on the yellow band [42]. Studies show that the yellow band is generated by a deep-level recombination. Also, they believe that this deep-level is a complex compound of V_{Ga} and carbon impurities [43,44]. Researchers believe that yellow band is transition from a deep donor to an acceptor [45,46]. There are many articles in literatures [47,48] reporting that the yellow band is generated by the transition induced by the defects. Theoretical calculations show that a yellow luminous center depends on V_{Ga} and oxygen donor. Because V_{Ga} forms complex compounds with its nearest neighbor O_N having lower formation energy, this is likely to be the

Table 1.3 Temperature dependence band-gap energy of gallium nitride

Reference	E_{g0} (eV)	α_c (eV/K)	T_0 (K)
[39]	3.471	-9.3×10^{-4}	772
[40]	3.489	7.32×10^{-4}	700
[41]	3.503	5.08×10^{-4}	−996

Table 1.4 Mechanical properties of the group III nitride materials

		Linear Elastic Constants (GPa)						
		C_{11}	C_{12}	C_{13}	C_{33}	C_{44}	B	
AlN	Experiment	345	125	120	395	118	201	[50]
		411	149	99	289	125	210	[51]
	Calculation	464	149	116	409	128	228	[52]
		398	140	127	382	96	218	[53]
		396	137	108	373	116	207	[54]
GaN	Experiment	296 ± 18	130 ± 10	158 ± 5	267 ± 17	24 ± 2	195	[55]
		390 ± 15	145 ± 20	106 ± 20	398 ± 20	105 ± 10	210	[56]
	Calculation	369	94	66	397	118	146	[57]
		396	144	64	476	91	172	[58]
InN	Experiment	190 ± 7	104 ± 3	121 ± 7	182 ± 6	10 ± 1	139	[55]
	Calculation	271	124	94	200	46	147	[59]

main reason for generating a yellow band. The luminous peak around 2.9 eV sometimes appears on undoped gallium nitride photoluminescence at room temperature [49]. This peak intensity is weak, often referred to as the blue band.

1.3.3 Mechanical and Thermal Properties

Because there is no appropriate substrate material matched with the group III nitride crystal materials, the current group III nitride materials are grown on sapphire, silicon carbide and silicon substrates through heteroepitaxy. The defect density up to $10^{10}\,cm^{-2}$ with orders of magnitude exists in the group III nitride material film obtained. High-density defects not only affect the light emission characteristics of group III nitride material, but also seriously affect the service life of the LED. How to reduce defect density in thin film materials of group III nitride materials becomes a major problem of LED epitaxial manufacturing. In addition, deformation and stress exist in the LED epitaxial growth processes. The loading processes such as mismatch among the materials, MOCVD flow field, temperature field and gas components will be accompanied by the emergence of stresses and strains. Different stress conditions in the multiple quantum well may increase or decrease the chip light extraction efficiency, cause defects and even damage, thus affecting the consistency, yield, and reliability of the products.

The impact of epitaxial materials' stress/strain on their properties requires detailed parameters of mechanical and thermal properties. Table 1.4 presents mechanical properties of the group III nitride materials. Table 1.5 presents typical thermal properties of group III nitride materials.

Thermal expansion coefficient is one of the most basic material properties. It not only affects some thermal properties of materials, such as thermal conductivity, specific heat capacity, and so on, but also affects the band-gap of materials. The value of coefficient of thermal expansion depends on materials' defect density, free charge concentration, stress/strain, and so on. Thermal expansion coefficient is particularly important to the epitaxial growth. There is the mismatch thermal expansion coefficient between the substrate and the epitaxial thin-film that leads to strain in the epitaxial film. People often use the linear thermal expansion coefficient to

Table 1.5 Thermal properties of group III nitride materials

Thermal Properties	Gallium Nitride (Wurtzite) [60–64]	Indium Nitride (Wurtzite) [60,65–68]	Aluminum Nitride (Wurtzite) [64,69–72]
Debye Temperature (K)	820	980	660
Specific Heat at 300 K ($Jg^{-1}{}^{\circ}C^{-1}$)	0.49	0.73	0.30
Temperature Coefficient of Specific Heat	$0.456 + 0.107 \times 10^{-3}T$ ($298 < T < 1773$)	$1.097 + 7.99 \times 10^{-5}T - 0.358 \times 10^{5}T^{-2}$ ($300 < T < 1800$)	$9.39 \times 10^{-5}T$ ($298 < T < 1273$)
Thermal Conductivity ($Wcm^{-1}{}^{\circ}C^{-1}$)	>2.1	2.85	0.45
Thermal Expansion, Linear ($^{\circ}C^{-1}$)	$a = 5.59 \times 10^{-6}$ $c = 3.17 \times 10^{-6}$	$a = 4.2 \times 10^{-6}$ $c = 5.3 \times 10^{-6}$	$a = 3.8 \times 10^{-6}$ $c = 2.9 \times 10^{-6}$

solve thermal and contingent problems. But sometimes it is difficult to obtain the accurate strain results by linear thermal expansion coefficient. Thus, the nonlinear thermal expansion coefficient is needed. Some researchers have studied nonlinear thermal expansion coefficient through experiment and theoretical research [73]. The relationship between the thermal expansion coefficient of gallium nitride material and temperature is shown in Figure 1.14 [74].

1.4 Industrial Chain of LED

High efficiency energy saving solid state lighting, known for long service life, has been listed in *Medium and Long Term Program for National Scientific and Technological Development Outline in China (2006–2020)* as the most preferential theme. During the

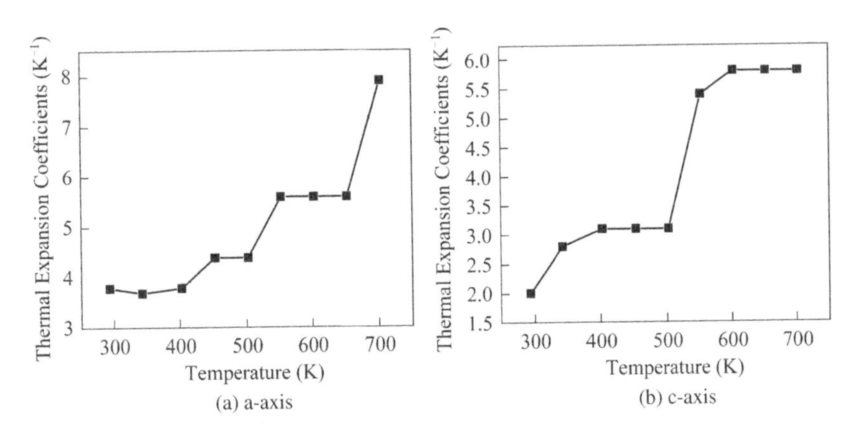

Figure 1.14 The relationship between the thermal expansion coefficient of gallium nitride material and temperature. (Reprinted with permission from M. Leszczynski, T. Suski, H. Teisseyre *et al.*, "Thermal expansion of gallium nitride," *Journal of Applied Physics*, **76**, 8, 4909–4911, 1994. © 1994 American Institute of Physics.)

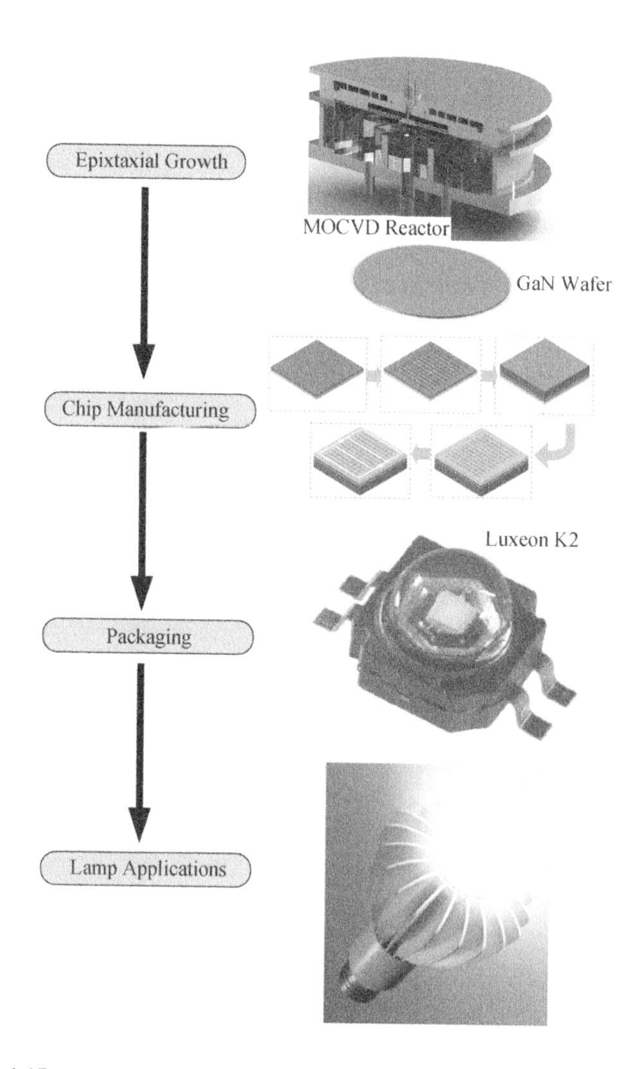

Figure 1.15 Schematic of the sequence in terms of major business classification.

Eleventh Five-Year Plan of China, semiconductor lighting has been treated as one of the high-tech industries with key national support. Many leading countries and areas such as USA, Japan, Taiwan, Korean, Netherland, and German also have active research plans for LED and lighting applications.

Industrial technology system of solid state lighting mainly includes four key technological fields which are shown in Figure 1.15 and Figure 1.16: epitaxy material technology, chip design and manufacturing, packaging materials and process technology and system integration technology and applications. Among the four fields, epitaxy material technologies and equipment consist of substrate materials such as GaN, SiC, Al_2O_3, Si, AlN, ZnO, MgO, and composite substrates and so on and certain epitaxial technologies; chip design and manufacturing consist of wafer preparation, micro-micro manufacturing such as wafer thinning by

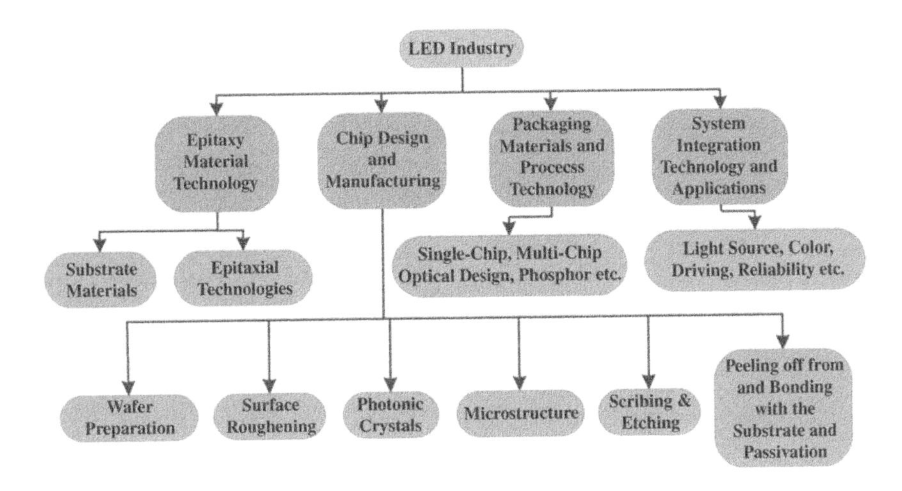

Figure 1.16 Industrial technology system of solid state lighting.

mechanical grinding and CMP (chimerical-mechanical-polishing), surface roughening/micro-structures by etching, photonic crystal and microlenses by nano-imprinting, wafer bonding/debonding, and substrate transfer; packaging technologies consist of single-chip packaging, multi-chip packaging, heat dissipation technology, secondary optics technology, phosphor preparation and coating technology and so on; system integration technology and applications consist of systematic module and lighting fixture structural design of lighting source, color control technology, powering circuit and lamp integration technology, and needed reliability technology and so on.

1.4.1 LED Upstream Industry

Upstream industry mainly refers to epitaxy and chip manufacturing. Due to the high development of epitaxy process, devices' key components such as emitting layer, cladding layer, buffer layer, and reflector and so on can be accomplished in the epitaxy process. As to chip micro-manufacturing, it mainly deals with the design of electrodes with appropriate current spreading, optical microstructures with high light extraction, segmentation and testing.

Presently, epitaxial methods for producing various brightness LEDs mainly include MOCVD, LPE, and VPE. The MOCVD method is used to produce red, yellow, and green LED epitaxial materials as well as blue, green, and ultraviolet epitaxial materials. The LEP method is used to produce ultra-high brightness red LED epitaxial materials and red and green LED materials with normal brightness. As for VPE, it is used to manufacture gallium-arsenide-phosphide epitaxial materials with ultra-high brightness gallium-phosphide as substrates. While zinc selenide white LED epitaxial materials, as a particular case, is manufactured by MBE which could also make the materials achieve ultra-high brightness level. However, these kinds of materials are not in mass production presently. Schematic of a typical epitaxial process is shown in Figure 1.17.

Enterprises producing LED chips are often called Chip Manufacturers for short. Their chief mission is to transfer epitaxial wafers into LED chips with positive and negative electrodes.

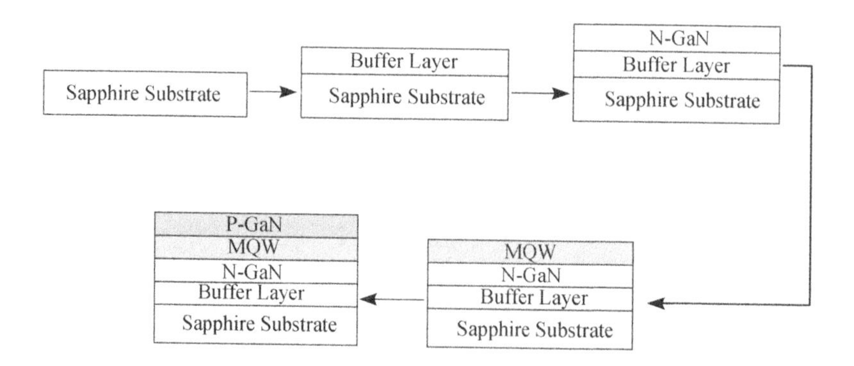

Figure 1.17 Schematic of the processes sequence of epitaxial growth.

This kind of LED chip can be divided into two categories according to different electrode positions: one is that with electrodes on the upper and lower sides such as indium-gallium-nitrogen chips grown on silicon carbide substrates, gallium-phosphide chips, gallium-aluminum-arsenide chips, and gallium-arsenide-phosphide chips; the other is that with electrodes on the same side such as epitaxial materials around sapphire substrates. Owing to substrate's being insulated, positive and negative electrodes have to be on the same side. As to its fabrication process, it includes metal film plating, lithography, chemical or iron etching, and scribing and so on. The size of chip ranges from $0.2 \times 0.2\,mm^2$ to $2.5 \times 2.5\,mm^2$. If divided according to the relative position of electrodes on devices when packaging, LED chips have flip chip type and conventional type. Figure 1.18 presents a typical chip manufacturing process.

1.4.2 LED Midstream Industry

LED midstream industry aims at packaging LED devices. The LED packaging industry is different from other semiconductor device packaging although all of them belong to semiconductor industry. A great variety of LED devices can be packaged not only according to their uses in different occasions but also on the basis of their color and shape.

Except for the power-type LED, other LED devices use almost the same packaging materials and technologies such as die attach with silver paste, oven curing, epoxy resin encapsulation packaging. Power packaging tends to use eutectic solder. CP Wong is working on nano-phased conductive epoxy with the hope to replace more expensive solders. A typical packaging process is shown in Figure 1.19.

The most commonly used power-type LED packaging structure is shown in Figure 1.20. Now, each procedure can work semi-automatically or fully automatically, which guarantees its efficacy and a high yield as well as good performance and optimized service life.

1.4.3 LED Downstream Industry

Downstream industry of LEDs refers to the industry formed after utilizing LED elements. There are many applications for LEDs.

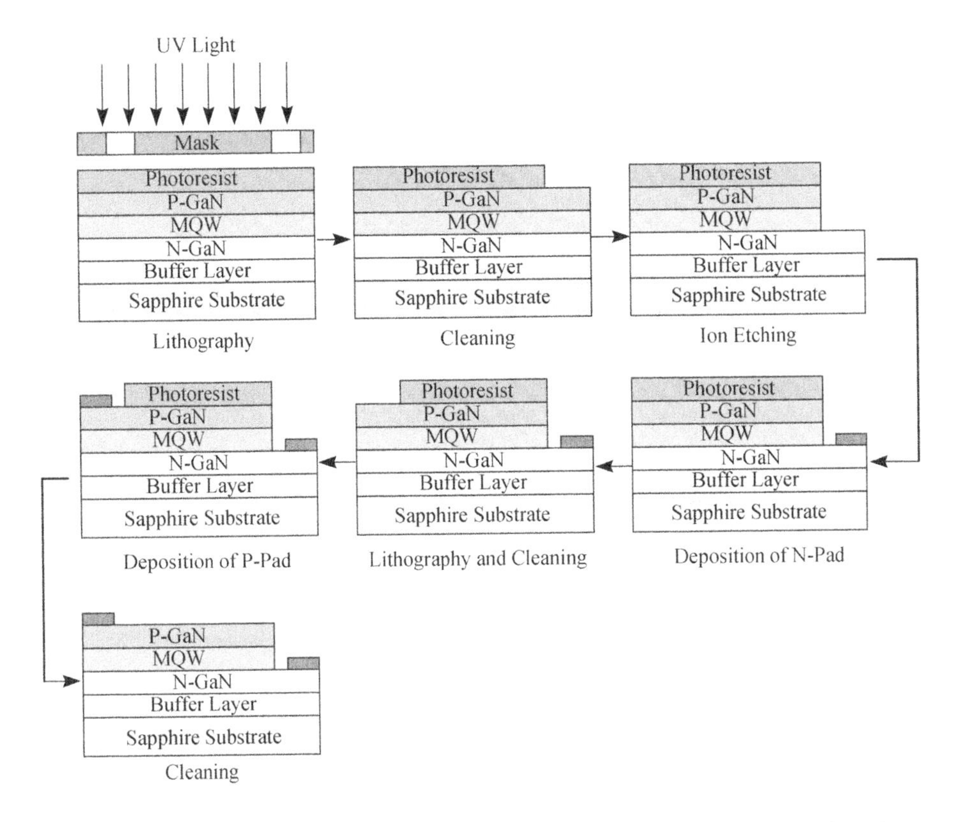

Figure 1.18 Schematic of a typical flow chart for processes in chip micro-manufacturing.

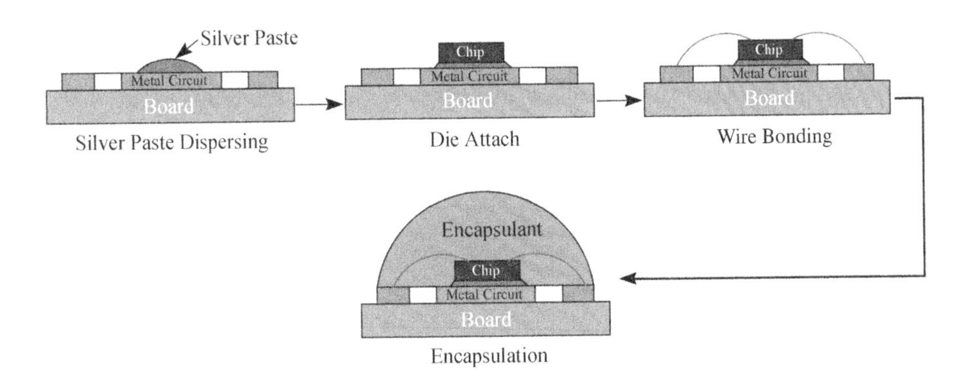

Figure 1.19 Schematic of packaging processes.

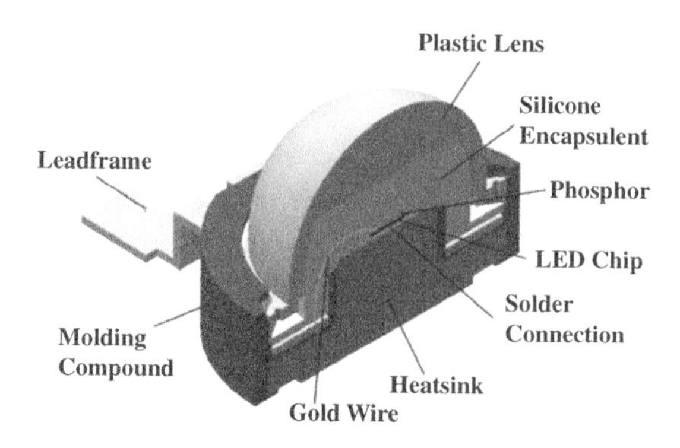

Figure 1.20 Most commonly used power-type LED packaging structure.

LED Display Screen

The LED display screen is a typical LED downstream product. At the beginning of the 1980s, monochromatic and polychromatic LED display screens became available. Screens such as word display screens, animation display screens, and some outdoor large-scale display screens like the CRT pixel screen and magnetic plate-turnover screen. In the early 1990s, in virtue of the development of electronic computer technologies and IC technologies, video technology of the LED display screen had been achieved and TV images could be directly presented on the screen. In the middle 1990s, because of the successful development and quick commercialization of InGaN ultra-high brightness blue and green LED, the quality of outdoor screens had been greatly increased. Large-scale color display screen has now become a rising industry with an annual output value of 3 billion Chinese Yuan (RMB). The advantages of the LED screen quickly extend to guidance screen, negotiable securities screen, exchange rate screen, and advertisement screen and so on.

LED Traffic Signal Lights

LED traffic signal lights as a new LED downstream industry emerged recently. Incandescent lamps were applied to traffic lights as their light sources in the past. The greatest weakness of incandescent lamps is the waste of electricity and short life, and the weakness remained unresolved for a long time. Since 1999, LED has been tried and adopted as the light source of traffic lights with quick development and wide applications. Due to the direct use of monochromatic light, LED lamps can save electricity by 90% as compared to incandescent ones. Till 2004, the first LED lamps had worked for more than five years. Presently, LED lamps are used in small and medium-sized cities. It is estimated that sales in the market will reach 1 billion Chinese Yuan annually in China.

Guide Light

Combined with solar cells, LED as guide light sources has been used for around 10 years and is extended to the field of channel signal lamps with various colors. LED beacons in the airport, projecting lights, and omnidirectional lamps are successfully developed. They are produced and applied widely due to their low-power feature, free maintenance, high brightness, good color purity, and distinct colors that enable signals to be easily recognized. It is difficult to

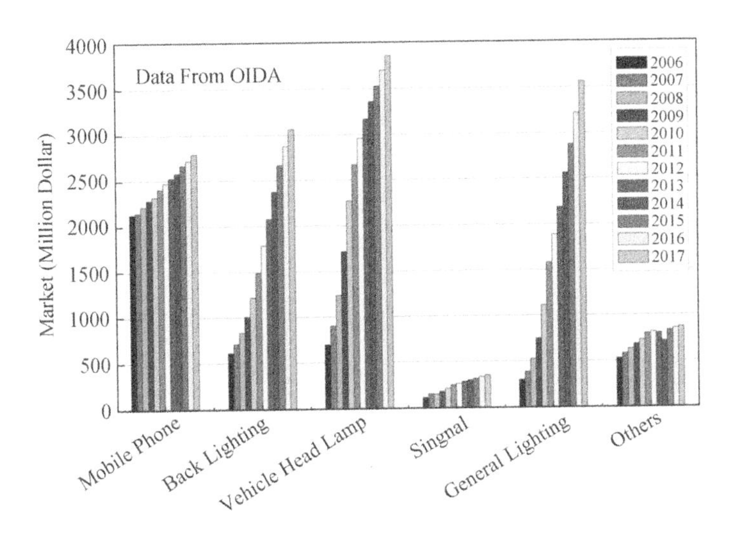

Figure 1.21 Market forecast of LED in several major business sectors.

develop railway lamps which require light intensity and perspectives different from other lamps. However, it is believed that they will be developed and applied in the future and that will be a huge market with big potential.

Liquid Crystal Backlight

Liquid crystal backlight is a developing industry. LED is applied to a LCD screen as its backlight, which upgrades mobile phone products. It is estimated that 3.5 billion LED chips are used each year and ultra-high brightness LED accounting for 30% of the total amount. Comparing with cold cathode fluorescent lamp (CCFL) backlight, LED is free of mercury, rich in colors, and compact in volume. LED backlight can display 105% color while CCFL can only display 65–75% color calculated by national television system committee (NTSC) standard. LED takes a leading role in the backlight of ordinary meters and household appliances, and its annual output has reached a considerable amount. With the new applications to large screen high definition television (HDTV), notebooks and desktop computer applications, LED for backlighting is one of the fastest growing industrial sectors, as shown in Figure 1.21.

LED Vehicle Light

LEDs have been widely used in cars, as shown in Figure 1.22 and the LED vehicle light industry has taken shape and is being developed rapidly. Red and yellow LEDs are quite suitable for brake lights, tail lights, and turn signals due to their low prices. They can also be used in instrument lighting and interior lighting of vehicles. Lumileds claimed LED has been used in nearly 15 million vehicles. Because there are 20 different types of car light source that can be replaced by LED, thus 200–400 LEDs will be needed just in one vehicle. It is possible that LEDs used in vehicle lighting will become a business sector with the greatest potential of all LED application industries and its output in the market will rise in one or two years.

Figure 1.22 Various applications of LEDs in vehicles. *(Color version of this figure is available online.)*

Landscape Decorative Lighting

Landscape decorative lighting industry has begun to take shape. In virtue of enhanced LED brightness, falling prices, long service life, low power, simpler driving and controlling procedures than neon lamps, and being capable of flashing and changing colors, LED is used to make monochromatic, polychromatic, and chromophoric light columns fitted with multi-colored luminescent units to decorate grand buildings, bridges, streets, entertainment places, and plazas. The LED gradually plays a leading role in the newly-built landscape lighting project in cities. Guardrail lamps, clearance lights, ball lamps, streamer lamps, underwater lights, and light bells are available in every lamp shop. This kind of lamps can also be used in household decoration. Beijing Olympic Games successfully demonstrated the powerfulness of LEDs and it is expected that Shanghai Expo 2010 will bring great motivation to the LED industry.

Special Lighting

Lamps requiring special lighting include lamps used either in low illumination lighting such as lawn lamps, courtyard lamps, and underground lamps, or in potable illumination lighting such as flashlights, headlights, and cap lamps. Figure 1.23 shows some typical special lighting scenes. The market for special lighting lamps in China has reached a considerable scale and formed regional industrial groups.

As to general illumination products, they are still in the early stages of development, but with great promise. Desk lamps, streamer lamps, and solar road lamps emerge in secession but the production scale is small. Large-scale application will be achieved after increasing luminous flux and lowering the price, plus with optimized designs in various technical sides this book is going to describe. At that moment, the industry will become the largest industry among the downstream industries. Recently, the Chinese government has lunched a new program,10 Cities 10,000 Lamps, with the goal of 10,000 lamps in each city for LED demonstration. In its first phase, the government is initiating in first 21 cities to promote various applications in road lighting, subway lighting, parking lot lighting, express train lighting, home lighting, and so on. Many more cities are applying to join this plan, which will help China to move much more aggressively in high power LED applications.

Figure 1.23 Various special lighting scenes. *(Color version of this figure is available online.)*

Those leading multinational companies such as Philips, Osram, and GE have been active in both traditional lighting and LED, while other pure LED players such as Cree, Nichia, and so on have been leading in the technology and product development of LEDs. A unique feature of these companies is that they are either vertically integrated, or have built a partnership, or have grown by acquisition of companies for the technology and/or market access. The following are a few recent acquisitions:

Philips Acquisition of Lumileds

Lumileds was founded in 1999, specialized in LED chip and LED packaging, which originated from HP, forming HP's optoelectronic department. When HP was divided into two business units, Philips acquired 47% share of Agilent with a value of 760 million Euros. This acquisition consolidated the determination of Philips' vertical integration, which should reflect in the market that Philips Lumileds did not market its bare blue chips and only provided the lighting users with high power modules.

Cree's Merger of Cotco

Before 2007, Cree did not have the packaging house in China and it needed to get access to mainland China's growing market. In April 2, 2007, Cree announced that it had completed the acquisition of privately held Cotco Luminant Device Limited. Cotco was a leading supplier of high brightness LEDs in China and had a strong distribution network in China. Through this acquisition of Cotco, Cree can access the solid state lighting market in China faster. In addition, the low cost manufacturing capabilities and the competitive and full power LED family of products of Cotco ensure that the packaging and manufacturing ability of Cree can be increased remarkably. Traditionally, most of Cree's revenue was from the sale of the LED chip. This acquisition will give Cree the chance to provide more value-added products in key markets of China.

Philips Acquisition of Color Kinetics

Philips and Color Kinetics reached agreement on June 19, 2007 and Philips acquired all the shares with a total deal of 592 million Euros. This move was expected to enhance the R&D capability of Philips in applications, as Color Kinetics had its IPs in various solutions. It was noted that Color Kinetics of Boston, USA, was only founded in April 1997 and was known for its various LED products and leading digital intelligent products.

Acquisitions and Mergers at Taiwan

Taiwan manufacturing specialists are used to positioning themselves competitively in the global market due to their success in IC and electronics OEM and ODM. The positive outlook has attracted a number of new players to participate in the industry. However, oversupply of LEDs with too many players in the industry has led to irrational price-cutting. IP infringement with those leading multinational companies forced them to use merger and acquisition for consolidation. A few important mergers and acquisitions were initiated since 2005. For example, Epistar and UEC enhanced their product offerings, technologies, and patent items to increase benefits. That was the major impetus behind the merger between Epistar and UEC. The merger was to help their status internationally. Another case in Taiwan included South Epitaxy's acquisition of Epitech.

On the other hand, due to the growing markets, many new companies are entering into the LED community and many venture capital (VC) firms are showing great interests in this green business. Therefore, many new companies are formed, including new companies in epitaxial growth and chip design and manufacturing. Even more packaging companies and application companies are born. For instance in China alone, it is estimated that more than 400 companies in packaging, more than 10 in LED chips, and 4000 in lighting applications have been found (www.LED-China.net). With the increasing international and domestic competition, merger and acquisition will occur. Collaborations in terms of Intellectual Property (IP) licenses will also occur.

System in Packaging (SiP) May Cause the Change of Industrial Chain

Traditionally, chip, packaging, and applications are three isolated sectors. However, with the development of SiP, these three sectors are closely interacted and even integrated. This may provide more advantages for those vertically integrated companies in competition. However, for those newly founded companies with their unique IPs, they may provide unique solutions in packaging and applications. SiP, if well developed in next few years to come, will provide very cost effective solutions to the markets described above and new applications that will be emerged later in future.

1.5 Summary

In this chapter, the historical evolution of lighting technology was reviewed in detail. Then the development of LEDs was examined from as early as 1907, to the early application of the semiconductor LEDs in the early 1960s, to the introduction of the blue LED in 1990s, and to the beginning of general lighting in 2000s. Basic physics of LEDs was also introduced in terms of materials involved covering the light-emitting layer, substrate, production method, and light color. Electrical, optical, and mechanical properties were briefly discussed with the goal of preparing the reader for the later chapters. Finally, the industrial chain of the LED was also discussed, mainly covering four key technological fields: epitaxy material technology, chip design and manufacturing, packaging materials and process technology, and system integration technology and applications. Examples from industrial sectors in upperstream, middlestream, and downstream were discussed and several cases of acquisition and merger were discussed, showing the parallel of technology and business in this growth stage of the industry.

References

[1] Slingo, W. and Brooker, A. (1900), *Electrical Engineering for Electric Light Artisans*, London, Longmans, Green and Co.

[2] Zhang, W. and Liang, C. (1997), "Progress in visible LED," *Semiconductor Information*, **43**(3): 1–9.

[3] Akira Fujimoto, Hideaki Watanabe, Masashi Takeuchi and Mikihiko Shimura, (1985) "LPE growth of lattice-matched InGaAsP on $GaAs_{0.69}P_{0.31}$ substrates and low thresholdcurrent density operation of visible InGaAsP DH lasers", *Japanese Journal of Applied Physics* **24**: L653–L656.

[4] Nishizawa, J., Itoh, K. and Okuno, Y. (1985), "LPE-AlGaAs and red LED," *Journal of Applied Physics* **57**: 2210–2214.

[5] Kuo, C., Fletcher, R. and Osentowaki, T. (1990), "High performance AlGaInP visible light-emitting diodes," *Applied Physics Letters* **57**: 2937–2939.

[6] Sugawara, H., Ishikawa, M. and Hatakoshi, G. (1991), "High–efficiency InGaAlP/GaAs visible light-emitting diodes," *Applied Physics Letters* **58**: 1010–1013.

[7] Sugawara, H., Itaya, K. and Hatakoshi, G. (1994), "Emission properties of InGaAlP visible light-emitting diodes employing a multi quantum-well active layer," *Japan Journal of Applied Physics* **33**: 5784.

[8] Yokoyama, H., Nishi, K. and Anan, T. (1990), "Enhanced spontaneous emission from GaAs quantum wells in monolithic micro cavities," *Applied Physics Letters* **57**(26): 2814–2816.

[9] Hunt, N.E., Schubert, E.F. and Logan, R.A. (1992), "Enhanced spectral power density and reduced line width at 1.3 μm in an InGaAsP quantum well resonant cavity light-emitting diode," *Applied Physics Letters* **61**: 2287–2289.

[10] Schubert, E., Wang, Y. and Cho, A. (1992), "Resonant cavity light-emitting diode," *Applied Physics Letters* **60**(8): 921–923.

[11] Schubert, E., Hunt, N. and Micovic, M. (1994), "Highly efficient light emitting diodes with micro cavities," *Science* **265**: 943–945.

[12] Blondelle, J., De Neve, H. and Borghs, G. (1996), "High efficiency (>20%) microcavity LEDs," *IEE Colloquium on Semiconductor Optical Micro cavity Devices and Photonic Band-gaps* 5: 12/1–12/6.

[13] Nakamura, S., Mukai, T. and Senoh, M. (1994), "Candela-class high-brightness InGaN/AlGaN double-heterostructure blue-light emitting diodes," *Applied Physics Letters* **64**(13): 1687–1689.

[14] Nakamura, S., Mukai, T. and Senoh, M. (1994), "High-brightness InGaN/AlGaN double-heterostructure blue-green light emitting diodes," *Journal of Applied Physics* **76**: 8189–8191.

[15] Nakamura, S., Senoh, M. and Iwasa, N. (1995), "High-brightness InGaN blue, green and yellow light-emitting diodes," *Japan Journal of Applied Physics* **34**: 797–799.

[16] Nakamura, S., Senoh, M. and Iwasa, N. (1995), "High-Power InGaN single-quantum-well- structure blue and violet light-emitting diodes," *Applied Physics Letters* **67**: 1868–1870.

[17] Nakamura, S., Senoh, M. and Iwasa, N. (1995), "Superbright green InGaN single-quantum- well-structure light-emitting diodes," *Japan Journal of Applied Physics* **34**: 1332–1335.

[18] Sheats, J.R., Antoniadis, H., Hueschen, M., Leonard, W., Miller, J. and Moon, R. (1996), "Organic electroluminescent devices," *Science* **273**: 5277.

[19] Maruska, H. P. and Tietjen, J. J. (1969), "The preparation and properties of vapor-deposited single-crystal-line GaN," *Applied Physics Letters* **15**: 327–329.

[20] Nakayama, N., Itoh, S., Ohata, T., Nakano, K., Okuyama, H., Ozawa, M., Ishibashi, A., Ikeda, M., and Mori, Y. (1993), "Room temperature continuous operation of blue-green laser diodes," *Electronic Letters* **29**: 1488–1489.

[21] Eason, D.B., Yu, Z., Hughes, W.C. (1995), "High Efficiency ZnCdSe/ZnSSe/ZnMgSSe Green Light Emitting Diodes," *Applied Physics Letters* **66**: 115–118.

[22] Morkoc, H., Strite, S., Gao, G. B., Lin, M. E., Sverdlov, B., and Burns, M. (1994), "II–VI ZnSe-based semiconductor device technologies," *Journal of Applied Physics*, **76**: 1363–1398.

[23] http://www.ecse.rpi.edu/~schubert/Course-ECSE-6290%20SDM-2/1%20P–N%20junction%20summary.pdf (2009/9/1)

[24] Xi, Y. and Schuberta, E. F. (2004), "Junction–temperature measurement in GaN ultraviolet light-emitting diodes using diode forward voltage method," *Applied Physics Letters*, **85**: 2163–2165.

[25] http://en.wikipedia.org/wiki/CIE_1931_color_space (2009/9/1)

[26] Nakamura, S., Makui, T. and Senoh, M. (1992), "In situ monitoring and Hall measurements of GaN grown with GaN buffer layers," *Journal of Applied Physics* **71**: 5543–5549.

[27] Kordos, P., Javorka, P., Morvic, M., Betko, J., Van, J.M., Wowchak, A.M., and Chow, P.P. (2000), "Conductivity and Hall-effect in highly resistive GaN layers," *Applied Physics Letters* **76**: 3762–3765.

[28] Kuznetsov, N.I., Nikolaev, A.E., Zubrilov, A.S., Melnik, Y.V. and Dimitriev, V.A. (1999), "Vapor phase epitaxy on SiC substrates," *Applied Physics Letters* **75**: 3138–3141.

[29] Perlin, P., Suski, T. and Teisseyre, H. (1995), "Towards the Identification of the Dominant Donor in GaN," *Physics Review Letters* **75**: 296–310.

[30] Pankove, J.I. and Berkeyheiser, J.E. (1974), "Properties of Zinc-doped GaN. II. photoconductivity," *Journal of Applied Physics* **45**: 3892–3896.

[31] Kaufmann, U., Kunzer, M., Kaier, M., Obloh, H., Ramakrishnan, A., Santic, B. and Schlotter, P. (1998), "Nature of the 2.8 eV photoluminescence band in Mg doped GaN," *Applied Physics Letters* **72**(11): 1326–1329.

[32] Tokunaga, H., Waki, I., Yamaguchi, A., Akutsu, N. and Matsumoto, K. (1998), "Growth condition dependence of Mg-doped GaN film grown by horizontal atmospheric MOCVD system with three layered laminar flow gas injection" *Journal of Crystal Growth*, **189/190**: 519–522.

[33] Amano, H., Kito, M., Hiramatsu, K. and Akasaki, I. (1989), "Light Genererating Carrier Recombination and Impurities in Wurtzite GaN/Al2O3 Grown by MOCVD," *Japan Journal of Applied Physics* **28**: 2112–2116.

[34] Nakamura, S., Iwasa, N., Senoh, M. and Mukai, T. (1992), "Thermal annealing effects on p-type Mg-doped GaN films," *Japan Journal of Applied Physics* **31**: 139–142.

[35] Littlejohn, M.A., Hauser, J.R. and Glisson, T.H. (1975), "Monte-Carlo calculation of velocity-field relationship for gallium nitride," *Applied Physics Letters* **26**: 625–628.

[36] Nakamura, S., Makui, T. and Senoh, M. (1992), "In situ monitoring and Hall measurements of GaN grown with GaN buffer layers," *Journal of Applied Physics* **71**: 5543–5547.

[37] Maruska, H.P. and Tietjen, J.J. (1969), "The preparation and properties of vapor-deposited single-crystal-line GaN," *Applied Physics Letters* **15**: 327–331.

[38] Monemar, B. (1974), "Fundamental energy gap of GaN from photoluminescence excitation spectra," *Physical Review B* **10**: 676–681.

[39] Piehugin, I.G. and Yaskov, D.A. (1970), "Preparation of gallium nitride," *Inorganic Materials* **6**: 1732–1738.

[40] Pankove, J.I., Berkeyheiser, J.E. and Maruska, H.P. (1970), "Luminescent properties of GaN," *Solid State Communications* **8**: 1051–1056.

[41] Nakamura, S., Mukai, T. and Seno, M. (1992), "Si- and Ge doped GaN films grown with GaN buffers layers," *Japan Journal of Applied Physics* **31**: 2883–2887.

[42] Li, S.T., Mo, C.L. and Wang, L. (2001), "The influence of Si-doping to the growth rate and yellow luminescence of GaN grown by MOCVD," *Journal of Luminescence* **93**: 321–326.

[43] Ogino, T. and Aoki, M. (1980), "Mechanism of yellow luminescence in GaN," *Japan Journal of Applied Physics* **19**: 2395–2399.

[44] Hofmann, D.M., Kovalev, D. and Steude, G. (1995), "Properties of the yellow luminescence in undoped GaN epitaxial layers," *Physical Review B* **52**: 16702.

[45] Kennedy, T.A., Glaser, E.R. and Freitas, J.A. (1995), "Native defects and dopants in GaN studied through photoluminescence and optically detected magnetic resonance," *Journal of Electronic Materials* **24**: 219–224.

[46] Glaser, E.R., Kennedy, T.A. and Doverspike, K. (1995), "Optically detected magnetic resonance of GaN films grown by organo metallic chemical-vapor deposition," *Physical Review B* **51**: 13326.

[47] Neugebauer, J. and van de Walle, C. (1996), "Gallium vacancies and the yellow luminescence in GaN," *Applied Physics Letters* **69**: 503–506.

[48] Mattila, T. and Nieminen, R.M. (1997), "Point-defect complexes and broad-band luminescence in GaN and AlN," *Physical Review B* **55**: 9571.

[49] Sasaki, T. and Zembutsu, S. (1987), "Substrate-orientation dependence of GaN single-crystal films grown by metal organic vapor-phase epitaxy," *Journal of Applied Physics* **61**: 2533–2537.

[50] Tsubouchi, K., Sugai, K. and Mikoshiba, N. (1981), New York, *IEEE Ultrasonics Symposium Proc.*, 1: 375–379.

[51] McNeil, L.E., Grimsditch, M. and French, R.H. (1993), "Vibrational spectroscopy of aluminum nitride," *Journal of the American Ceramic Society* **76**: 1132–1136.

[52] Ruiz, E., Alvarez, S. and Alemany, P. (1994), "Electronic structure and properties of AlN," *Physical Review B* **49**: 7115–7123.

[53] Kim, K., Lambrecht, R.L. and Segall, B. (1996), "Elastic constants and related properties of tetrahedrally bonded BN, AlN, GaN, and InN," *Physical Review B* **53**: 16310.

[54] Wright, A.F. (1997), "Elastic properties of zinc-blende and wurtzite AlN, GaN, and InN," *Journal of Applied Physics* **82** (6): 2833–2839.

[55] Sheleg, A.U. and Savastenko, V.A. (1979), "Determination of elastic constants of hexagonal crystals from measured values of dynamic atomic displacements," *Izv. Akad. Nauk SSSR. Neorganicheskie Materials* **15**: 1598–1602.

[56] Polian, A., Grimsditch, M. and Grzegory, I. (1996), "Elastic constants of gallium nitride," *Journal of Applied Physics* **79**: 3343–3344.

[57] Azuhata, T., Sota, T. and Suzuki, K. (1996), "Elastic constants of III–V compound semiconductors: modification of Keyes'relation," *Journal of Physics: Condensed Matter* **8**: 3111.

[58] Kim, K., Lambrecht, R.L. and Segall, B. (1994), "Electronic structure of GaN with strain and phonon distortions," *Physical Review B* **50**: 1502–1505.

[59] Wang, K. and Reeber, R.R. (2001), "Thermal expansion and elastic properties of InN," *Applied Physics Letters* **79**: 1602–1604.

[60] Nipko, J.C., Loong, C.K., Balkas, C.M. and Davis, R.F. (1998) "Phonon density of states of bulk gallium nitride," *Applied Physics Letters* **73**: 34–36.

[61] Barin, I., Knacke, O. and Kubaschewski, O. (1977), "Thermochemical properties of inorganic substances," *Springer-Verlag, Berlin-Heidelberg-New York.*

[62] Florescu, D. I., Asnin, V. M., Pollak, F. H., Molnar, R. J. and Wood, C. E. C. (2000), "High spatial resolution thermal conductivity and Raman spectroscopy investigation of hydride vapor phase epitaxy grown n-GaN/ sapphire (0001): Doping dependence," *Journal of Applied Physics* **88**: 3295–300.

[63] Kotchetkov, D., Zou, J., Balandin, A.A., Florescu, D.I. and Pollak, F.H. (2001), "Effect of dislocations on thermal conductivity of GaN layers," *Applied Physics Letters* **79**: 4316–4318.

[64] Harima, H. (2002) "Properties of GaN and related compounds studied by means of Raman scattering," *Journal of Physics: Condensed Matter* **14**: R967–R993.

[65] Davydov, V. Y., Emtsev, V. V., Goncharuk, I. N., Smirnov, A. N., Petrikov, V. D., Mamutin, V.V., Vekshin, V. A., Ivanov, S. V., Smirnov, M. B. and Inushina, T. (1999), "Experimental and theoretical studies of phonons in hexagonal InN," *Applied Physics Letters* **75**: 3297–3299.

[66] Krukowski, S., Witek, A., Adamczyk, J., Jun, J., Bockowski, M., Grzegory, I., Lucznik, B., Nowak, G., Wroblewski, M., Presz, A., Gierlotka, S., Stelmach, S., Palosz, B., Porowski, S. and Zinn, P. (1998), "Thermal properties of indium nitride," *Journal of Physics and Chemistry of Solids* **59**: 289–295.

[67] Sheleg, A. U. and Savastenko, V. A. (1979), "Determination of elastic constants of hexagonal crystals from measured values of dynamic atomic displacements," *Izv. Akad. Nauk SSSR. Neorg. Mater.* **15**: 1598–1602.

[68] Wang, K. andReeber, R.R. (2001), "Thermal expansion and elastic properties of InN," *Applied Physics Letters* **79**: 1602–1604.

[69] Pankove, J. I. and Schade, H. (1974), "Photoemission from GaN," *Applied Physics Letters* **25**: 3–55.

[70] Koshchenko, V. I., Grinberg, Y. K. and Demidenko, A. F. (1984), "Thermodynamic properties of AlN (5–2700 K), GaP (5–1500 K), and BP (5–800 K)," *Inorganic Materials*, **20**: 111787–90.

[71] Kim, J. G., Frenkel, A.C., Liu, T. and Park, R. M. (1994), "Growth by molecular beam epitaxy and electrical characterization of Si-doped zinc blende GaN films deposited on beta -SiC coated (001) Si substrates," *Applied Physics Letters* **65**: 91–93.

[72] Slack, G. A., Tanzilli, R.A., Pohl, R.O. and Vandersande, J.W. (1987), "The intrinsic thermal conductivity of AlN," *Journal of Physics and Chemistry of Solids*, **48**: 641–647.

[73] Kirchner, V., Heinke, H. and, Hommel, D. (2000), "Thermal expansion of bulk and homoepitaxial GaN," *Applied Physics Letters* **77**(10): 1434–1436.

[74] Leszczynski, M., Suski, T. and Teisseyre, H. (1994), "Thermal expansion of gallium nitride," *Journal of Applied Physics* **76**(8): 4909–4911.

2

Fundamentals and Development Trends of High Power LED Packaging

2.1 Brief Introduction to Electronic Packaging

LED packaging has many similarities with electronic packaging. We will first present the fundamentals of electronic packaging before entering into the relevant topics on high power LED packaging.

2.1.1 About Electronic Packaging and Its Evolution

Electronic packaging is defined as a system assembly that does not include a chip or chips in terms of the system in packaging (SiP) definition. In general, an electronic packaging consists of four major functions: powering, signal distribution, mechanical protection, and thermal management. There are many types of electronic packaging, depending on the number of input/outputs (IOs) and applications. Electronic packaging can be in a single module or in a system and has been evolving rapidly. Many excellent books in packaging have been published [1,2] but there is no intention of reviewing them here. However, it is worthwhile to point out the rich development of electronic packaging by presenting a roadmap of single chip integrated circuit (IC) packaging development in Figure 2.1. Figure 2.2 presents the latest development in IC packaging: through silicon via (TSV) based three-dimensional (3D) packaging.

Due to the introductory nature of this section, we only focus on a popular technology, flip-chip, to show the evolution of the materials and processes involved. Figure 2.3 shows an IC flip-chip packaging on an FR4 board, demonstrating a two-level packaging, in a typical control board for an engine control system.

Figure 2.4 shows a cross-sectional view of a flip-chip packaging, showing clearly the local features of flip-chip interconnects. In this figure, the passivation layer provides the protection for the silicon chip which is done on the wafer level. Solder mask is a polymer used to protect

---

LED Packaging for Lighting Applications: Design, Manufacturing and Testing, First Edition. Sheng Liu and Xiaobing Luo.
© 2011 Chemical Industry Press. All rights reserved. Published 2011 by John Wiley & Sons (Asia) Pte Ltd.

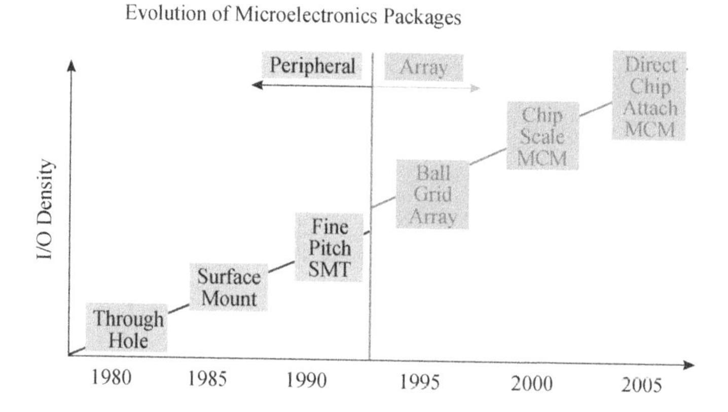

Figure 2.1 Single chip packaging evolution. (Courtesy of Daniel X.Q. Shi, The Hong Kong Applied Science and Technology Research Institute Company Limited.)

the surface conductors on the printed board from moisture, and is also used to mask off the outer areas of the board where solder is not required and provides an excellent protective coating over fine conductors on the printed board. Underfill is a highly filled polymer to provide an additional support to the solder ball interconnect. It is noted that material selections and combinations need precision engineering and process mechanics modeling.

Figure 2.5 shows a flowchart for typical flip-chip packaging processes, presenting how packaging is done by dicing, die attaching, soldering, conventional underfilling, and so on.

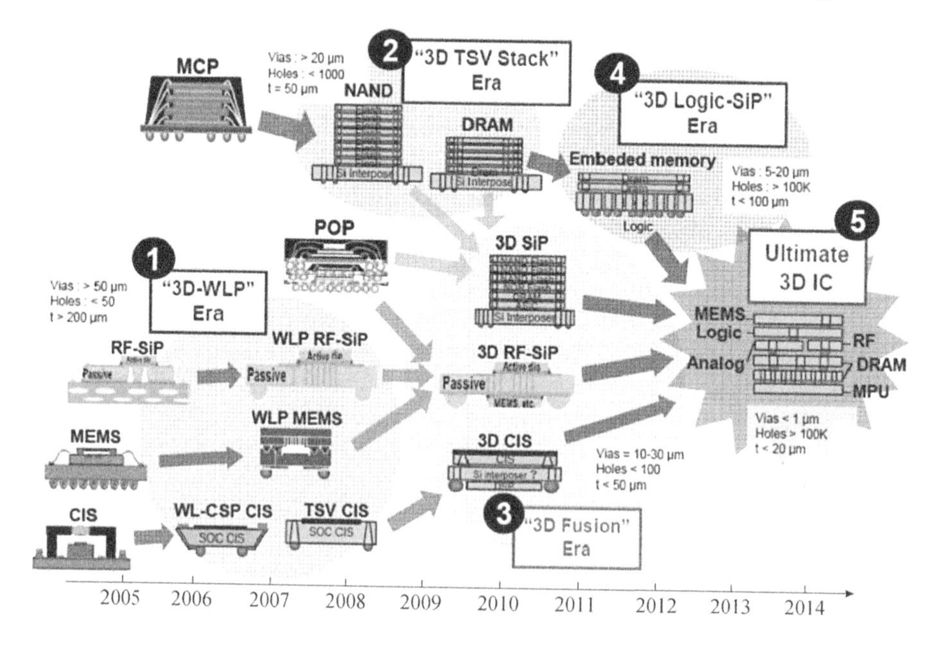

Figure 2.2 Roadmap for three-dimensional interconnect and packaging by Yole Development, 2007. (Courtesy of Daniel X.Q. Shi, The Hong Kong Applied Science and Technology Research Institute Company Limited.)

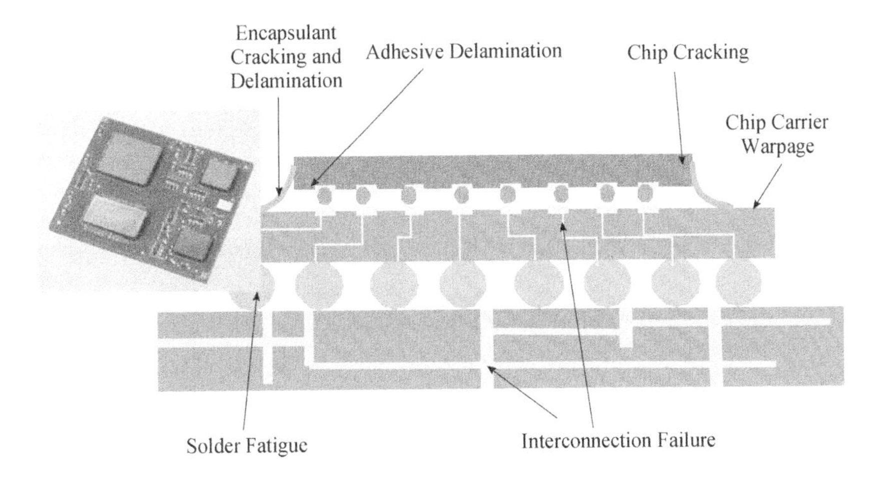

Figure 2.3 An IC flip-chip packaging on FR4 board.

In the first level flip-chip packaging, tiny solder balls link the electrodes on the active side of a chip and those pads on the substrate (FR5) to form the first chip-substrate level interconnects. Those solder balls from the FR5 to the FR4 form the second level interconnects. Packaging does not handle chip-level aluminum or copper interconnects. Due to the various materials with significant local and global hydro-thermal mismatch (that is, different thermal coefficients of expansion and moisture coefficients of expansion), and the various processes which will also involve those materials and subassemblies experiencing various temperature, moisture, and mechanical loading (pressure, handling, and so on), deformation will be induced. Warpage and co-planarity will become common issues in both device level, which may affect the subsequent assembly processes such as soldering, and yield could be a serious issue. Stresses can also be induced in bulk materials, along interfaces and at corners, which will make the materials and interfaces fail, or result in initiate voids, and partial or full debonding along interfaces. In the

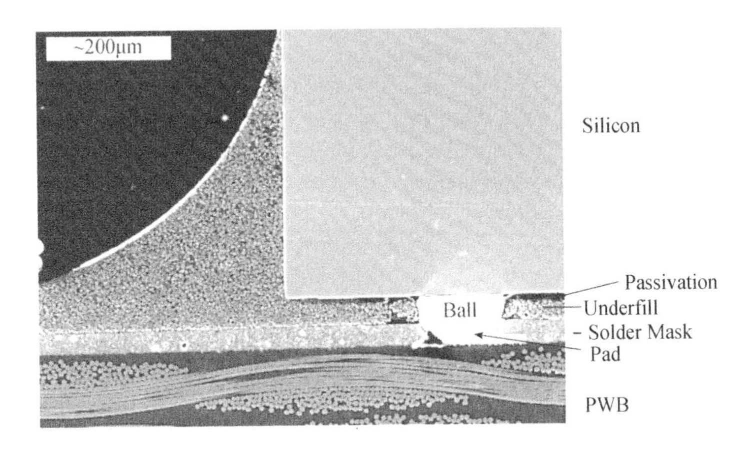

Figure 2.4 A local view in SEM of a typical flip-chip packaging.

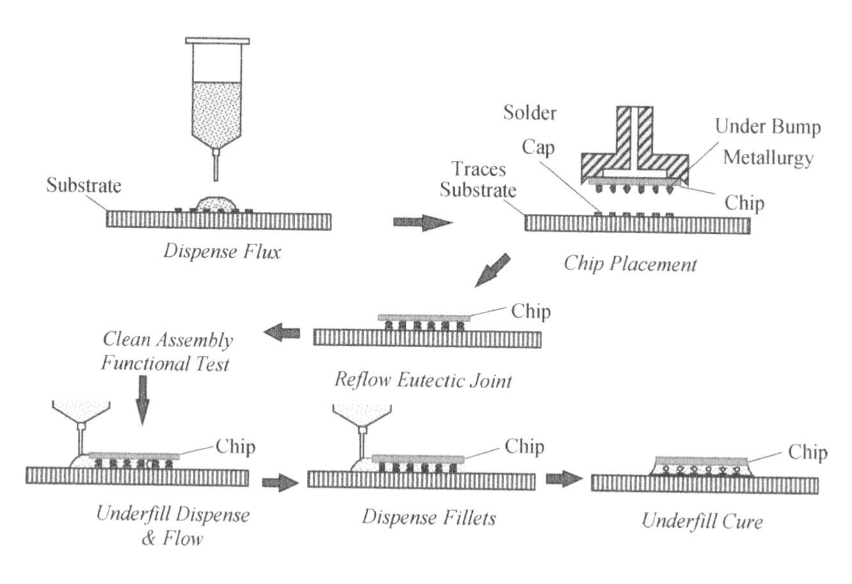

Figure 2.5 A conventional flip-chip packaging processes sequence. (Courtesy of C.P. Wong, Georgia Institute of Technology, USA.)

early stage of processes development, contaminations easily occur, which will make delaminations exist after the molding, die attaching, laminating, and so on. Moisture will easily penetrate into the bulk and interfaces, failure can easily occur during the reflow process, causing a famous pop-corning problem. Figure 2.3 also presents some typical failure modes such as die cracking, adhesive delamination, encapsulant cracking and delamination, soldering fatigue, chip carrier warpage, and substrate level interconnect failure.

In addition to the processes involved in making IC packaging, design issues have to be addressed. For instance, electrical design deals with maintaining signal integrity, one of the major bottlenecks for enabling reliable systems. In addition to deformation and the stresses mentioned above, electromigration by electron wind and thermal stress gradient is another bottleneck. Thermal management also becomes an issue, as the chips get very hot, which will degrade electrical property, radio frequency (RF) property, and optical property, naturally resulting in a coupling between them. With the increasing frequency in RF circuits, coupling for multiple physical variables exists and co-design for all these variables is essential for complex systems. It is also equally important that the designs need validation by tools, some of which are in place, while others may still need development. All these issues have been discussed in a number of books and one such book is by Suhir *et al.* [2].

2.1.2 *Wafer Level Packaging, More than Moore, and SiP*

In our definition above, packaging is defined as a system itself and it may be used as a device, a module, or a system. In the concept of SoC (System on Chip), it is desirable to package all the functions on the chip and the industry is also driving the feature size on the chip to be in nano scale, as shown in the integrated technology roadmap for semi conductors [3]. However, due to the many constraints on the system, it is impossible to integrate all the functions on the chip due

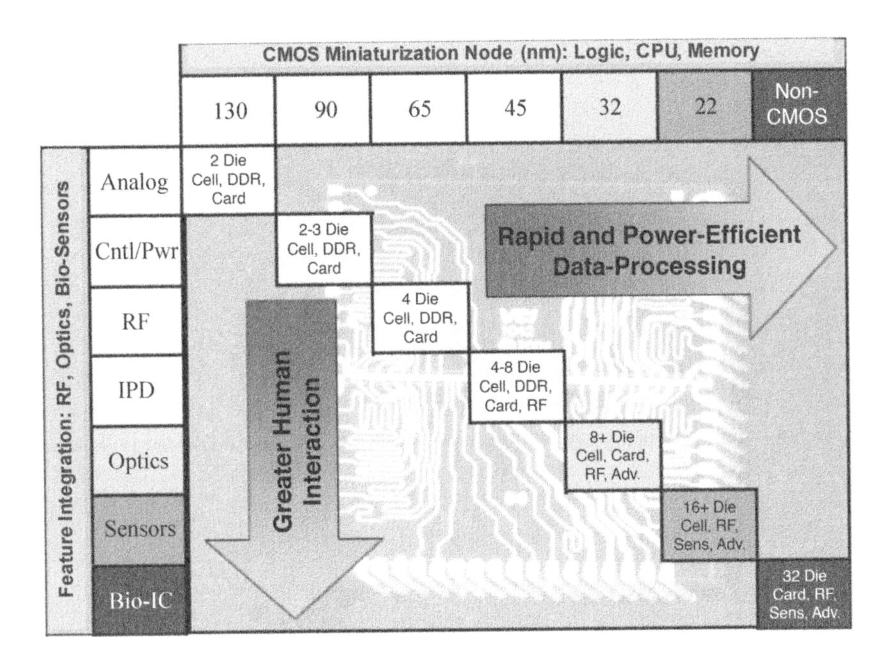

Figure 2.6 More Than Moore and More Moore from VCI. (Courtesy of Daniel X.Q. Shi, The Hong Kong Applied Science and Technology Research Institute Company Limited.)

to different materials, different processes, and even different wafer sizes involved. The concept of More Than Moore was then proposed, which is naturally integrating various functions in the packaging level, and is another interpretation of SiP, as shown in Figure 2.6.

It is naturally desirable to package a system on the wafer scale to reduce the number of processes. Definition of wafer level packaging is one in which the chip and packaging are fabricated and tested on the wafer level prior to singulation. SiP is more general than wafer level packaging. Significant advances in electrical, mechanical, and thermal performance can be enhanced with a much more reduced size in final modules and systems. For instance, one can achieve the smallest IC package size as it is really chip scale packaging (CSP), lowest cost per IO because the interconnects are all done at the wafer level in one set of parallel steps, lowest cost of electrical testing as this is done at the wafer level, and enhanced electrical performance due to the short interconnects. Traditionally, millions of steps for regular IC packaging could be reduced from IC packaging level to wafer level packaging to no more than a dozen steps. With the emerging of TSV coupled with mechanical grinding, CMP (chemical polishing), and wafer bonding, SiP indeed will perform best and the rate of integration is even more significantly enhanced.

2.2 LED Chips

2.2.1 Current Spreading Efficiency

Modeling and simulation of current spreading and injection of the GaN-based LED chip is relatively new. Hyunsoo Kim *et al.* used resistive network coordinating with a constant voltage drop model of the diode to study the transverse current spreading of a GaN-based LED, and pointed out that the phenomenon of current crowding exists near the N and P pads and the distance of current transverse spreading is similar to an exponential function [4]. Huapu Pan *et al.*, after simplifying the LED model based on the basic equation of an electrostatic field, established the LED's current spreading model. They proposed parameters and standards of quantitative evaluation for its characteristics, and pointed out that current spreading is more uniform and series resistance is smaller for an LED using interdigitated mesa structure compared to that using conventional mesa structure [5]. Study associated with simulation software SimuLED, developed by STR Company, on the current spreading in the LED chip is fundamental and comprehensive, covering the energy band structure, the calculation of internal quantum efficiency, the distribution of current density, the heat conduction equation, and the temperature changes, which simulates a variety of electro-optical thermal situations of LED chip [6].

Theories behind SimuLED software are briefly introduced here. The first module of this software is SiLENSe, used to calculate the energy band structure, carrier distribution, recombination situation, and relationship between voltage and current, and so on of a one-dimensional structure. The main physical equation involved is as follows.

Poisson equation:

$$\frac{d}{dz}\left(P_z^0 - \varepsilon_0\varepsilon_{33}\frac{d\varphi}{dz}\right) = q\left(N_D^+ - N_A^- + p - n\right) \tag{2.1}$$

where P_z^0 is the spontaneous radiation polarization vector, ε_0 is the vacuum dielectric constant, ε_{33} is the corresponding static dielectric constant, N_D^+ is the number of ionized donors, N_A^- is the number of ionized acceptors, p and n are the numbers of holes and electrons respectively.

Under the steady-state condition, the flow of electrons $\vec{J}_n$ and holes $\vec{J}_p$ is determined by the equation of continuity:

$$\nabla \cdot \vec{J}_n - qR = 0, \nabla \cdot \vec{J}_p + qR = 0 \tag{2.2}$$

where R is the recombination rate of non-equilibrium carrier; the current densities of the flow of electrons and holes are respectively:

$$\vec{j}_n = -q\vec{J}_n, \vec{j}_p = +q\vec{J}_p \tag{2.3}$$

Carrier's recombination rate is:

$$R = R^{SR} + R^{dis} + R^{A} + R^{rad} \tag{2.4}$$

Radiative recombination rate of non-equilibrium carrier is:

$$R^{rad} = B \cdot np \cdot \left[1 - \exp\left(-\frac{F_n - F_p}{kT}\right)\right] \tag{2.5}$$

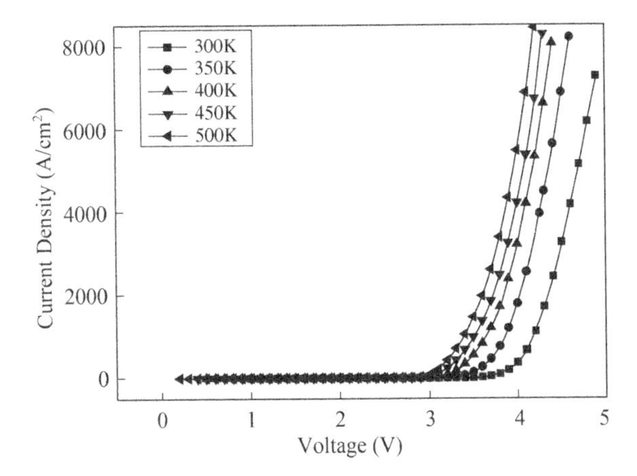

Figure 2.7 Relationship between voltage and current density of LED one-dimensional structure at different temperatures. (Courtesy of Semiconductor Technology Research Inc.)

where B is the temperature-dependent recombination constant, R^{SR} is the Shockley- Rayleigh nonradiative recombination rate induced by point defects, R^{dis} is nonradiative recombination rate induced by line defects, R^{A} is the Auger nonradiative recombination rate.

The internal quantum efficiency is:

$$IQE = R^{rad}/R \tag{2.6}$$

The first software module's objective is to obtain the relationship between the voltage and the current of the one-dimensional structure, as shown in Figure 2.7, as well as the relationship between current density and internal quantum efficiency as shown in Figure 2.8. The data of the following two diagrams are from examples provided by the software.

Figure 2.7 is a typical curve of the relationship between the voltage and the current of the diode. When the voltage is low, the diode is not functional and the current is very low. When the voltage increases to a certain level, the LED is functional, and then a very small change in voltage can lead to a large change in current. In addition, the entire curve shifts to the left when the diode working temperature rises. That is, at the same voltage, the current will be greater, indicating when the working temperature rises, carriers have more energy and move more actively.

Figure 2.8 shows the relationship between the current density and the internal quantum efficiency of the diode. It can be seen that the internal quantum efficiency of the diode will increase with the increase of current density, then, as it reaches a maximum value, it will drop with the increase of the current density. In addition, when the working temperature of the diode rises, the point of the maximum internal quantum efficiency moves in the direction whereby the current density is greater, and the maximum internal quantum efficiency becomes smaller which also shows that with rising temperature, the non-radiative recombination becomes more noticeable, reducing the internal quantum efficiency.

As can be seen from Table 2.1, the current density is relatively small when the internal quantum efficiency of the diode is maximum. When the diode works at 300 K, the current

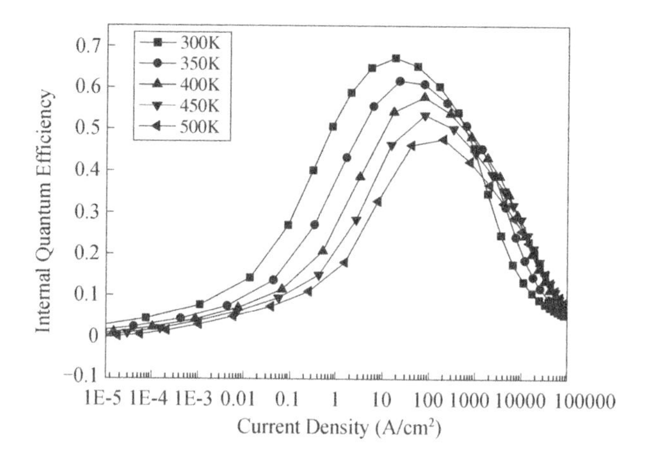

Figure 2.8 Relationship between current density and internal quantum efficiency of LED one-dimensional structure at different temperatures. (Courtesy of Semiconductor Technology Research Inc.)

density of the maximum internal quantum efficiency is 18 A/cm^2, that is, a current of 180 mA can be injected into a chip of 1 mm × 1 mm, however, a current of 350 mA is usually injected into the chip of the same size in normal working condition. Therefore, usually the LED chip works under a relatively large current, which means that the corresponding internal quantum efficiency will be reduced, that is why, in order to maximize the use of LEDs, we should make the current of LED chips be distributed as uniformly as possible and make the LED work at a low temperature.

The second software module is a mixed three-dimensional model based on the calculation results of the first module. With the relationship between the voltage and current as well as the relationship between the internal quantum efficiency and the current density of a one-dimensional structure, together with certain boundary conditions and electrostatic field equation, we can simulate the shape of the specific mesa structure, ohmic contact, and the current distribution and recombination situation of three-dimensional chips in the chip structure. In addition, the software couples related heat conduction equation and sets certain heat conduction boundary conditions and the heat conduction coefficient of the material, then temperature distribution inside the chip can be calculated.

Table 2.1 Maximum IQE and corresponding current density of the diode working at different temperatures

Temperature (K)	Maximum IQE	Current Density of Maximum IQE (A/cm^2)
300	0.67217	18.049
350	0.6173	22.203
400	0.57884	75.297
450	0.53594	77.837
500	0.4775	200.53

The basic equation of a three-dimensional model:

$$\overset{v}{j} = (\hat{\sigma}/q)\nabla F \tag{2.7}$$

where $\hat{\sigma} = q\hat{\mu}N_i^{\pm}$ is the conductivity tensor of material, where q is the electron, $\hat{\mu}$ is the mobility tensor, $N_i^{\pm}(i = D, A)$ is the ionization rate of dopant concentration, F is the quasi-Fermi level corresponding to the carrier, and $\overset{v}{j}$ is the current density.

Heat conduction equation:

$$-\nabla(\lambda\nabla T) = q \tag{2.8}$$

where T is the temperature, λ is the heat conductivity, and q is the local heat source dependent on the current density.

Based on the above basic equations, we will be able to calculate the current and voltage distribution within the LED chip of a three-dimensional structure, recombination situation of the active layer, and the temperature distribution of the entire chip. At this point, the current spreading model of LED chips has been completed.

2.2.2 *Internal Quantum Efficiency*

Internal quantum efficiency (IQE) is the ratio of electron hole pairs involved in radiation recombination to the injected electron hole pairs. The current injected in LED is divided into current component involved in radiation recombination, current component involved in non-radiation recombination and leakage current caused by the structure. Leakage current can be reduced through the optimal design of the structure; and when low current is injected, the leakage current can be reduced to a very low degree where its impact can be neglected; both the quantum-confined Stark effect and carrier localized effect affect radiation recombination. Therefore, the methods which can effectively improve the radiation recombination and reduce the non-radiation recombination can improve the IQE of LED. As an important parameter of LED devices, reflecting the overall quality of epitaxial growth, IQE indicates the overall quality of epitaxial growth; thus it is especially important that it is measured. IQE is closely related to the injected current and temperature, and the small current is injected while measuring IQE.

One method is to indirectly obtain IQE by measuring the external quantum efficiency of the LED [7,8]. Measuring the external quantum efficiency of the LED and modeling its light extraction efficiency, the ratio of the former to the later is the IQE of the LED. However, the light extraction efficiency is largely affected by the geometrical structure of the chip, refractive indexes of the materials and various losses. The value obtained by modeling is only an estimation, and therefore not very accurate. Therefore, there exists a big variation for IQE obtained by this method and the true value.

Internal quantum efficiency at the room temperature can be obtained by the ratio of resonant excitation PL efficiency at the room temperature to that at a low temperature [8–11]. This is because with the decrease of operating temperature, the non-radiation recombination centers are gradually frozen out and lose the activity; therefore, IQE gradually increases and becomes stable when the temperature has dropped to a point where the non-radiation recombination

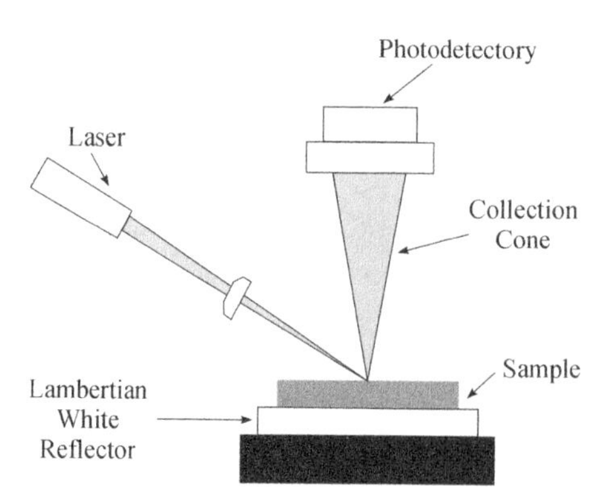

Figure 2.9 Experimental setup for photoluminescence measurements [10]. (Reprinted with permission from M. Boroditsky, I. Gontijo, M. Jackson *et al.*, "Surface recombination measurements on III–V candidate materials for nanostructure light-emitting diodes," *Journal of Applied Physics*, **87**, 7, 3497–3504, 2000. © 2000 American Institute of Physics.)

centers lose the activity completely. This method assumes that the internal quantum efficiency at low temperature is approximately 100%; this approximation is reasonable if excited by a low temperature and a low power [8]. Figure 2.9 is its experimental setup; during the low temperature PL testing, the sample is placed in the cryostat.

Martinez [12] adopted the technology of heat-pulse measurements and time-resolved photoluminescence to obtain low-temperature IQE. Martinez has proven that the technology of heat-pulse can detect the phonon in the process of non-radiation recombination, and it can also be identified in the PL spectrum. Therefore, the PL spectrum can be used to determine the relative intensity of non-radiation recombination and radiation recombination. Figure 2.10 is a schematic of its setup. Gap isolates the phonon generated by photoluminescence and obtains optical response through the PL intensity measured by the bolometer. If there is no PL photon

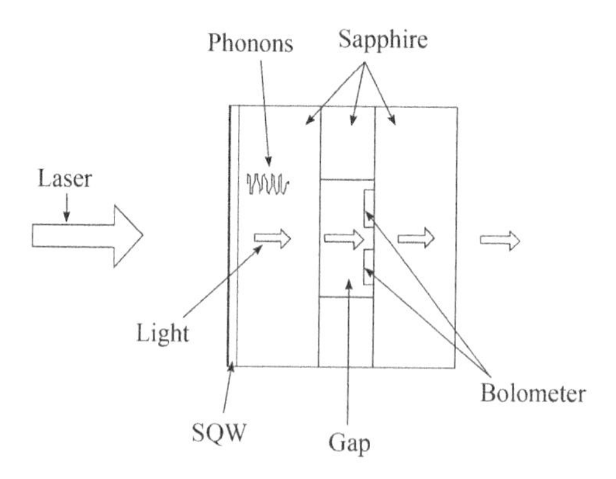

Figure 2.10 Schematic of the sample designed to isolate the optical response from the bolometer signals [12]. (Reprinted with permission from C.E. Martinez, N.M. Stanton, A.J. Kent *et al.*, "Determination of relative internal quantum efficiency in InGaN/GaN quantum wells," *Journal of Applied Physics*, **98**, 5, 053509, 2005. © 2005 American Institute of Physics.)

response or optical response measured by gap, the internal quantum efficiency is calculated with the phonon signal and the optical signal.

Non-radiation recombination of part of the injected carriers caused by lattice defects and impurities leads to lower internal quantum efficiency. Therefore, it is necessary to guarantee the quality of crystals in the process of LED material epitaxial growth and chip manufacturing and reduce the defect density. Threading dislocations are largely generated during the heteroepitaxial growth process of GaN because of a lack of homogeneous substrates. These dislocations will lead to non-radiation recombination centers and become leakage current channels, which have a great impact on the lifetime and performance of devices. The commonly used methods to reduce the dislocations are low temperature buffer layer technology and epitaxial lateral over-growth technology. The low temperature buffer layer technology is a two-step growth method which is firstly to grow a thin GaN or AlN buffer layer at a low temperature on the substrate (500–600 °C), and secondly to grow GaN at a high temperature (1100 °C or so). The epitaxial lateral growth technology is firstly to grow a GaN layer about 1–2 μm thick on the sapphire substrate, secondly to deposit a SiO_2 layer about 0.1 μm thick, thirdly to etch the SiO_2 stripe (mask), and lastly to regrow a GaN layer on it. Zang [13] and others have reported a growth technology of nanoepitaxial lateral overgrown (NELO), which makes a nanopore array on the SiO_2 film, and then GaN is grown on it. This method has effectively reduced the defect density, and improved the internal quantum efficiency. In addition, local SiO_2 nano-array structure features can be used as the light scattering points, thereby increasing the efficiency of light extraction. Wuu [14] and some other researchers have combined epitaxial lateral overgrowth (ELOG) and patterned sapphire substrate (PSS) to make UV-LED and achieved good results.

Wu *et al.* [15] have doped silicon in the barrier layer of In GaN/GaN multi-quantum well (MQW) structure to improve its internal quantum efficiency and thermal stability. As Si atoms occupy Ga vacancies, indium composition in multi-quantum wells is reduced. Xie *et al.* [16] have studied and showed that P-type doping in the barrier layer of InGaN/GaN MQW can reduce the efficiency droop, that is, under high current injection, the reduction in internal quantum efficiency can be mitigated.

2.2.3 *High Light Extraction Efficiency*

Light extraction efficiency is a key factor in influencing LED luminous efficiency. Owing to significant difference in refractive indexes for semiconductor materials and air, total reflection induced optical loss and other optical loss like Fresnel diffraction will be produced at the interface (Figure 2.11), which leads to the fact that a very small amount of light can be radiated out of the LED-emitting surface. For instance, with the refractive index being 3.5 and the total reflection angle being 16°, light extraction efficiency is only 2%. In addition, it is the fact that light is absorbed by lattice defects and different substrate materials, and is also absorbed at electrode contacts and the active layer, all contributing to the low light extraction efficiency.

After the invention of the red LED, various methods have been employed to improve the light extraction efficiency of the LED. The main methods having been tried are as follows: distributed Bragg reflector (DBR), resonant cavity, geometrically shaped chips, surface roughening, patterned substrate and photonic crystal, and so on. All those methods mentioned previously are still at the stage of experimental study but with new results obtained continuously.

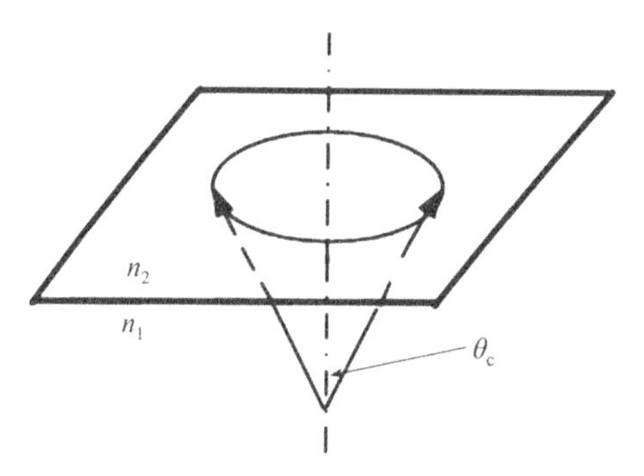

Figure 2.11 Schematic of light-emitting escape cone.

(i) Distributed Bragg Reflector (DBR)

By inserting a DBR between the active layer and the substrate, light radiating towards the substrate can be reflected back to the surface or the side, whereby light absorbed by the substrate decreases and extraction increases.

A DBR structure is composed of alternatively superimposed materials of high-refractive and low-refractive indices matching the substrate lattice (Figure 2.12). The optical path of each layer is a quarter wavelength as long as its emission wavelength is defined and its geometrical thickness can be calculated by applying the following formula [17]:

$$\begin{aligned} h_H &= \lambda_0/4n_H \cos\theta_H \\ h_L &= \lambda_0/4n_L \cos\theta_L \end{aligned} \tag{2.9}$$

where λ_0 represents the central wavelength radiating out from the active layer; n_H, n_L represent the refractive index of the materials respectively; θ_H, θ_L are two corresponding incident angles.

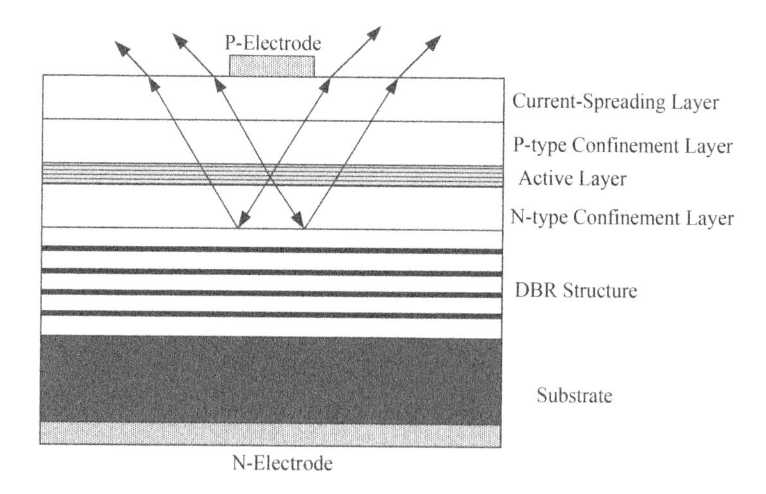

Figure 2.12 Schematic of LED chip with DBR.

Kato *et al.* [18] first applied and further developed a way to improve GaAs/AlGaAs LED grown on absorption type GaAs substrate (light extraction efficiency assisted by DBR was four times higher than the one without the assistance of DBR). Sugawara *et al.* [19] applied AlInP/AlGaInP distributed Bragg reflector to high-brightness AlGaInlP green light-emitting diodes. Chen *et al.* [20] adopted both DBR and charge asymmetric resonance tunneling structures in developing high-efficiency InGaN/GaN MQW green light-emitting diodes. This led to a rise in external quantum efficiency reaching 11.25% under 20 mA injection current. Nakada *et al.* [21] improved the external quantum efficiency of blue LED from 0.16% to 0.23% by using GaN/AlGaN DBR grown on sapphire substrate through MOCVD.

DBR reflectivity is directly related to the contrast of refractive index of the materials and the period numbers. The larger the numbers and the greater the difference in refractive index will be, the greater the peak value of DBR reflection. As shown in Equation 2.9, DBR only works in the limited range of incident angle with a specific wavelength, and cannot achieve good isotropic optical transmission with a wide spectrum of LED. Moreover, the growth process of the Bragg reflecting layer is complex and the cost is high.

(ii) Resonant Cavity

Cavity LED is structured on the theory of the Fabry-Perot interferometer. Based on the theory, the active layer is placed in a planar micro-cavity (Fabry-Perot cavity) composed of two parallel mirrors with a distance in the size of wavelength. Only a resonant fundamental wave and higher order harmonics can propagate along the optic axis of the micro-cavity. At the resonant wavelength, the emission intensity increases significantly. And if the active layer is placed in the locations of antinodes, the emission intensity can be doubled.

The first resonant cavity LED, developed by Schubert *et al.* [22], was formed by inserting GaAs/AlGaAs between a semi-transparent Ag/CdSnOx reflector (reflectivity ≈ 0.9) and a AlAs/AlGaAs distributed Bragg reflector (reflectivity ≈ 0.99) (Figure 2.13). The depth of the sandwich structure was λ approximately and GaAs active layer was on an antinode. As a result, the emission intensity is 1.7 times greater than that of ordinary LEDs. In order to improve the

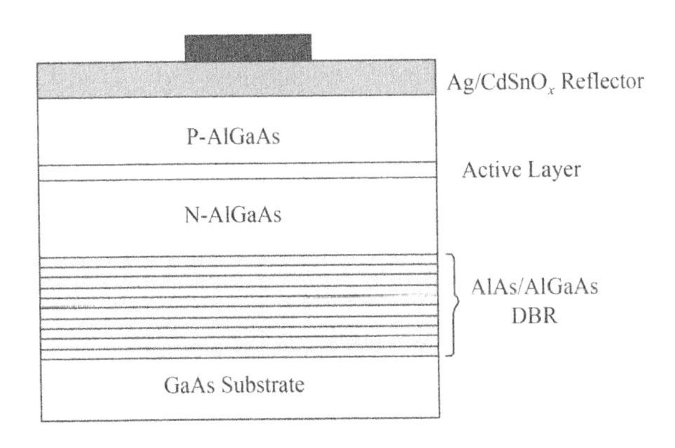

Figure 2.13 Schematic of resonant cavity LED chip structure [22]. (Reprinted with permission from E. F. Schubert, Y.-H. Wang, A.Y. Cho *et al.*, "Resonant cavity light-emitting diode," *Applied Physics Letters*, **60**, 8, 921–923, 1992. © 1992 American Institute of Physics.)

performance of the resonant cavity LED, Dill *et al.* [23] adopted the tuning resonant cavity mode to make external quantum efficiency reach 14.8%. Neve *et al.* [24] utilized a photonic cycle and guided mode propagating transversally between the mirrors to make external quantum efficiency reach 20%. Benisty *et al.* [25–26] analyzed light emission of Fabry-Perot modes, leaky modes, and guided modes based on the theories and simulation, and analyzed as well the performance of the practical planar micro-cavity by introducing photonic cycling effect.

(iii) Geometrically Shaped Chips

Most LEDs are of plain rectangle structure. Due to restrictions of the geometry, it is difficult to obtain high light extraction efficiency for plain rectangle LEDs. Therefore, the method of using non-rectangle and non-plain structures for extracting light has been under consideration.

In 2000, Schmid *et al.* [27] introduced an LED chip structure with lateral tapers whose external quantum efficiency was up to around 45%. InGaAs quantum well structure was grown by MBE techniques. Lithography process and chemically assisted ion beam etching were utilized to make it disk-like with gradually decreasing thickness from center to edge, and then copper reflector was evaporated onto its surface.

Lee and Song *et al.* [28] suggested a polyhedron chip: top and bottom surfaces are parallel, cross-section is parallelogram or triangle and the side is inclined. Because each internal reflection decreases the incident angle by α, the photons propagating to horizontal surface are bound to emit. Likewise, if the side is inclined, most photons can emit to whatever direction they travel (Figure 2.14). By adopting such a structure, the efficiency of light extraction rises by 120% compared with that of rectangle structure.

Krames *et al.* [29] introduced a truncated-inverted-pyramid AlGaInP/GaP LED structure, by whose shape the light reflected from the side to the top or vise versa was reoriented into the small incident angle (Figure 2.15). When the wavelength is 611 nm, its highest luminous efficiency is more than 100 lm/W. And when the wavelength is 652 nm, the external quantum efficiency is up to 50%.

As for GaN-based LED grown on sapphire substrate, it is difficult to make an inclined plane due to its great hardness. But recently, there have been some researchers who have successfully applied this method to the GaN base LED chip. Kao's research team improved the light output efficiency by 70% by etching sidewalls of the GaN epitaxial layer inclined at an angle

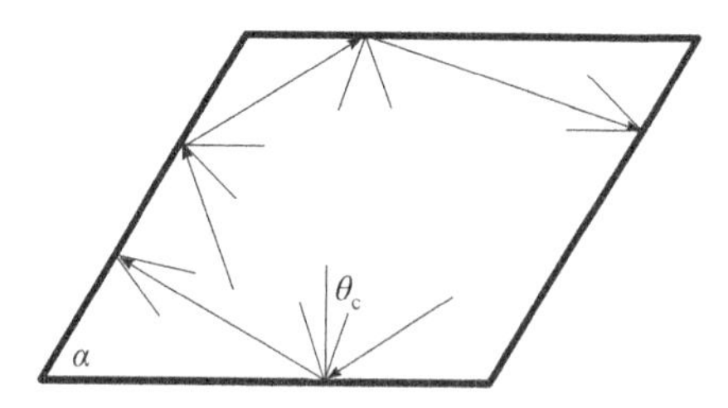

Figure 2.14 Light propagation path in cross-section of parallelogram [28]. (Reprinted with permission from S.J. Lee and S.W. Song, "Efficiency improvement in light-emitting diodes based on geometrically deformed chips," SPIE Conference on Light-Emitting Diodes: Research, Manufacturing and Applications III, *Proceedings of SPIE Vol. 3621*, 237–248, 1999. © 1999 SPIE.)

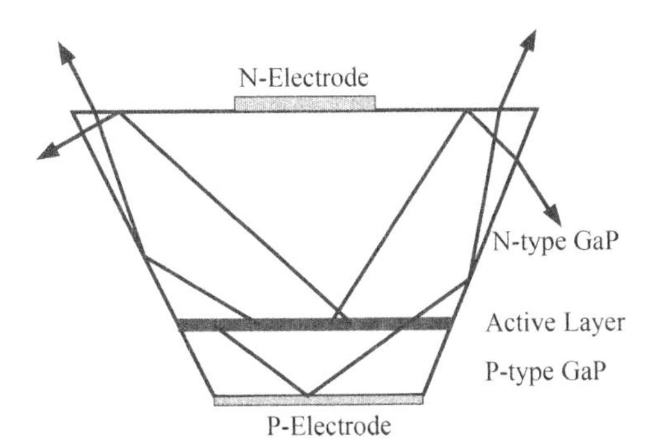

Figure 2.15 Schematic of truncated-inverted-pyramid LED structure [29]. (Reprinted with permission from M.R. Krames, M. Ochiai-Holcomb, C. Carter-Coman and E.I. Chen, "High-power truncated-inverted- pyramid (AlxGa1-x) 0.5In0.5P/GaP light-emitting diodes exhibiting >50% external quantum efficiency," *Applied Physics Letters*, **75**, 16, 2365–2367, 1999. © 1999 American Institute of Physics.)

of 22° [30]. While Chang *et al.* etched GaN sidewalls of LED chip to make sidewalls textured, by which way the light efficiency of the chip was easily enhanced by 10% [31].

(iv) Surface Roughening

Surface roughening is mainly applied to the transparent current spreading layers of P-side-up chips and N-type layers of vertical chips. By roughening indium tin oxide (ITO) layer, Yao *et al.* raised the light extraction efficiency by 70% compared to a non-roughening chip [32]. As for the vertical chip, its sufficient space ensures surface roughening because its uppermost layer is usually N-GaN whose thickness is normally up to 3 μm, and has also attracted lots of research efforts done in this area. Lee *et al.* proposed in their research that as to the vertical chip, the lower surface of P-GaN layer was first roughened to be a diffuse reflecting surface and then bonded be to Si substrate; meanwhile the upper surface of N-GaN layer was roughened to be a diffuse transmitting surface. In this way, the external quantum efficiency reached around 40%, a result indicating an efficiency enhancement by 136% was achieved compared to that of the conventional chips [33]. Horng *et al.* also greatly enhanced the light extraction efficiency by roughening both the P-GaN and N-GaN. Compared to the P-GaN roughened chip, the light extraction efficiency of achip with two roughened sufraces had a further rise of 77% [34].

(v) Patterned Substrate

Most methods of enhancing the efficiency of LED external quantum have been based on the principle of enlarging the angle of incidence. But light absorption by lattice defects can also decrease the external quantum efficiency. In GaN-based LEDs, the density of dislocation defects is very great because of the big difference of thermal expansion coefficients and lattice constants between GaN epitaxial layer and sapphire substrate. In order to solve this problem, a patterned substrate is proposed (Figure 2.16). Research efforts have shown that this method can effectively reduce the density of dislocation defects and increase the escape of photon emitted in the active layer due to the change in geometry of some structures, and finally enhance the LED external quantum efficiency.

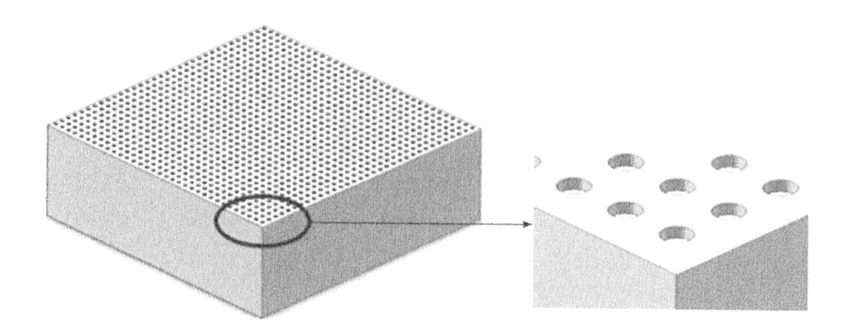

Figure 2.16 Schematic of patterned substrate structure.

Lee *et al.* adopted patterned substrate technique in their study and the external quantum efficiency increased by 15% [35]. Wuu *et al.* applied patterned substrate technique to the near-ultraviolet LED, and found that the luminescence intensity of patterned substrate LED was enhanced by 63% compared to the conventional LED [36]. Wang *et al.* also designed a patterned substrate structure that raised light extraction efficiency by 30% [37].

(vi) Photonic Crystal

It has been predicted that the photonic crystal structure could enhance the light extraction efficiency most, to even more than 90%, although no practical results have been achieved [38]. Photonic crystal is a dielectric structure whose refractive index periodically varies in space, with the variation period having the same order of magnitude as the light wavelength. The fundamental feature of photonic crystal is that it has a photonic band-gap which can control the spontaneous emission. When the spontaneous emission frequency of atoms in photonic crystals is just in the band gap, the number of the photon state at this frequency is zero, and the probability of spontaneous emission is also zero, so that the spontaneous emission is fully suppressed. To strengthen spontaneous emission, the number of the photon states should be increased. Kim and his colleagues used electron beam lithography to make two-dimensional photonic crystal patterns on the surface of GaN. When the pattern period was 500 nm, the output optical power doubled [39]. Truong *et al.* also mentioned in their latest paper that they succeeded in raising the light emitting power of an LED chip by 80% by utilizing the nano-imprinting technique, and the experiment result was verified in the simulation [40]. Kang *et al.* have been researching the photonic crystal and achieved some good results in proposing an integrating method of laser lift-off of sapphire and patterned GaN, which made the light extraction efficiency of the LED chip improve by 40–100% [41].

2.3 Types and Functions of LED Packaging

LED packaging technology was first developed with the invention of the early red LED. The LED in the early stage was mostly used in the field of indicators because of its low luminance. The input electrical power of this type LED is generally lower than 0.1 watt, which is consequently called a low power LED. With the emergence of the blue LED, especially the discovery of the technology that uses the blue light to stimulate YAG phosphor to produce white

LED, the method that increases the size of LED chips and the drive current to enhance the brightness of LEDs has become an important direction of LED packaging. This promotes the development of high power LED, for which the input current is in general more than 100 mA. However, high power packages in the market are those with electrical power being normally larger than 1 W and the chip size being larger than 1 mm × 1 mm.

2.3.1 Low Power LED Packaging

As shown in Figure 2.17, the low power LED currently refers to the LED packaged with epoxy resin. The chip is firstly bonded onto the leadframe with solder or silver paste, and the top of the chip is bonded on the other part of the leadframe with gold wire or aluminum wire. Then, the whole leadframe is immersed into the mould filled with epoxy resin. After the curing of epoxy resin by raising the temperature, the leadframe is demoulded to obtain the fully packaged LED. In the leadframe of the LED, there is generally a small reflector that is used to make the lights emitting from the side of the chip converge to the central portion. Because the low power LED is mostly used in indicators and it has no special requirements on intensity distribution, the most common shape of the epoxy resin lens is hemispherical. Hemispherical lens can make the lights emitting from the chips converge close to the axes of the lens, which can significantly improve the indication effect of LED. Since the epoxy resin materials in low power packaging are very easy to be yellowed under the ultraviolet radiation and the light transmittance of the material can be decreased due to the influence of high temperature, the optical attenuation of the low power LED is severe after long time operation. The elastic modulus of the epoxy resin materials is relatively high, therefore, under the condition of high temperature and high moisture, the gold wire or aluminum wire at the top of the chip may fracture under high stress, resulting in failure in the LED packaging. The other reason for the short life of the low power LED is the small size of the metal leadframe. The metal material is generally copper or aluminum. Since the heat generated by LED chips cannot be effectively conducted out through the shelf, the accumulated thermal deformation and thermal stress may lead to the failure of the LED.

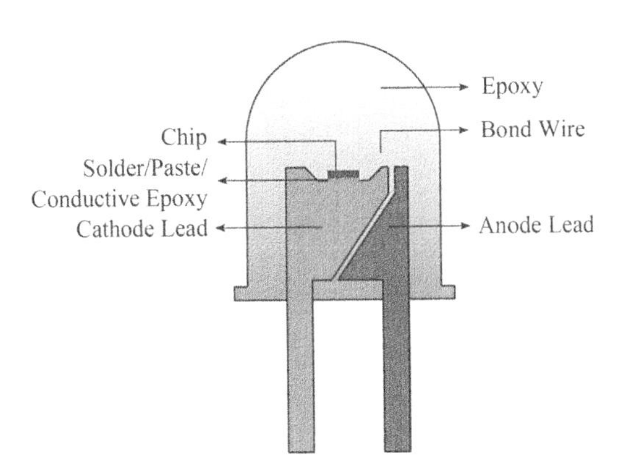

Figure 2.17 Schematic of low power LED packaging.

2.3.2 High Power LED Packaging

High power LED generally refers to the LED module for which the single chip size is not less than 1 mm × 1 mm and the drive current of which is at least 350 mA, although some people prefer to lower down this current to be 100 mA. The emergence of the high power LED proposes great challenges for the traditional low power packaging. Because the traditional low power LED has no serious problems in heat dissipation, LED packaging technology has made little progress in the past decades. With the invention of high power LEDs, the problem of LED heat dissipation has become more serious. If the heat of the chip cannot be conducted to the environment rapidly, the junction temperature of the chip will rise quickly. The over high temperature may lead to a large decrease of the electron hole recombination efficiency in the active layer of the chip and affect the internal quantum efficiency. Furthermore, the mismatch of the thermal expansion coefficient between the chip and the leadframe may lead to a significant increase in the internal stress of the chip. In 1998, Lumileds proposed the leadframe plastic packaging structure (Luxeon) used in high power LED packaging, as shown in Figure 2.18.

In the Luxeon structure, firstly, the chip is bonded through solder onto the large size metal heat slug. The material for the heat slug is generally copper with high thermal conductivity. Secondly, through gold wire bonding, the electrodes of the chip are respectively connected to the metal frame on both sides. Then, finally embedding the lens made by polycarbonate into the plastic frame, and filling the lens with silicone gel to protect the chip and gold wire. Since the large size copper heat slug is adopted, the thermal resistance of the LED with Luxeon packaging is reduced significantly, which consequently improves the heat dissipation ability in LED packaging. Luxeon packaging changes the LED intensity distribution mainly through changing the exterior shape of the lens made by polycarbonate. Figure 2.19 shows the shape of three major types of lenses.

Although Luxeon packaging has solved the problem of heat dissipation in the traditional low power packaging, a large volume metal slug leads to an increase in the volume and weight of the whole LED packaging modules. All these are unfavorable in reducing the size/profile and weight of LED lamps and improving the product portability, especially of the road lights and tunnel lights for which the total electrical power is over 100 W. Therefore, based on Luxeon packaging, Osram has proposed the GoldenDragon-series that are based on leadframe packaging. Cree has proposed the XLamp-series that are based on the technology of the chip on ceramic board. After the Luxeon-series, Lumileds also proposed the Rebel-series based on the technology of the chip on ceramic substrate. Figure 2.20 shows the four types of packages mentioned above.

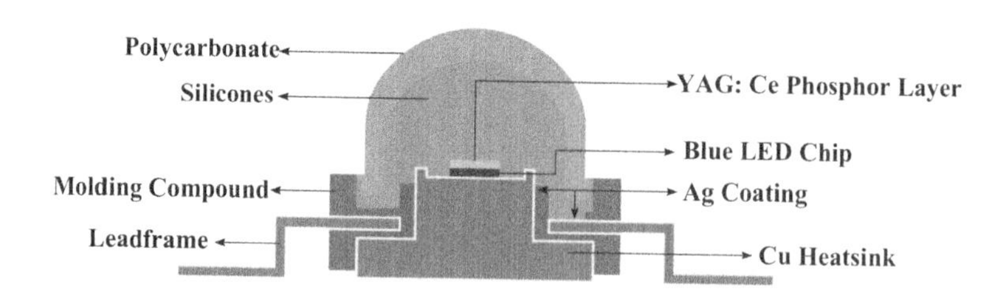

Figure 2.18 Schematic of a high power LED packaging.

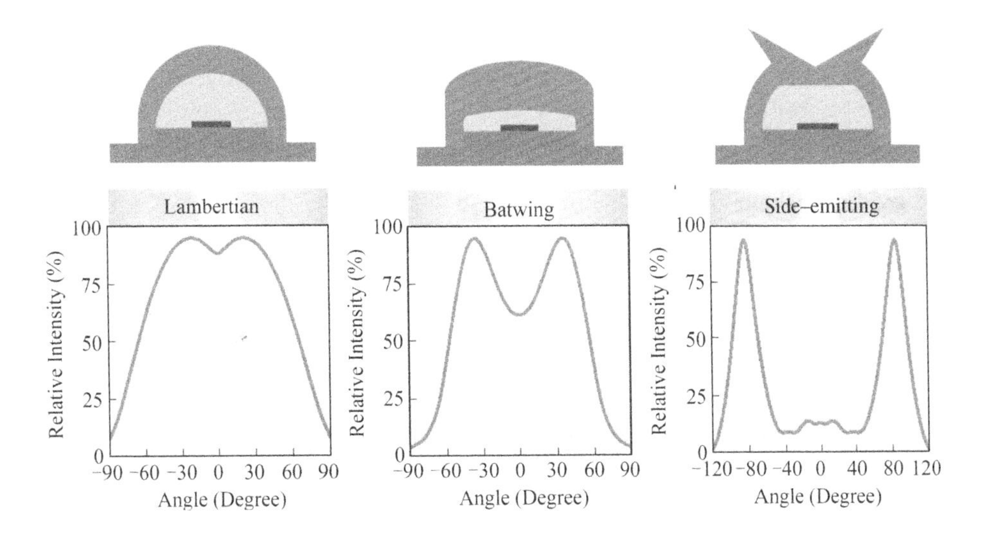

Figure 2.19 Lens used for a high power LED packaging.

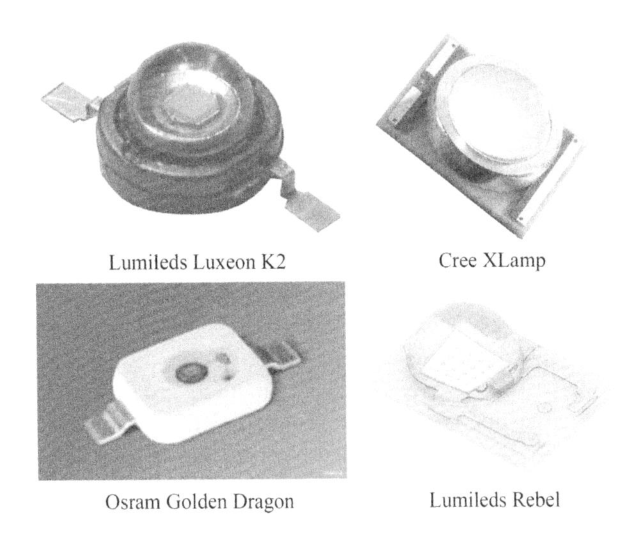

Figure 2.20 Typical high power LED package products.

2.4 Key Factors and System Design of High Power LED Packaging

Since the first light-emitting diode (LED) was invented by Holonyak and Bevacqua in 1962 [42], LEDs have made remarkable progress in the past four decades with the rapid development of epitaxy growth [43], chip design and manufacture [44], packaging structures, processes, and materials. In the early 1990s, Nakamura successfully grew blue and green GaN-based LEDs on sapphire substrate [45,46]. In 1995, as part of the method whereby YAG was applied to the blue LED chip, Ce phosphor to generate white light was discovered and first enabled the white LED to be commercially available [47,48]. White LEDs have superior characteristics such as high efficiency, small size, long life, durability, low power consumption, high reliability, and so on [49]. The market for white LED is growing rapidly in various applications such as large size flat panel backlighting, road lighting, vehicle forward lamp, museum illumination and residential illumination [50–52]. It has been widely accepted that solid state lighting, in terms of white LEDs, will be the fourth illumination source to substitute incandescent lamps, fluorescent lamps, and high pressure sodium lamps [53].

The emerged and emerging illumination applications require that white LEDs provide higher performance and reduce the cost in an acceptable range for consumers. Nowadays, the highest luminous efficiency of commercially available high power white LEDs is close to 120 lm/W at room temperature and pulse mode, which is adequate for the requirements of most applications, by considering the features of LEDs including the direction lighting. However, the performance/price ratio (lm/$) of LEDs is almost ten times lower than that of traditional lamps. This restricts the penetration of white LEDs in general illumination. Further research efforts are needed to improve the performance and cut down the cost.

In 2008, Philips Lumileds and Osram Opto Semi-conductors unveiled impressive lab results with luminous efficiency of 140 lm/W using 1 mm^2 chip at a drive current of 350 mA [54]. Packaging is considered to be critical to maintaining this lab record realized for volume production in industry stably and in low cost. The theoretical viewpoint is that the LED can work as long as one hundred thousand hours given perfect conditions. However, the LED chip is very sensitive to damage in terms of electrostatic discharge (ESD), moisture, high temperature, chemical oxidation, shocking, and so on. Packaging protects the LED chip from damage. In addition, packaging is responsible for enhancing the light extraction to provide high luminous flux, dissipating generated heat from the chip to increase the reliability and life and controlling the color for specific requirements [55]. The trend of the high performance/price ratio of the white LED proposes more rigorous requirements on packaging.

The main requirements on white LED packaging are high luminous efficiency, high color rendering index (CRI), adjustable correlated color temperature (CCT), excellent color stability, low thermal resistance, rapid thermal dissipation, high reliability, and low cost. It is not a simple task for packaging to satisfy all these requirements simultaneously. There normally exist some contradictions such as high luminous efficiency and low CCT, high performance, and low cost [56]. Warm white LED with low CCT needs the addition of longer wavelength phosphor, which will increase the conversion loss and thereby reduce the luminous flux. Advanced packaging materials and techniques are essential for high performance LEDs, whereas low cost limits the materials selection. Diamond has super thermal conductivity (2000 W/mK), but the high cost makes it impossible to use for thermal dissipation. Packaging design should choose balanced considerations in these contradictions and guarantee the most favorable targets.

White LED packaging is a complicated technology of both fundamental science and engineering natures, as shown in Figure 2.21. It mainly concerns optics, thermal science, materials, mechanics, electronics, packaging processes, and equipment. These key factors determine the final performance of white LEDs. Generally, materials, processes and equipment are the basis for the development of white LED packaging. It is the advanced materials, processes and equipment that make the high quality white LEDs realizable. For example, the emergence of silicones ensures that the high power LEDs are used longer than traditional LEDs with epoxy resin encapsulants. The adoption of the ceramic board as the

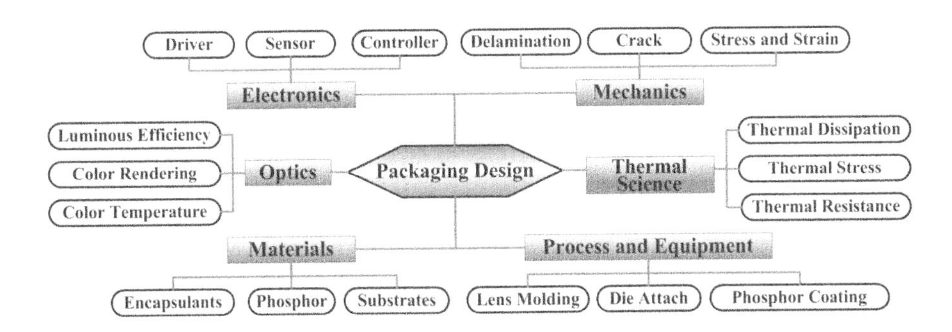

Figure 2.21 Schematic for packaging with various design issues. As a complicated system, packaging design cannot focus on one issue and should address these issues simultaneously and provide an integrated system with balanced performance.

substrate of chip-on-board technology ensures that the heat of the chip is efficiently transferred and the size of LED modules is significantly reduced. Bonding chips with solder also makes the thermal resistance between the chip and the substrate reduce dramatically and therefore increases the operation temperature of white LEDs, before the successful implementation of nano phased conductive adhesive with both high thermal conductivity and handling. Higher junction temperature means that LEDs can be used in harsher environments and prolongs the life of LEDs. Automatic equipment is consistent and repeatable. This is very important for LEDs, because the cost of white LEDs is still higher than traditional lighting sources. Continuous reduction of LED cost in the following years can accelerate the penetration of white LEDs into illumination applications.

Based on the improvements in materials, processes, and equipment, the optimization of optics, thermal science, mechanics, and electronics can be realized. However, the optimization of these key factors is still the subject of further studies and more attention need to be paid to them. The optimization requires fundamental understanding of how the varieties of materials, processes, and equipment affect the final performance of LEDs. Additionally, the reliability issues caused by the operation are also not yet completely known. This is challenging work for packaging design, especially when the LEDs are used in harsh lighting applications. In the following text, simple explainations will be given and more detailed discussions can be found in the following chapters.

The optics is related with the luminous efficiency, correlated color temperature (CCT), and color rendering index (CRI). Luminous efficiency is the main concern in optics, since it is the direct criteria that determines whether the LED has the potential to save more energy than other lighting sources. To enhance the luminous efficiency, the general methods used in LED packaging include conformal phosphor coating, free form lens, high transparent and high refractive index encapsulants, and optimized reflectors. Phosphor materials are also important in the improvement of luminous efficiency. Since the luminous efficiency of white LEDs has exceeded 100 lm/W, which makes white LEDs usable in various applications, the correlated color temperature and color rendering index are becoming more important in the optics. These two factors can provide humans with more comfortable and more real visions in illumination. In applications such as residential lighting and hotel lighting, the applications require warm, white light (2500 K–4000 K) to make people feel more comfortable. Other applications such as medical lighting and museum illumination require high CRI (>90) to exhibit the natural color of objects [57]. However, to provide high quality white LEDs with warm white and high Ra, the products fabricated by phosphor-converted LEDs should sacrifice the luminous efficiency due to the addition of longer wavelength converters such as red phosphor.

Thermal dissipation has been a serious issue with the invention of high power LEDs. Constrained by the internal and external quantum efficiencies, a non-radiative process in active layer converts most of the electrical power to heat. Increased drive current means that there is almost 70 W/cm^2 for a 1W LED with a 1 mm^2 area, which is higher than conventional microprocessor chips. Generated heat will increase the junction temperature significantly. Higher temperature may damage the PN junction, lower luminous efficiency, increase forward voltage, cause wavelength shift, reduce lifetime, and affect quantum efficiency of phosphor [58]. Degradation of materials is mostly caused by long term heating. Elevated temperature can induce thermal stresses in packaging components due to the mismatch of coefficient of thermal expansion (CTE). Since the thickness of the active layer and the P-GaN of the chip are only dozens of nanometers, they are very sensitive to thermal stresses. Significant deformation or warpage may be also induced. For most commercial LEDs, junction temperatures cannot exceed some values such as 120 °C for most vendors and 150 °C, 180 °C for a few exceptionally well packaged products. High thermal stresses caused by an over-heated LED may lead to voids, cracks, delaminations, and premature failures. Therefore, rapidly removing heat from the chip and keeping the junction temperature below a certain limit are crucial for the maintenance of LED performance.

There are two paths for heat dissipation. One is conducting heat through upper phosphors and encapsulants, the other is conducting heat through chip attached materials. Since encapsulants are mostly polymer and both encapsulants and phosphors are heat insulated, all of the heat must be conducted through the materials. Considering the super heat flux density of chips, these materials should not only present high thermal conduction, but also rapidly spread the heat to areas with effective configurations. Applying high conductive metal as the heat slug is the first solution of high power LED packaging. Luxeon products are the representatives in these types of LED packaging. Figure 2.22 is the schematic of this product. Then, the chip-on-ceramic board technique is proposed to minimize the size of LED packaging. Through adopting the ceramic board, the CTE mismatch issue is solved and can also conduct the heat efficiently. Other techniques proposed for the solution of high power LED packaging include directed bonded cooper (DBC), metal core printed circuit board (MCPCB), and composite materials [59,60].

Reliability is also critical in LED packaging. Since the first high power white LED was invented, great efforts have been put into the study to fabricate high reliability LEDs with a lifetime as long as 50,000–100,000 hours. Although white LEDs have presented longer life than traditional lamps, the relatively higher cost and requirements on environmental protection and energy saving propose more rigorous expectations on reliability. However, the influencing mechanisms of manufacturing on the reliability and operation are not substantially understood. This restricts the development of LEDs since we cannot predict the lifetime and failure of LED devices accurately. Unexpected failure may induce serious impacts. A representative incident was the Luxeon recall of Lumileds, which was caused by non-conforming epoxy material [61]. Potential issues affecting the reliability mainly include deformation, voids, delamination, cracks, impurities, moisture, temperature, and current. These issues are generally attributed to

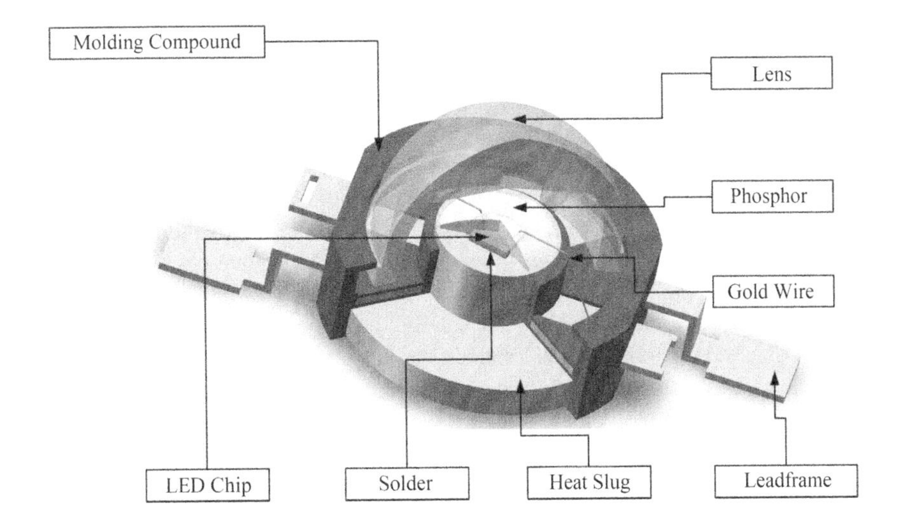

Figure 2.22 Schematic illustration of Luxeon packaging.

improper choice of materials [62], incorrect handling and processes, non-optimized structures, excessive usage under harsh conditions, and so on.

In the manufacturing process, voids normally exist in die attach materials and thermal interface materials. In the soldering and curing process, complicated physical and chemical reaction will happen and may release gas or vapor. In the annealing process, melted materials should cool down and shrink. Different composite alloys have various shrinkage rates and generate internal stresses in materials. This induces the initial crack and voids as shown in Figure 2.23 [63]. These small voids can block thermal conduction. In the following thermal

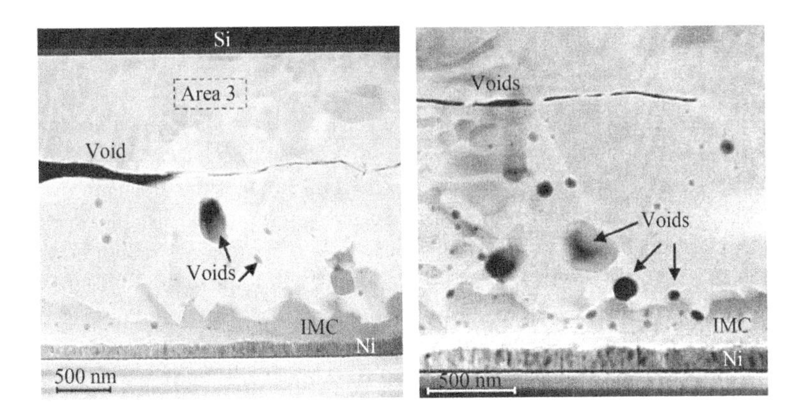

Figure 2.23 SEM images for die attach materials [63]. Small voids and initial cracks are inevitable in the curing process. IMC denotes the intermediate metallic compound. (Reprinted from *Materials Science and Engineering*: *A*, **441**, J.-W. Park, Y.-B. Yoon, S.-H. Shin and S.-H. Choi, "Joint structure in high brightness light emitting diode (HB LED) packages," 357–361, 2006, with permission from Elsevier.)

cycles such as encapsulants curing and operations, the voids and initial crack may expand and result in breakage and delamination in the die attaching interface.

Thermal load is the primary factor affecting the lifetime of LEDs. Heat accelerates the moisture absorption rate of materials and speeds up the erosion reactions of impurities and metals. Cracks and delaminations may grow and lead to degradation and even failure of components under thermal cycles. Generally, heat is coupled with drive current. Increasing current generates more heat and induces worse electrical stress and thermal stress. These coupling effects of heat and current on the reliability of LEDs have been studied by many experiments [64–71]. The main impacts of high current and elevated temperature include increase of thermal resistance and forward voltage, reduction of thermal conductivity and transmittance of silicones and QE of phosphors, carbonization of packaging components and phosphors, wavelength shift of chip and phosphors, current crowding, materials degradation, and so on. It is believed that pulsed current with high duty cycles is beneficial for the improvement of reliability instead of direct current. The operating current of LEDs should be carefully set to avoid over-loaded electrical and thermal stresses according to different environments.

Since the LED packages should be subjected to mechanical, thermal, and environmental loadings during manufacturing processes and services, special attention should be paid to the adhesion strength in the interfaces to maintain LEDs in a safe state. It has been proven that the weak adhesion strength makes the delamination grow faster and it cannot resist multi-shocks or reflows. Precondition can improve the adhesion strength, but the materials should be carefully chosen to avoid penetration of impurities and harmful ions.

It should be noted that reliability issues cannot be permanently solved. The only solution is minimizing the probabilities of failures by keeping decay of components within acceptable ranges. For a long time, structures of LED packaging were designed for the purposes of easy operations and simplified processes and reduced cost. However, this induces adverse results of LEDs since people find that the performance is not as stable as declared. Typical cases are failures of road lights in many cities of China in their early attempts to implement high power road lights in primary and secondary streets. Therefore, we believe that applying advanced technologies by a small increase of cost to improve the reliability is really essential for the future of LEDs.

Finally, it should be noted that these key factors are correlated and they determine the final cost and performance of white LEDs. Designers cannot focus on one aspect without considering the impacts of other aspects. Light extraction and thermal dissipation are normally interplayed. The more the light is extracted, the less the heat should be dissipated. Low junction temperature will further increase the initial optical power and make the LED brighter. On the other hand, if less light is extracted or the junction temperature is too high, there will be a harmfully optical-thermal cycle. This cycle can damage the chip and cause degradation of materials. Other reliability issues such as delamination and cracks may also emerge and shorten the lifetime. Since the cost of LEDs will be higher than traditional lamps in the next 2–4 years, designers should firstly try to improve the performance of LEDs. The correlation of these packaging issues demand that white LEDs should be co-designed as an optical-thermal-reliability system. Construction of a system level design platform is therefore important for LED packaging.

An excellent packaging design should not only focus on the issues caused by individual packages, but also should address the issues caused by the operation of LEDs.

Designing and manufacturing a single LED package without specifically considering the impacts of environment and usage may induce unexpected results such as elevated temperature, poor weld ability, and installation damage. Typical early examples include the failure of Cree XLamp's lens and surface contamination of Seoul Semicon's silicone encapsulants. To reduce the cost of lamps or road lights, manufactures prefer to increase the drive current to more than 500 mA so as to provide high lumen output, which will accelerate LED failure. Secondly, the optical lens is sensitive to the position of the LED, which may be varied by the reflow soldering. Therefore, designers should carefully choose materials and package configurations to avoid manufacturing "defective" products. The packaging design tolerance should also be higher to make the LED durable in more rigorous conditions. System solutions for specific applications are essential for future packaging design.

2.5 Development Trends and Roadmap

Although the LED packaging industry has finished significant technology accumulation, prototype developments, IP development, and so on, we still have room for new developments due to the demand of 10× cost reduction in the next 4 to 5 years and it is worthwhile to predict the development trends and roadmap [72].

When the LED goes to general lighting, large size backlighting, and automotive headlamps, we still see a lot of room for innovations for more compact, high performance/cost products. In particular in those developing countries, we still see the absence of the mature technologies for high power LED packaging. We still have observed poor reliability of the existing products. We are still far away from having practical products in terms of reliability and standards. Current LED lighting fixtures are still too heavy and we are still in need to integrate technology which can cover efficient optics, compact thermal management modules, reliable powering unit, and mechanical protection. In this section, we will propose a brief prediction of packaging technology in terms of various aspects.

2.5.1 Technology Needs

In order to further reduce the cost of LED packaging modules and systems, we will need a system approach to co-consider the reliability and durability, secondary optics and first level packaging optics, thermal management, and cost issues. DfX, coined for design for reliability and durability, manufacturability, assembly, testing, cost, and so on, is actually both a framework and a methodology for concurrent engineering of new products. It is challenging to build the platform and execute it in the design of LED packaging for applications. Even for multi-national companies, many issues need handling before one can really use it to full scale for the actual LED packaging and application design. However, even the platform is partially completed, the benefits can be significant. For academic institutions, a few leading universities in packaging have been pushing the establishment of this platform for the past 19 years, such as Georgia Tech, University of Maryland at College Park, Wayne State University, Huazhong University of Science and Technology, HK ASTRI, and so on. In the following sub-sections, we will briefly discuss the development needs for optical structures, color control technology, phosphor layer materials and processes, thermal management technology, substrate materials, and reliability.

(i) Optical Structures

Currently, simple shape reflectors, spherical and simple shape lens are commonly used in commercial packaging and lighting fixture products. Research and developments will likely target new nano-micro-lens array, integrated reflectors and lens optics, freeform reflectors, and lenses.

(ii) Color Control Technology

With the wide applications of LED packaging in various sectors, it is desirable to develop methods to control the phosphor density, phosphor layer profile, multi-layer phosphor, RGB color, and multi-chip forward current control.

(iii) Phosphor Layer Materials and Processes

Currently, leading companies have used phosphors mixed with silicone. The research community is working on multi-wavelength phosphor layer, nanometer phosphor layer, phosphor in nano-phased ceramics which has much high refractive index. In addition to direct coating, conformal coating, new methods are being developed for direct molding, electrostatic spray, and so on. Efforts are being developed in the authors' group in terms of coating the spherical shaped phosphor layer in volume which may result in a uniform color temperature.

(iv) Thermal Management Technology

Currently, both epoxy and eutectic die attach solder are used for the LED chip to substrate bonding in chip-substrate level packaging. Eutectic die attach solder is effective as it provides low thermal resistance. Epoxy provides a low cost solution due to the easy operation and inexpensive equipment. But its thermal performance counts on the thickness control to reduce the thermal resistance. One new development is in nano materials enhanced polymer bonding which may replace solders [73]. More efficient stripe fin design, microjet fluid cooling, micro vapor network chamber based substrate, micro heat pipe substrate, micro channel substrate, nano thermal interface materials, and integrated substrate and heat sink technology are on-going research efforts in the community. Substrates with micro structures will help reduce the spreading resistance and environment resistance. Thermal management technology in active cooling needs more reliability evaluation.

(v) Substrate Materials

Currently, metal core printed circuit board (MCPCB) is used. Low temperature co-fired ceramics substrate is one candidate for high power LED packaging. Direct bonded copper (DBC) has been proven to be useful for high power lasers and is being tested for LED array packaging. Multi-layer MCPCB, multi-layer DBC, microstructured substrate integrated with drive circuits, and even distributed powering unit are new research directions.

(vi) Reliability

Model based reliability approach for both modules and systems needs further development in terms of refined material databases, components and module database, sensors based loading

Table 2.2 Technical specifications of packaging as a function of years

Year	2008	2010	2012	2014	2016	2018	2020
Luminous Efficacy (lm/W)	100	130	150	170	180	190	200
Color Rendering Index	80	83	85	88	>90	>90	>90
Color Uniformity	>70%	%75%	>75%	>75%	>80%	>80%	>85%
Thermal Resistance (° C/W)	8	<6	<5	<4	<4	<3	<3
Life Time (h)	30000	40000	50000	70000	90000	90000	100000
Packaging Efficiency	80%	84%	88%	90%	92%	94%	>95%
Lumens Cost (¥/klm)	120	60	30	20	10	8	5

database, multi-physical variables multi-scale based CAE modeling for both processes modeling and reliability qualification, model based simplified testing vehicles for material and material-pair selections, failure mechanisms and degradation models for coupled optical/color, thermal, stress, moisture, and electrical performance. The system level reliability covers the health management and will be coupled to the wireless sensing modules. More on reliability will be discussed in Chapter 5.

2.5.2 *Packaging Types*

As described in Chapter 2.1, IC packaging has been evolving from simple dual in-line packaging to ball grid array (BGA) to flip-chip packaging (FC) to chip scale packaging (CSP) and finally to wafer level packaging (WLP) based CSP. The driving forces behind the evolution are minimization, low cost, high performance, and robust reliability. This has been also true for the development of LED packaging so far in the past 40 years. The LED packaging has evolved from the simple low power 3 mm and 5 mm packaging to current XLamp and Rebel type packages. According to what was proposed for the purpose of discussion, Table 2.2 and Figure 2.24 present the roadmap for performance specifications.

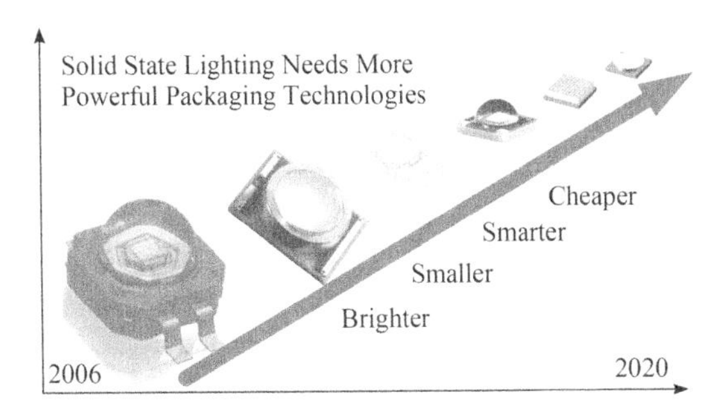

Figure 2.24 SSL needs more compact packaging.

(i) Application Specific LED Packaging

In IC designs, we have so called application specific integrated circuits (ASIC) for each application of microelectronics. By analogy, we define application specific LED packaging (ASLP) or just ASP to meet the needs of LEDs for particular applications such as road lights, in-house lamps, automotive headlights, tunnel lights, and so on. Bulk secondary lenses or reflector covers have been in use, and if optimally designed, can work well applications such as road lights. However, the bulk secondary optics makes the metallic parts heavy (fins, mechanical components), which adds to the cost so as not to be able to compete with traditional lighting in terms of the initial installation. It is also desirable to use an LED array so that the extended light source is used and it is challenging to design compact lens to meet the needs of both individual LED and LED array. It is even more desirable to integrate the secondary optics and first level packaging spherical lens, which will be discussed in detail in Chapters 3 and 6.

(ii) Wafer Level Packaging

Traditionally, packaging is done after one obtains a bare die or dice and package them in a module, which is defined to be the device level packaging. In wafer level packaging, processes tend to be done more in the wafer level. Typical processes include wafer thinning, film deposition, wafer bonding, wafer debonding or lift off for substrate transfer, TSV (through silicon via), and so on. Wafer thinning can be done by mechanical grinding, CMP (chemical-mechanical-polishing), wet etching or dry etching, in sequence, but with an optimization combination of these processes to make sure sub-surface damage can be minimized. Wafer level packaging can be from wafer on wafer, and chip on wafer. Details can be found in a recent book [1]. For LED wafer level packaging, a simple example is a simple white light concept from wafer level. Phosphor silicone can be spin coated or screen printed on the sapphire wafer, or bonded with a glass phosphor or nano phase ceramics wafer. Schematic processes are presented in Figure 2.25. If on-chip lens

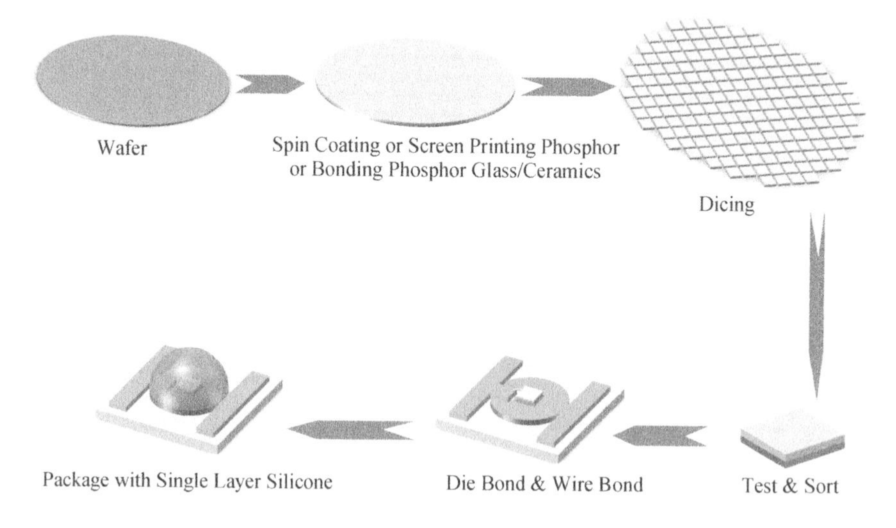

Figure 2.25 A wafer level packaging for the direct white light.

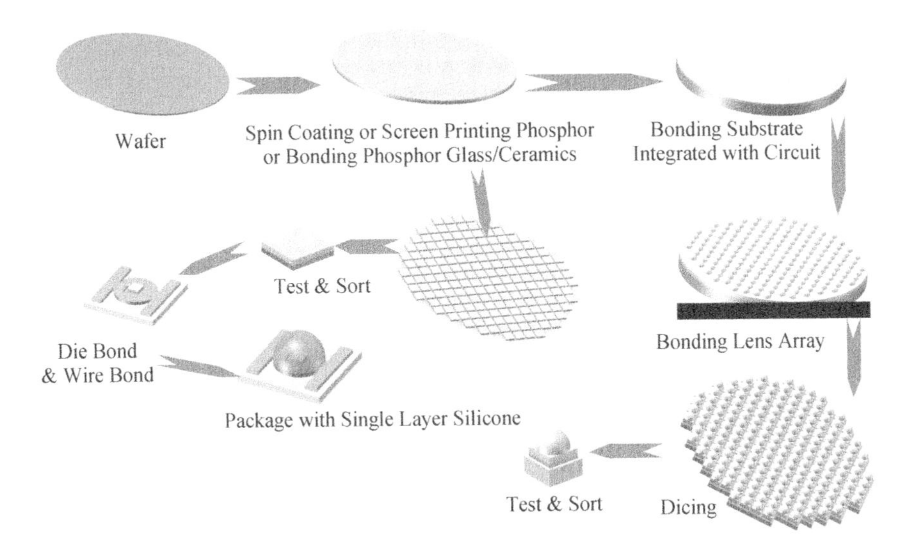

Figure 2.26 A wafer level packaging for white light LED with integrated optics.

(optical chip) or lens array are considered. We can use chip on wafer or wafer on wafer concept. Schematic processes are also presented in Figure 2.26 for wafer on wafer concept [74–77]. It is noted that non-spherical lens can be designed for specific applications. We also need to point out that we can make individual LED chips by dicing, individual lens or lens array, and use chip on wafer concept to realize the wafer level integration.

(iii) System in Packaging

System in packaging (SiP) was revolutionary in that it covered the whole system as a packaging. For LED in particular, systems can be a full street lamp, or a tunnel lamp, or a small in-house

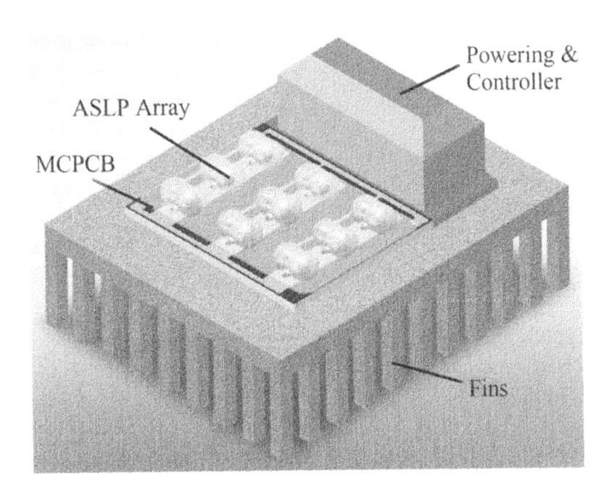

Figure 2.27 Schematic of an integrated LED packaging module with lens, cooling function, powering and controllers.

lighting screen, or a large screen backlighting system integrated with LCD, and so on. Wafer level packaging can be part of the SiP. In an LED SiP, we may perceive how to integrate the major functions for a system into a three dimensional packaging (3D). These functions can be integrated optics, powering, signal distribution, thermal management, and sensors based health monitoring and wireless monitoring. One such example is shown in Figure 2.27 for a typical street lamp module. There are thousands of companies out there in traditional lighting and they need the LED technology. However, most of them are short of knowledge and technology of secondary optics, thermal management, and powering. By the SiP modules proposed here, it would be very easy for those companies to switch to the LED community so as to minimize the pains caused to them by the transferring from traditional lighting to SSL.

2.6 Summary

In this chapter, with the belief that LED packaging shares many similar points to electronic packaging and with the hope that the cost breakdown will continue in LED packaging, we first presented the fundamentals of electronic packaging from the conventional leadframe based plastic packaging, to chip scale packaging, to wafer level integration, and to the system in packaging (SiP) which is based on More Than Moore, the so-called second law of microelectronics [1]. The last concept is important, as in solid state lighting, More Than Lighting will be important for the LED, as the LED will generate many more opportunities due to their unique features. Then, we discussed the LED chips in terms of current spreading efficiency, internal quantum efficiency, and high light extraction efficiency. Types and functions of LED packaging are described in details in terms of conventional low power packaging, and high power packaging. Factors and system design of high power packaging are seriously discussed, with particular focus on the cross-disciplinary co-design approach to the design of LED packaging. As a complicated system, packaging design cannot focus on one issue and should address these issues simultaneously and provide an integrated system with balanced performance. Reliability issues are briefly discussed. Finally, development trends and a roadmap are provided, with particular interest in the integration from a simple application specific LED packaging (ASLP), to more advanced level wafer level integration. It is believed that silicon based packaging will be rapidly used in LED packaging.

References

[1] Tummala, R.R. and Swaminathan, M. (2008), *Introduction to System-on-Package (SOP)*, McGraw-Hill.

[2] Suhir, E., Lee, Y.C., Wong, C.P. (2007), *Physical Design-Reliability and Packaging*, Springer.

[3] ITRS (2006), *International Technology Roadmap for Semiconductors, USA*.

[4] Kim, H., Park, S.J. and Hwang, H. (2007), "Lateral current transport path, a model for GaN-based light-emitting diodes: applications to practical devices designs," *Applied Physics Letters* **81**(7): 1326–1328.

[5] Pan, H.P., Huang, L.W. and Li, R. (2007), "Finite element model of GaN based LED and the optimization of the mesa structure," *Chinese Journal of Luminescence* **28**(1): 114–118. (In Chinese)

[6] Bulashevich, K.A., Mymrin, V.F., Karpov, S.Y., Zhmakin, I.A., Zhmakin, A.I. (2006), "Simulation of visible and ultra-violet group- nitride light emitting diodes," *Journal of Computational Physics* **213**: 214–238.

[7] Lastras, A. (1978), "Internal quantum efficiency measurement for GaAs light-emitting diodes," *Journal of Applied Physics* **49**(6): 3565–3570.

[8] Hangleiter, A., Fuhrmann, D., Grewe, M., Hitzel, F., Klewer, G., Lahmann, S., Netzel, C., Riedel, N. and Rossow, U. (2004), "Towards understanding the emission efficiency of nitride quantum wells," *Physica Status Solidi* (a) **201**(12): 2808–2813.

[9] Jackson, M., Boroditsky, M. and Yablonovitch, E. (1997), "Measurement of internal quantum efficiency and surface recombination velocity in InGaN structures," *10th Lasers and Electro-Optics Society Annual Meeting*, San Francisco, CA, USA, IEEE, 378–379.

[10] Boroditsky, M., Gontijo, I., Jackson, M., Vrijen, R., Yablonovitch, E., Krauss, T., Cheng, C. C., Scherer, A. and Krames, M. (2000), "Surface recombination measurements on III–V candidate materials for nanostructure light-emitting diodes," *Journal of Applied Physics* **87**(7): 3497–3504.

[11] Laubsch, A., Sabathil, M., Bruederl, G., Wagner, J., Strassburg, M., Baur, E., Braun, H., Schwarz, U. T., Lell, A., Lutgen, S., Linder, N., Oberschmid, R. and Hahn, B. (2007), "Measurement of the Internal Quantum Efficiency of InGaN Quantum Wells," *Proceedings of SPIE*, San Joes, CA, USA, SPIE, 6486,64860J.

[12] Martinez, C.E., Stanton, N.M., Kent, A.J., Graham, D.M., Dawson, P., Kappers, M.J. and Humphreys, C.J. (2005), "Determination of relative internal quantum efficiency in InGaN/GaN quantum wells," *Journal of Applied Physics* **98**(5): 053509.

[13] Zang, K.Y., Chua, S.J., Teng, J.H., Ang, N.S.S., Yong, A.M. and Chow, S.Y. (2008), "Nanoepitaxy to improve the efficiency of InGaN light-emitting diodes," *Applied Physics Letters* **92**(24): 243126.

[14] Wuu, D.S., Wang, W.K., Wen, K.S., Huang, S.C., Lin, S.H., Huang, S.Y. and Lin, C.F. (2006), "Defect reduction and efficiency improvement of near-ultraviolet emitters via laterally overgrown GaN on a GaN/patterned sapphire template," *Applied Physics Letters* **89**(16): 161105.

[15] Wu, G.M., Chung, T. J., Nee, T.E., Kuo, D.C., Chen, W.J., Ke, C.C., Hung, C.W., Wang, J.C. and Chen, N.C. (2007), "Improvements of quantum efficiency and thermal stability by using Si delta doping in blue InGaN/GaN multiple quantum well light-emitting diodes," *Physica Status Solidi* (c) **4**(7): 2797–2801.

[16] Xie, J.Q., Ni, X.F., Fan, Q., Shimada, R., Özgür, Ü. and Morkoç, H. (2008), "On the efficiency droop in InGaN multiple quantum well blue light emitting diodes and its reduction with p-doped quantum well barriers," *Applied Physics Letters* **93**(12): 121107.

[17] Sugawara, H., Itaya, K. and Hatakoshi, G. (1993), "Characteristics of a distributed Bragg reflector for the visible-light spectral region using InGaAlP and GaAs: comparison of transparent- and loss-type structures," *Journal of Applied Physics* **74**(5): 3189–3193.

[18] Kato, T., Susawa, H., Hirotani, M., Saka, T., Ohashi, Y., Shichi, E. and Shibata, S. (1991), "GaAs/GaAlAs surface emitting IR LED with Bragg reflector grown by MOCVD," *Journal of Crystal Growth* **107**(1–4): 832–835.

[19] Sugawara, H. and ltaya, K. (1992), "High-brightness lnGaAlP green light-emitting diodes," *Applied Physics Letters* **61**(15): 1775–1777.

[20] Chen, C.H., Chang, S.J., Su, Y.K., Chi, G.C., Sheu, J.K. and Chen, J.F. (2002), "High-efficiency InGaN/GaN MQW green light-emitting diodes with CART and DBR structures," *IEEE Journal of Selected Topics in Quantum Electronics* **8**(2): 284–288.

[21] Nakada, N., Nakaji, M., Ishikawa, H., Egawa, T., Umeno, M. and Jimbo, T. (2000), "Improved characteristics of InGaN multiple-quantum-well light-emitting diode by GaN/AlGaN distributed Bragg reflector grown on sapphire," *Applied Physics Letters* **76**(14): 1804–1806.

[22] Schubert, E.F., Wang, Y.H., Cho, A.Y., Tu, L.W. and Zydzik, G. J. (1992), "Resonant cavity light-emitting diode," *Applied Physics Letters* **60**(8): 921–923.

[23] Dill, C., Stanley, R.P., Oesterle, U. (1998), "Effect of detuning on the angular emission pattern of high-efficiency microcavity light-emitting diodes," *Applied Physics Letters* **73**(26): 3812–3814.

[24] Neve, D.H., Blondelle, J., Van Daele, P., Demeester, P., Borghs, G. and Baets, R. (1997), "Recycling of guided mode light emission in planar microcavity light emitting diodes," *Applied Physics Letters* **70**(7): 799–801.

[25] Benisty, H., De Neve, H., Weisbuch, C. (1998), "Impact of planar microcavity effects on light extraction—part I: basic concepts and analytical trends," *IEEE Journal of Quantum Electronics* **34**(9): 1612–1631.

[26] Benisty, H., De Neve, H., Weisbuch, C. (1998), "Impact of planar microcavity effects on light extraction—part II: selected exact simulations and role of photon recycling," *IEEE Journal of Quantum Electronics* **34**(9): 1632–1643.

[27] Schmid, W., Eberhard, F., Jager, R., King, R., Miller, M., Joos, J. and Ebeling, K.J. (2000), "45% quantum efficiency light emitting diodes with radial outcoupling taper." *Proceedings of SPIE*, San Joes, CA, USA, SPIE, 3938,90.

[28] Lee, S.J. and Song, S.W. (1999), "Efficiency improvement in light emitting diodes based on geometrically deformed chips," *Proceedings of SPIE*, San Joes, CA, USA, SPIE, 3621,237.

[29] Krames, M.R., Ochiai-Holcomb, M., Carter-Coman, C. and Chen, E.I. (1999), "High-power truncated-inverted-pyramid (AlxGa1-x) 0.5In0.5P/GaP light-emitting diodes exhibiting >50% external quantum efficiency," *Applied Physics Letters* **75**(16): 2365–2367.

[30] Kao, C.C., Kuo, H.C., Huang, H.W., Chu, J.T., Peng, Y.C., Hsieh, Y.L., Luo, C.Y., Wang, S.C., Yu, C.C. and Lin, C.F. (2005), "Light-output enhancement in a nitride-based light-emitting diode with 22° undercut sidewalls," *IEEE Photonics Technology Letters* **17**(1): 19–21.

[31] Chang, C.S., Chang, S.J. and Su, Y.K. (2004), "Nitride-based LEDs with textured side walls," *IEEE Photonics Technology Letters* **16**(3): 750–752.

[32] Yao, Y., Jin, C.X., Dong, Z.J., Sun, Z. and Huang, S.M. (2007), "Improvement of GaN-based light-emitting diodes using surface-textured indium-tin-oxide transparent ohmic contacts," *Chinese Journal of Liquid Crystals and Displays* **22**(3): 273–277. (In Chinese)

[33] Lee, Y.J., Kuo, H.C., Lu, T.C. and Wang, S.C. (2006), "High light-extraction GaN-based vertical LEDs with double diffuse surfaces," *IEEE Journal of Quantum Electronics* **42**(12): 1196–1201.

[34] Horng, R.H., Zheng, X.H., Hsieh, C.Y. and Wuu, D.S. (2008), "Light extraction enhancement of InGaN light-emitting diode by roughening both undoped micropillar-structure GaN and P-GaN as well as employing an omnidirectional reflector," *Applied Physics Letters* **93**(2): 021125–021125.

[35] Lee, Y.J., Hwang, J.M., Hsu, T.C., Hsieh, M.H., Jou, M.J., Lee, B.J., Lu, T.C., Kuo, H.C. and Wang, S.C. (2006), "Enhancing the output power of GaN-based LEDs grown on wet-etched patterned sapphire substrates," *Photonics Technology Letters* **18**(10): 1152–1154.

[36] Wuu, D.S., Wang, W.K., Shih, W.C., Horng, R.H., Lee, C.E., Lin, W.Y. and Fang, J.S. (2005), "Enhanced output power of near-ultraviolet InGaN-GaN LEDs grown on patterned sapphire substrates," *Photonics Technology Letters* **17**(2): 288–290.

[37] Wang, C.C., Ku, H., Liu, C.C., Chong, K.K., Hung, C.I., Wang, Y.H. and Houng, M.P. (2007), "Enhancement of the light output performance for GaN-based light-emitting diodes by bottom pillar structure," *Applied Physics Letters* **91**(12): 121109–121109.

[38] Fan, S., Villeneuve, P.R., Joannopoulos, J.D. and Schubert, E.F. (1997), "Photonic crystal light-emitting diodes," *Proceedings of SPIE*: 67–73.

[39] Kim, D.H., Cho, C.O., Roh, Y.G., Jeon, H., Park, Y.S., Cho, J., Im, J.S., Sone, C., Park, Y., Choi, W.J. and Park, Q. H. (2005), "Enhanced light extraction from GaN-based light-emitting diodes with holographically generated two-dimensional photonic crystal patterns," *Applied Physics Letters* **87**(20): 203508–203508.

[40] Truong, T.A., Campos, L.M., Matioli, E., Meinel, I., Hawker, C.J., Weisbuch, C. and Petroff, P.M. (2009), "Light extraction from GaN-based light emitting diode structures with a noninvasive two-dimensional photonic crystal patterns," *Applied Physics Letter* **94**(2): 023101–023101.

[41] Kang, X.N., Zhang, B., Dai, T., Bao, K., Wei, W. and Zhang, G.Y. (2008), "Improvement of light extraction from GaN-based LED with surface photonic lattices," *5th China International Forum on Solid State Lighting*.

[42] Holonyak, J.N. and Bevacqua, S.F. (1962), "Coherent (visible) light emission from $Ga(As_{1-x}P_x)$ junctions," *Applied Physics Letters* **1**(4): 82–83.

[43] Dupuis, R.D. and Krames, M.R. (2008), "History, development, and applications of high-brightness visible light-emitting diodes," *IEEE Journal of Lightwave Technology* **26**(9): 1154–1171.

[44] Lee, Y.J., Lu, T.C., Kuo, H.C. and Wang, S.C. (2007), "High brightness GaN-based light-emitting diodes," *IEEE Journal of Display Technology* **3**(2): 118–125.

[45] Nakamura, S., Mukai, T. and Senoh, M. (1991), "High-power GaN PN junction blue-light-emitting diodes," *Japanese Journal of Applied Physics* **30**(12A): L1998–L2001.

[46] Nakamura, S., Senoh, M. and Mukai, T. (1993), "High-power InGaN/GaN double-heterostructure violet light emitting diodes," *Applied Physics Letters* **62**(19): 2390–2392.

[47] Nakamura, S., Pearton, S. and Fasol, G. (1997), *The Blue Laser Diode: GaN Based Light Emitters and Lasers*. Berlin, Springer.

[48] Schlotter, P., Schmidt, R. and Schneider, J. (1997), "Luminescence conversion of blue light emitting diodes," *Applied Physics A: Materials Science & Processing* **64**(4): 417–418.

[49] Evans, D.L. (1997), "High-luminance LEDs replace incandescent lamps in new applications," *Light-Emitting Diodes: Research, Manufacturing, and Applications*, San Jose, CA, USA, SPIE, 142–153.

[50] Steranka, F.M., Bhat, J.C., Collins, D., Cook, L., Craford, M.G., Fletcher, R., Gardner, N., Grillot, P., Goetz, W., Keuper, M., Khare, R., Kim, A., Krames, M., Harbers, G., Ludowise, M., Martin, P.S., Misra, M., Mueller, G., Mueller-Mach, R., Rudaz, S., Shen, Y.C., Steigerwald, D., Stockman, S., Subramanya, S., Trottier T. and Wierer J.J. (2002), "High power LEDs-technology status and market applications," *Physica Status Solidi* (a) **194**(2): 380–388.

[51] Craford, M.G. (2005), "LEDs for solid state lighting and other emerging applications: status, trends, and challenges," *Fifth International Conference on Solid State Lighting*, San Diego, CA, USA, SPIE, 594101-10.

[52] Lee, Y.J., Kuo, H.C., Lu, T.C., Wang, S.C., Ng, K.W., Lau, K.M., Yang, Z.P., Chang, A.P. and Lin, S.Y. (2008), "Study of GaN-Based light-emitting diodes grown on chemical wet-etching-patterned sapphire substrate with V-shaped pits roughening surfaces," *Journal of Lightwave Technology* **26**(11): 1455–1463.

[53] OIDA. (2002), "Light emitting diodes (LEDs) for general iIllumination, an OIDA technology roadmap update 2002," from *http://lighting.sandia.gov/lightingdocs/OIDA_SSL_LED_Roadmap_Full.pdf*.

[54] LEDs Magazine (2008), "LED chips set new R&D records," from *http://www.ledsmagazine.com/features/5/10/5*.

[55] Haque, S., Steigerwald, D., Rudaz, S., Steward, B., Bhat, J., Collins, D., Wall, F., Subramanya, S., Elpedes, C. and Elizondo, P. (2000), "Packaging challenges of high-power LEDs for solid state lighting," from *www.lumileds.com/pdfs/techpaperspres/manuscript_IMAPS_2003.PDF*

[56] Hahn, B., Weimar, A., Peter, M. and Baur, J. (2008), "High-power InGaN LEDs: present status and future prospects," *Light-Emitting Diodes: Research, Manufacturing, and Applications XII*, San Jose, CA, USA, SPIE, 691004-8.

[57] Braune, B., Brunner, H., Strauss, J. and Petersen, K. (2005), "Light conversion in opto semiconductor devices: from the development of luminous materials to products with customized colors," *Optoelectronic Devices: Physics, Fabrication, and Application II*, Boston, MA, USA, SPIE, 60130D-8.

[58] Arik, M., Becker, C.A., Weaver, S.E. and Petroski, J. (2004), "Thermal management of LEDs: package to system," *Third International Conference on Solid State Lighting*, San Diego, CA, USA, SPIE, 64–75.

[59] Carl, H.Z. (2004), "New material options for light-emitting diode packaging," *Light-Emitting Diodes: Research, Manufacture, and Applications VIII*, Bellingham, WA, SPIE, 173–182.

[60] Zweben, C. (2008), "Advances in LED packaging and thermal management materials," *Light-Emitting Diodes: Research, Manufacturing, and Applications XII*, San Jose, CA, USA, SPIE, 691018-11.

[61] LEDs Magazine (2008), "Lumileds recalls some Luxeons, halts production line," from *http://ledsmagazine.com/news/5/1/19*

[62] Hsu, Y.C., Lin, Y.K., Chen, M.H., Tsai, C.C., Kuang, J.H., Huang, S.B., Hu, H.L., Su, Y.I. and Cheng, W.H. (2008), "Failure mechanisms associated with lens shape of high-power LED modules in aging test," *IEEE Transactions on Electron Devices* **55**(2): 689–694.

[63] Park, J.W., Yoon, Y.B., Shin, S.H. and Choi, S.H. (2006), "Joint structure in high brightness light emitting diode (HB LED) packages," *Materials Science and Engineering: A* **441**(1–2): 357–361.

[64] Jayasinghe, L., Dong, T.M. and Narendran, N. (2007), "Is the thermal resistance coefficient of high-power LEDs constant?," *Seventh International Conference on Solid State Lighting*, San Diego, CA, USA, SPIE, 666911-6.

[65] Meneghini, M., Trevisanello, L., Sanna, C., Mura, G., Vanzi, M., Meneghesso, G. and Zanoni, E. (2007), "High temperature electro-optical degradation of InGaN/GaN HBLEDs," *Microelectronics Reliability* **47**(9–11): 1625–1629.

[66] Su, Y.K., Chen, K.C. and Lin, C.L. (2007), "Ultra high power light-emitting diodes with electroplating technology," *2007 IEEE Conference on Electron Devices and Solid-State Circuits*. Tainan, Taiwan, IEEE, 19–22.

[67] Trevisanello, L.R., Meneghini, M., Mura, G., Sanna, C., Buso, S., Spiazzi, G., Vanzi, M., Meneghesso, G. and Zanoni, E. (2007), "Thermal stability analysis of high brightness LED during high temperature and electrical aging," *Seventh International Conference on Solid State Lighting*, San Diego, CA, USA, SPIE, 666913-10.

[68] Biber, C. (2008), "LED light emission as a function of thermal conditions," *2008 24th Annual IEEE Semiconductor Thermal Measurement and Management Symposium*, 180–184.

[69] Buso, S., Spiazzi, G., Meneghini, M. and Meneghesso, G. (2008), "Performance degradation of high-brightness light emitting diodes under DC and pulsed bias," *IEEE Transactions on Device and Materials Reliability* **8**(2): 312–322.

[70] Nam, H. (2008), "Failure analysis matrix of light emitting diodes for general lighting applications," *15th International Symposium on the Physical and Failure Analysis of Integrated Circuits*, 1–4.

[71] Trevisanello, L., Meneghini, M., Mura, G., Vanzi, M., Pavesi, M., Meneghesso, G. and Zanoni, E. (2008), "Accelerated life test of high brightness light emitting diodes," *IEEE Transactions on Device and Materials Reliability*, **8**(2): 304–311.

[72] Liu, S., Wang, K., Liu, Z.Y., Chen, M.X. and Luo, X.B. (2008), "Roadmap for LED packaging development: A personal View," *5th China International Exhibition & Forum on Solid State Lighting*, Shenzhen, China.

[73] Li, Y., Moon, K.S. and Wong, C.P. (2005), "Microelectronics without lead," 10.1126/*Science*.1110168.

[74] Liu, S., Chen, M.X., Luo, X.B. and Gan, Z.Y. (2006), Method to fabricate white light-emitting diode, *China Patent Application*, 200610029858.6.

[75] Liu, S., Chen, S.Y., Luo, X.B. and Chen, M.X. (2006), Method to fabricate high brightness white light-emitting diode, *China Patent Application*, 200610029856.7.

[76] Liu, S., Chen, M.X., Luo, X.B. and Gan, Z.Y. (2006), Light-emitting diode packaging structure and method, *China Patent Application*, 200610029859.0.

[77] Liu, S., Chen, M.X., Luo, X.B. and Gan, Z.Y. (2006), Method to fabricate light-emitting diode by spin coating and lithography, *China Patent Application*, 200610029857.1.

3

Optical Design of High Power LED Packaging Module

3.1 Properties of LED Light

3.1.1 Light Frequency and Wavelength

Light is a common natural phenomenon, through which the world can be perceived by humans, and through which various kinds of information can be obtained. The study about the nature of light can be traced back to the seventeenth century. Human's understanding of light has changed from the initial theory of particles into the wave theory. However, at the very beginning, people just simply believed that light was a mechanical wave. In the 1860s, Maxwell, basing his summary on previous work, proposed that light was a kind of electromagnetic wave in a certain frequency range. Later, the electromagnetic attribute of light was confirmed by a large number of experiments. The electromagnetic wave that human eyes can feel within the frequency range is known as visible light. The frequency range of visible light is 380–780 nm. Within this range, human visual sense of different colors can be caused by the electromagnetic waves. Usually, the approximate corresponding relationship between colors and wavelengths is as shown in Figure 3.1.

By the end of the nineteenth century and the early twentieth century, with the birth of the theory of relativity and quantum theory, the human's understanding of light was further deepened. Einstein proposed a quantum theory of light, which was the physical basis of the light-emitting of the LED. However, it cannot be generalized that light is particles or fluctuations. On certain occasions, light shows the characteristics of particles; but on other occasions, it shows the characteristics of fluctuations; it has the quality of "wave corpuscle duality". It is only through the different natures which the light shows under different conditions that people can interpret the light; in the same way, the appropriate theory must be adopted to describe the light under different conditions. For example, in the interpretation of the light-emitting principle of LED chips, the quantum theory of light should be adopted; while on the lighting optical design of LED, often the ray theory is adopted.

LED Packaging for Lighting Applications: Design, Manufacturing and Testing, First Edition. Sheng Liu and Xiaobing Luo.
© 2011 Chemical Industry Press. All rights reserved. Published 2011 by John Wiley & Sons (Asia) Pte Ltd.

Visible Light			
Colour	Wavelength λ (nm)	Frequency v (THz)	Energy E (eV)
Red	$622<\lambda<780$	$384<v<482$	$1.59<E<1.99$
Orange	$597<\lambda<622$	$482<v<502$	$1.99<E<2.08$
Yellow	$577<\lambda<597$	$502<v<520$	$2.08<E<2.15$
Green	$492<\lambda<577$	$520<v<609$	$2.15<E<2.52$
Blue	$455<\lambda<492$	$609<v<659$	$2.52<E<2.72$
Purple	$380<\lambda<455$	$659<v<789$	$2.72<E<3.26$

Figure 3.1 Approximate corresponding relationship between colors and wavelengths [1]. *(Color version of this figure is available online.)*

The light-emitting principle of the LED chip is that in the active region, the spontaneous radiation recombination of the electron-hole pairs produces light of different wavelengths. The relationship between the wavelength of light waves and frequency is shown in Equation 3.1.

$$\lambda = \frac{c}{v} \tag{3.1}$$

While the relationship between photon energy and frequency is as follows:

$$E = hv \tag{3.2}$$

Based on the two equations above, the relationship between wavelength and energy is shown in Equation 3.3,

$$\lambda = \frac{c}{v} = \frac{hc}{E} \approx \frac{1240}{E}(nm) \tag{3.3}$$

where c is the light velocity, h is Plank constant, v is photon frequency. And the photon energy caused by radiation is approximately equivalent to the band-gap energy E_g of materials in active region. Therefore, the light wavelength of the LED is as follows:

$$\lambda \approx \frac{1240}{E_g}(nm) \tag{3.4}$$

The bandgap of materials in different active regions are different. Nowadays the light emitting from LED chip can be a monochromatic light ranging from red to ultraviolet.

3.1.2 Spectral Distribution

As one type of electromagnetic wave, light is provided with energy. Usually, the quantity of radiation is used to express the size of the radiant energy of a light source or objects irradiated. The energy of radiation per unit time is expressed as the radiant flux Φ_e, the unit is watt. However a beam of light does not contain only one wavelength (frequency); it often includes light of continuous wavelengths (frequency) within a certain range. Even the light emitted by an LED chip is not strictly a monochromatic light. The optical wavelength (frequency) is also continuously distributed in a very small range. The spectral power distribution function $S(\lambda)$ is adopted to describe the distribution of light in the range of wavelength or frequency. $S(\lambda)$ is defined through the following equation:

$$\Phi_e = \int_0^{+\infty} S(\lambda)d\lambda \tag{3.5}$$

In order to describe the light source intuitively, the relative spectral power distribution curve is usually adopted to indicate the spectral distribution. Figure 3.2 is the spectral distribution curve of a vendor phosphor-converted white LED packaging module in the market; the peak spectral wavelength is 565 nm.

3.1.3 Flux of Light

It has been pointed out in Section 3.1.2 that the radiant flux Φ_e can be used to measure the radiant energy power of the light source or objects irradiated. The radiation can contain light of any frequencies. The situation of different optical receivers in response to the lights of different frequencies must also be taken into consideration. Usually, optical receivers have different

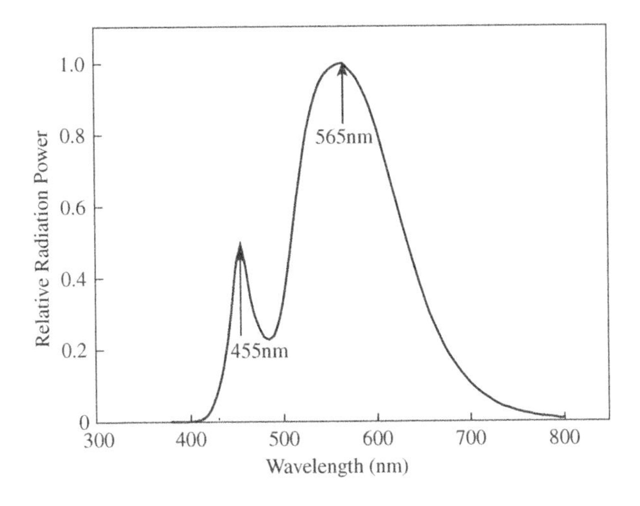

Figure 3.2 Spectral distribution curve of some LED.

responses to lights of different frequencies. Therefore, when considering a particular optical receiver, it is not enough to only use radiant flux.

In the same way, for the human eye, they are provided with different selectivity to lights of different frequencies. Experiments show that the human eye is most sensitive to the yellow light whose wavelength is 555 nm. In other words, under the condition of same radiant flux, compared with lights of other wavelengths, the yellow light (555 nm) has the strongest stimulation on the human eyes. Therefore, the radiant flux cannot be used to describe the degrees of stimulation of light source or irradiated objects imposed on the human eye. It is necessary to introduce a new physical quantity—luminous flux, whose unit is lm (lumen). It is prescribed that the corresponding luminous flux of the yellow light with wavelength of 555 nm is 683 lm when the radiant flux is 1 W; and for the rest of the lights with any other wavelengths, their luminous fluxes are all less than 683 lm when the radiant flux is 1 W. Therefore, if the physiologically luminous efficiency of the human eyea to the light with wavelength of 555 nm is prescribed as 1, then the physiologically luminous efficiencies of human eyes to lights with other wavelengths are all less than 1. The physiologically luminous efficiency capability of the human eye to lights with different wavelengths is called visual sensitivity function.

Studies have shown that there are different visual sensitivity functions of the human eye in different light environments, such as visual sensitivity functions of photopic vision and scotopic vision (CIE DS 010.2–2001 Photometry—The CIE system of physical photometry, 2001). In 1924, the Commission Internationale de l'Eclairage (CIE) recommended the visual sensitivity function of photopic vision $V(\lambda)$; in 1951, CIE recommended the visual sensitivity function of scotopic vision $V'(\lambda)$. The curves of visual sensitivity function are shown in Figure 3.3.

The luminous flux is a physical quantity which is used to measure the size of radiation that the the human eye perceives the light source or objects irradiated, so that the size of luminous flux is not only related to the absolute radiant flux of light source or objects of illumination, but

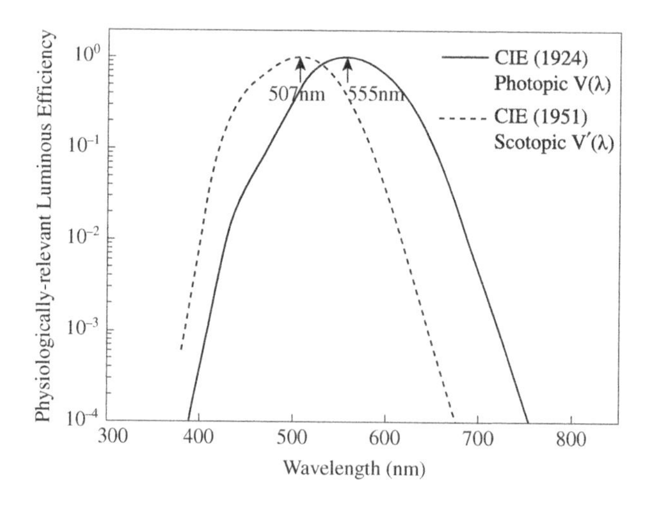

Figure 3.3 Curves of visual sensitivity function recommended by CIE.

also related to the visual sensitivity function of human eyes as well as the spectrum. The luminous flux can be expressed in the following equation:

$$\Phi_v = 683(lm/W)\int V(\lambda)S(\lambda)d\lambda \tag{3.6}$$

3.1.4 Lumen Efficiency

As introduced previously, we know that not all of the electromagnetic waves can be felt by the human eye; even in the range of visible light, the sensitivities of the human eye to lights with different wavelengths are also different. Consequently, in lighting applications, different light sources with different luminescent spectrums are provided with different lighting effects; a light source with a large radiant flux is not necessarily having a large luminous flux. Take the incandescent lamp as an example; its radiation spectrum contains a large number of infrared rays not visible to the human eye, so the luminous flux is very low; while the white light LED, the emission spectrum is mainly concentrated in the range of visible spectrum. Therefore, although the absolute value of the radiant flux of the LED is not large, it has a large luminous flux; and this is one of the reasons for LED's energy-saving. Hereon, the lumen efficiency can be adopted to measure the efficiency of the radiation spectrum of the light source. The unit of lumen efficiency is lm/W and the equation is as follows:

$$k = \frac{\Phi_v}{\Phi_e} = 683(lm/W)\frac{\int_{380}^{780} V(\lambda)S(\lambda)\mathrm{d}\lambda}{\int_0^{+\infty} S(\lambda)\mathrm{d}\lambda} \tag{3.7}$$

where k means the luminous flux generated by 1 watt radiant flux. As mentioned above, the lumen efficiency of the light with 555 nm wavelength can reach as high as 683 lm/W, the lumen efficiency of light with other wavelengths are all less than 683 lm/W. The lumen efficiency, a concept different from the luminous flux of light source, is not related to the absolute radiated power of light source, only related to the spectral distribution. Different light sources are provided with different spectral distributions, so the lumen efficiencies are not the same; light sources with high lumen efficiency are often more energy-efficient. As the light source is driven by electricity, and the power efficiency η is often low, so that the light-emitting efficiency k' with 1 W electric power is lower than k, and k' can be expressed in the following equation:

$$k' = \eta k = \frac{\Phi_v}{\Phi_e} = 683(lm/W)\eta\frac{\int_{380}^{780} V(\lambda)S(\lambda)\mathrm{d}\lambda}{\int_0^{+\infty} S(\lambda)\mathrm{d}\lambda} \tag{3.8}$$

Table 3.1 is a list of lumens efficiencies of some typical light sources:

3.1.5 Luminous Intensity, Illuminance and Luminance

In various directions of the space, the luminous flux of most light sources or objects irradiated are not the same. In order to describe the light emitting from the radiation body in all directions in the space, it is necessary to introduce the physical quantity of luminous intensity $\boldsymbol{I}_v$.

Table 3.1 Lumens efficiencies of some typical light sources

Light Source	Incandescent Bulb	HFED	High Pressure Mercury Lamp	High Pressure Sodium Lamp	Low Pressure Sodium Lamp	Metal Halide Lamp	LED
k' (lm/W)	<25	40–80	30–50	60–120	100–175	60–80	60–120

The luminous intensity $I_v = \mathrm{d}\Phi_v/\mathrm{d}\omega(lm/Sr)$ is the luminous flux of the electric light source in a unit solid angle of certain direction. The unit of luminous intensity is cd or lm/sr; a basic unit of international system of units (SI). 1 cd is equivalent to the luminous flux of a monochromatic point light source in a unit solid angle, of which the radiant flux is 1/683 W and the light-emitting wavelength is 555 nm. The relationship between the luminous intensity and the unit of luminous flux is as follows:

$$1\ lm = 1\ cd \times sr \tag{3.9}$$

where *cd* is pronounced candela, and *sr* is pronounced steradian.

Since the light intensities of most light sources in various directions of the space are not equal, in most cases they cannot be given analytically. In order to achieve a complete light-emitting description of the light source in the space, it is necessary to measure and provide the light intensities of the light source from various angles in the space.

Actually, the real point source of light does not exist, but the luminous intensity can still be used to describe the light emitting of the light source in the space. At this time, it is necessary to observe the luminous intensity of the light source in a place which is much greater in size than the light source. The principle used to measure the luminous intensity of a light source is shown in Figure 3.4a. After obtaining the luminous intensity of the light source in the space, a convenient method to describe the luminous intensity of a light source is the luminous intensity distribution curve, which, in a coordinate system, can provide absolute or relative luminous intensity values (Figure 3.4b) of the light source on different planes (the xz plane in Figure 3.4a).

The luminous intensity of most light sources in the space is not axi-symmetric. In order to obtain an accurate luminous intensity of the light source, luminous intensity distribution curves on a number of planes need to be measured.

When the light emitted by the light source is shining on the surface, the optical energy transfers to the object's surface. In order to show the power distribution of the optical energy on the object of the illumination's surface, the concept of illuminance ***E*** is introduced. Its definition is the luminous flux per unit area, expressed in Equation 3.10.

$$E = \frac{\mathrm{d}\Phi_v}{\mathrm{d}S} \tag{3.10}$$

The unit of illuminance is lux (lx). In the lighting area, the range of illuminance is prescribed on certain occasions. Table 3.2 is a list of illuminance ranges in some typical occasions.

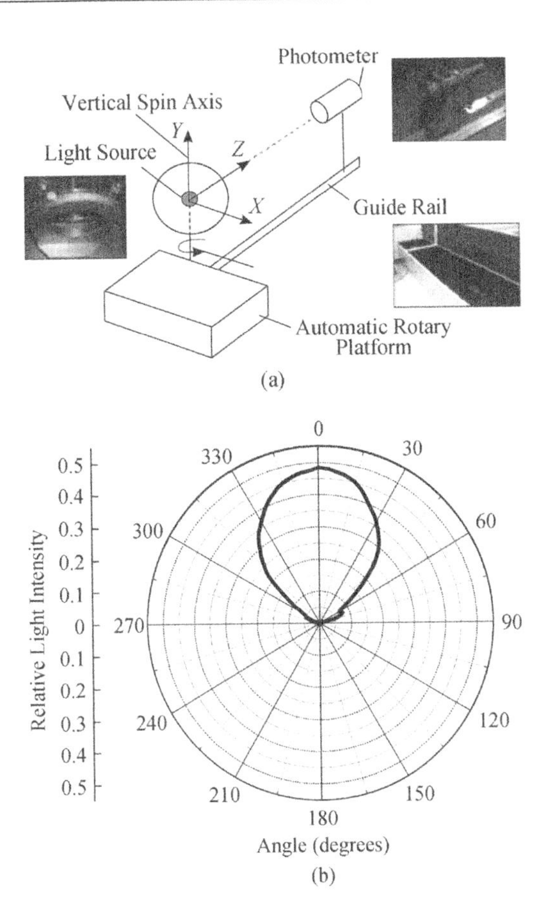

Figure 3.4 (a) The luminous intensity test system; (b) Cree Xlamp LED luminous intensity distribution curve.

Table 3.2 Illuminance ranges in some typical occasions [2]

Illumination Condition	Illuminance Range (lx)
Moonless Night	3×10^{-4}
Full Moon	0.5
Street Lighting	10–30
Reading & Writing	300
Home Lighting	30–300
Office Light	100–1000
Surgery Lighting	10000
Photostudio Lighting	10000
Direct Sunlight	100000

(Reproduced with permission from E.F. Schubert, *Light-Emitting Diodes*, 2nd ed., Cambridge University Press, Cambridge. © 2006.)

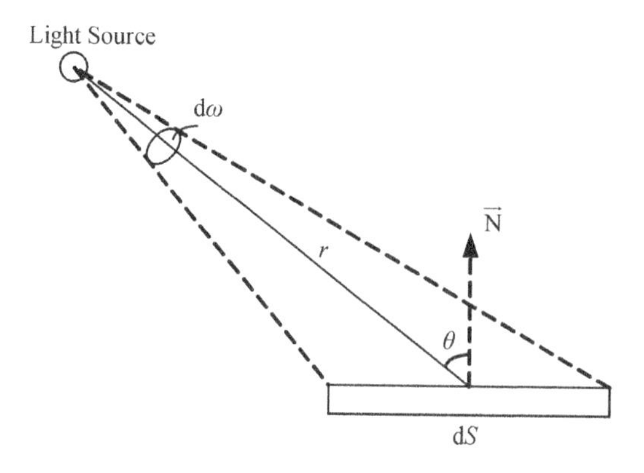

Figure 3.5 Location of the object to the light source.

The illuminance on the object's surface was defined both by the luminous intensity distribution of light sources and the location of the object to the light source. It is shown in Figure 3.5:

$$d\omega = \frac{dS\cos\theta}{r^2} \tag{3.11}$$

while $d\Phi_v = I_v d\omega = I\frac{dS\cos\theta}{r^2}$, therefore, the illuminance for the dS area is:

$$E = \frac{d\Phi}{dS} = \frac{I\cos\theta}{r^2} \tag{3.12}$$

As shown in Figure 3.5, the illuminance ***E*** on the microfacet ds which is away from the point light source is directly proportional to the luminous intensity of the light source and is inversely proportional to the square of the distance r. When the surface is perpendicular to r, the illuminance can reach the largest value.

Although the illuminance ***E*** can show the optical energy distribution on the surface of the irradiated object, it cannot show the light and shade of the light source or the object irradiated observed by the human eye. The human eye does not necessarily feel brighter in the area with high illuminance than that in the area with low illuminance. The light and shade that the human eye can feel is also related to the location of the human eye to the radiation object. Therefore, another physical quantity is introduced: luminance L, its unit is cd/m^2. As shown in Figure 3.6, obtain a microfacet dS on the surface of the radiator, suppose the luminous flux in the solid angle $d\omega$ is $d\Phi_{vi}$, where the angle between this microfacet and the normal N is i, then the luminous intensity in the direction $\vec{\mathbf{I}}$ is:

$$I_{vi} = \frac{d\Phi_{vi}}{d\omega} \tag{3.13}$$

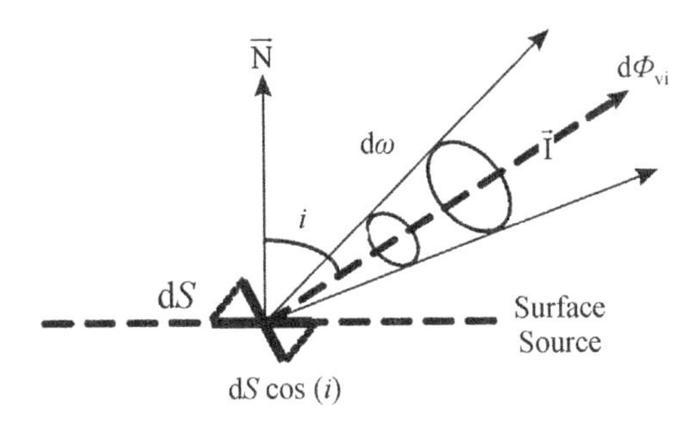

Figure 3.6 Luminance of the microfacet dS in the direction $\vec{\mathbf{I}}$ [3].

The luminance of the microfacet dS in the direction $\vec{\mathbf{I}}$ is defined as the ratio of the luminous intensity I_i of the microfacet in the direction $\vec{\mathbf{I}}$ to the projected area dScos$i$ of the microfacet on the plane which is perpendicular to the direction $\vec{\mathbf{I}}$:

$$L_i = \frac{I_{vi}}{dS\cos i} = \frac{d\Phi_{vi}}{\cos i dS d\omega} \tag{3.14}$$

Normally, the luminance of the radiator in all directions in the space is not the same, that is, from different angles, the light and shade that the human eye feels is not the same. However, there is a kind of special radiator of which the luminance is the same in all directions in the space-Lambert radiator. As shown in Figure 3.7, the luminous intensity of the Lambert radiator is:

$$I_{vi} = I_0 \cos i \tag{3.15}$$

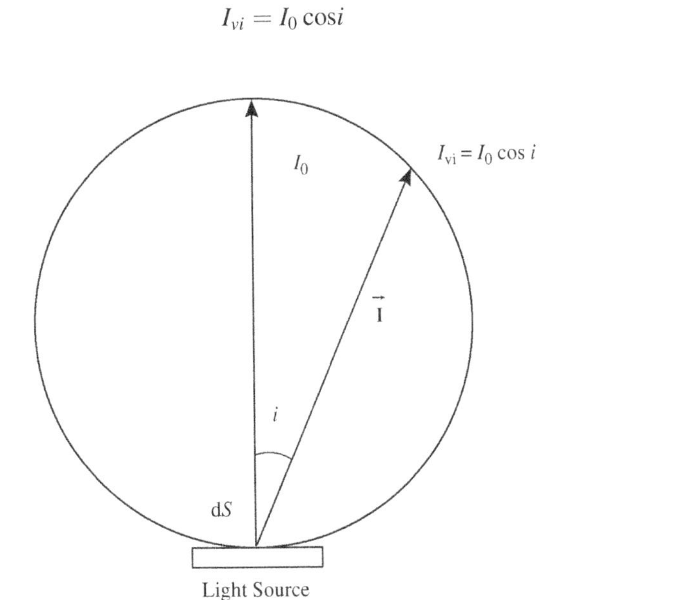

Figure 3.7 Luminous intensity distribution of Lambert radiator.

Based on the two equations above, then the luminance of the Lambertian radiator in direction $\vec{\mathbf{I}}$ is:

$$L_i = \frac{I_0}{\mathrm{d}S} \tag{3.16}$$

As $\frac{I_0}{\mathrm{d}S}$ is constant, make $L_i = \frac{I_0}{\mathrm{d}S} = L_0$, then the luminance of the Lambert radiator in all directions is equal, not related to the angle i.

3.1.6 Color Temperature, Correlated Color Temperature and Color Rendering Index

In addition to photometry, there is also a feel of color of human eyes to light. Therefore, some physical quantities about light colors need to be introduced to describe the feelings of the human eye to the light colors. The experimental results show that the majority of colors can be expressed through not more than three kinds of colors. These three primary colors can be red (R), green (G), and blue (B). Therefore, other colors can be defined according to these three colors. However, some colors cannot be expressed by the positive values of these three colors, especially those ones which are close to monochromatic color. Hereon, it is necessary to adopt three supposed color values (X, Y, Z) to express them. These color values (X, Y, Z) are the function values obtained by integrating spectral power distribution $S(\lambda)$ with standard color matching functions $\bar{x}(\lambda), \bar{y}(\lambda), \bar{z}(\lambda)$, which are shown in Equation 3.17. Figure 3.8 provides the modified versions of the color matching function of CIE 1931–the modified versions of Judd (1951) and Vos (1978).

$$\begin{aligned} X &= \int \bar{x}(\lambda) S(\lambda) \mathrm{d}\lambda \\ Y &= \int \bar{y}(\lambda) S(\lambda) \mathrm{d}\lambda \\ Z &= \int \bar{z}(\lambda) S(\lambda) \mathrm{d}\lambda \end{aligned} \tag{3.17}$$

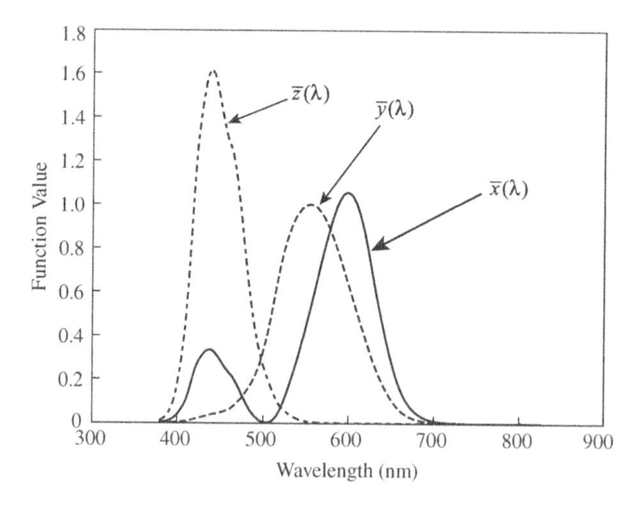

Figure 3.8 Color matching function.

In order to achieve a convenient presentation, the concept of color coordinates (x, y) is introduced and the relationship between the color coordinates and color values is as follows:

$$x = \frac{X}{X+Y+Z}$$
$$y = \frac{Y}{X+Y+Z} \tag{3.18}$$
$$z = 1 - x - y = \frac{Z}{X+Y+Z}$$

Therefore, the color can be presented by the coordinate value of the point in the plane.

Figure 3.9 is a chromaticity figure of 1931 CIE. In the figure, the curves contain a variety of monochromatic color coordinates (x, y); the area surrounded by the curves presents the color coordinates of other colors. The trichromatic RGB is also on the curves. The area surrounded by the triangle with the three points (R, G, B) as its three vertexes are the colors which may consist of trichromatic RGB.

In Figure 3.9, there is a curve in the area surrounded by the curves. This curve is called the Planckian Locus, which represents the color coordinates of the blackbody radiation spectrum under different temperatures. When the color coordinate of a light source is at a point of the Planckian Locus, the blackbody temperature corresponding to the point on Planckian Locus can be used to describe the color; it is the color temperature (CT). Figure 3.10 presents the colors corresponding to different color temperatures. If the color coordinates of the light source is not the same as that of any point on the Planckian Locus, then the temperature of the blackbody which can cause proximate color feeling of the light source to the human eye is adopted, and this temperature is the correlated color temperature (CCT).

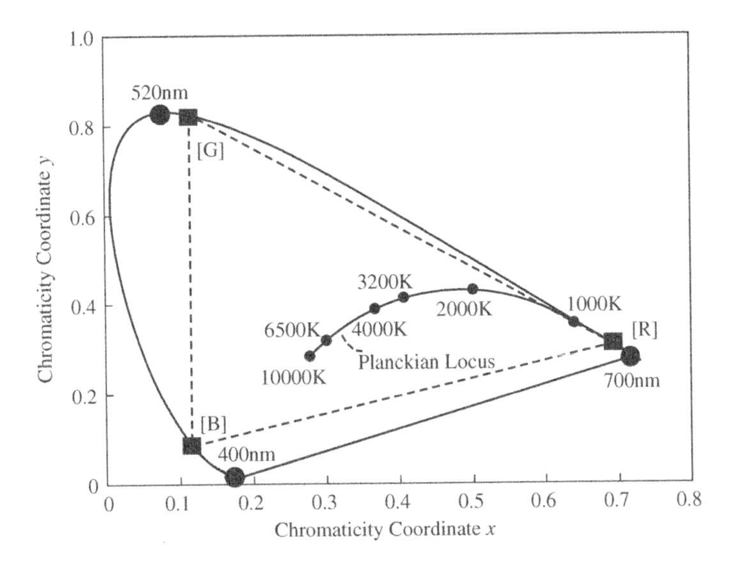

Figure 3.9 1931CIE colorimetric diagram.

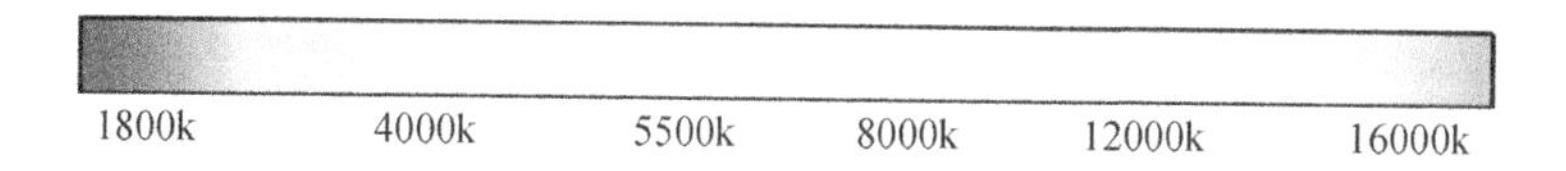

Figure 3.10 Colors corresponding to different color temperatures. *(Color version of this figure is available online.)*

By means of color coordinates, it can be very convenient to calculate the color coordinates of several mixed-light sources [1]. Supposing there are n light sources, the color coordinate of each light source is (x_i, y_i), the radiant flux is Φ_{ei}, then the color coordinates after mixing is:

$$x = \frac{\sum_{i=1}^{n} x_i \Phi_{\mathrm{ei}}}{\sum_{i=1}^{n} \Phi_{\mathrm{ei}}} \qquad y = \frac{\sum_{i=1}^{n} y_i \Phi_{\mathrm{ei}}}{\sum_{i=1}^{n} \Phi_{\mathrm{ei}}} \tag{3.19}$$

According to the response of the human eye to different spectrums, the color coordinates and color temperature of the light source are defined. However, due to visual reasons, the light sources of a different spectrum may be provided with the same color coordinates. Then the light sources with different spectrums but with the same color coordinates are called metameric sources [1]. As each object irradiated has a reflectance spectrum, the same object irradiated by metameric source may show different colors, resulting in a shift of the color coordinate. In the lighting area, the degree of shift of the color coordinate reflects the quality of light source. A standard light source is usually defined, and the color of the object irradiated by the standard light source is called the real color of the object. Then compare it with the shift of color coordinates of other light sources, and define the shift of this color coordinate as the color rendering index Ra. In this way, the ability that the light source has to restore the real color of the object can be described quantitatively. The color-rendering index of a light source can reach 100 at most. The higher the color-rendering index is, the stronger the ability that the light source has to restore the real color of the object. Some methods to calculate the color rendering index of the light source are introduced as follows. The studies have shown that the color density that the human eye can distinguish in the figure of 1931CIE is non-homogeneous [1]. The chromaticity diagram 1931CIE mentioned above is not suitable for evaluating the shift of color coordinates. Therefore, the uniform chromaticity scale (UCS) diagram is introduced. The coordinates can be obtained through the following change:

$$u = \frac{4X}{X + 15Y + 3Z} = \frac{4x}{-2x + 12y + 3}, \quad v = \frac{6X}{X + 15Y + 3Z} = \frac{6y}{-2x + 12y + 3} \tag{3.20}$$

Figure 3.11 is the uniform chromaticity scale (UCS) diagram after the coordinate transformation.

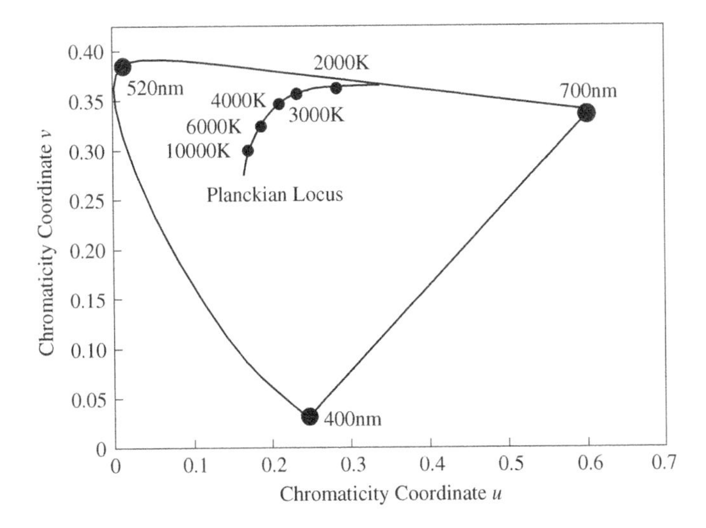

Figure 3.11 UCS chromaticity diagram.

As the reflectance spectrums of different objects are not the same, in the evaluation of the color-rendering index of the light source, the reflectance spectrums of eight standard samples (1964CIE) are generally chosen as references.

The evaluation method is to compare the colorimetric shifts of the standard light source to these eight samples, with the tested light source of the eight samples, then get the average value and obtain the color rendering index of the tested light source. When evaluating the color rendering index of the light source, Planckian black body is usually used as the standard light source and the color rendering index of the standard light source Ra is assumed to be equal to 100 (Ra = 100), for the daylight illumination is often very similar to the blackbody radiation source.

Supposing that the spectrum of the standard light source is $S_r(\lambda)$, the reflectance spectrum of eight samples irradiated by a standard light source is $S_r(\lambda)\rho_i(\lambda)(i = 1, \ldots, 8)$, the spectrum of the tested light source is $S_K(\lambda)$ and that the reflectance spectrums of eight samples irradiated by tested light source is $S_k(\lambda)\rho_i(\lambda)(i = 1, \ldots, 8)$, then the color rendering index of the tested light source Ra can be calculated through the steps shown in Figure 3.12.

In the above chart, the parameters c and d are defined by the following equations:

$$c = \frac{4 - u - 10v}{v}$$
$$d = \frac{1.708v + 0.404 - 1.481u}{v} \tag{3.21}$$

W is defined by the following equation:

$$W = 25Y^{\frac{1}{3}} - 17 \tag{3.22}$$

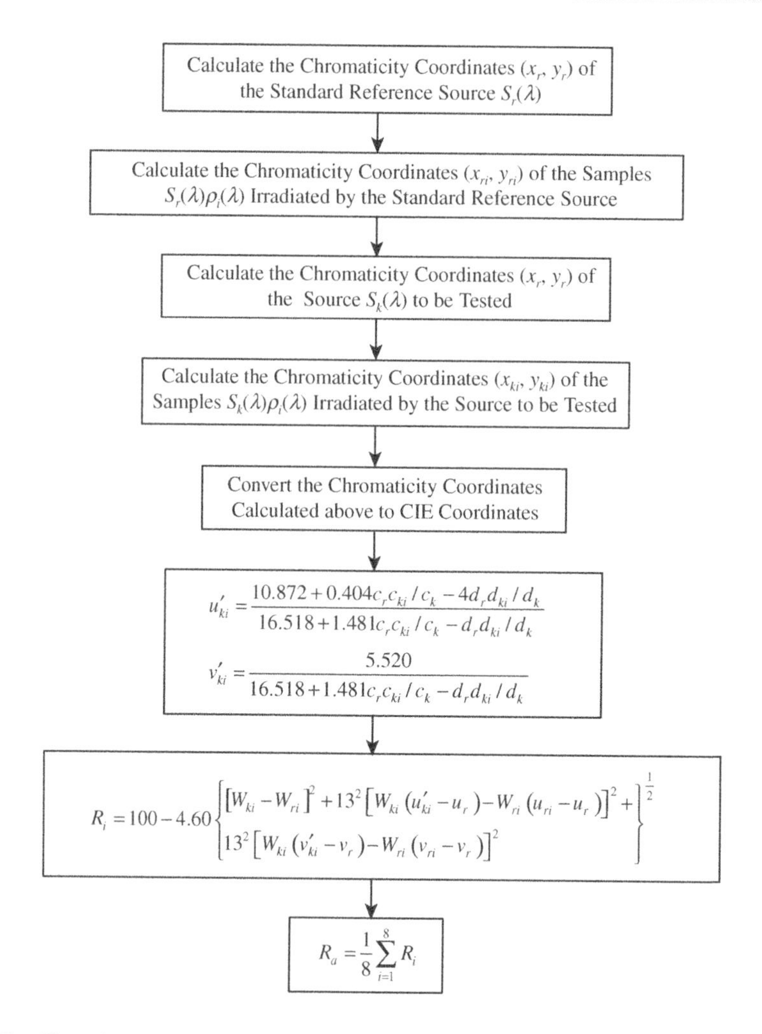

Figure 3.12 Flow chart of calculating the color rendering index [1]. (Adapted from A. Zukauskas, M.S. Shur and R. Gaska, *Introduction to Solid-State Lighting*, Wiley-Interscience, 2002.)

In the process of calculating Y, $S_r(\lambda)$, $S_k(\lambda)$ should be multiplied by a factor, making $Y_r = Y_k = 100$.

3.1.7 White Light LED

In lighting applications, white light LED is usually used as the light source. However, the LED chip can only generate monochromatic light with a very narrow spectral range or ultraviolet light that the human eye cannot see. Therefore, it is necessary to change the light generated by the chip into white light. There are three methods to generate white light: mixing the lights with a variety of colors to generate white light in module level; using phosphor conversion to generate white light; and utilizing an LED with multiple-cavity white light chip.

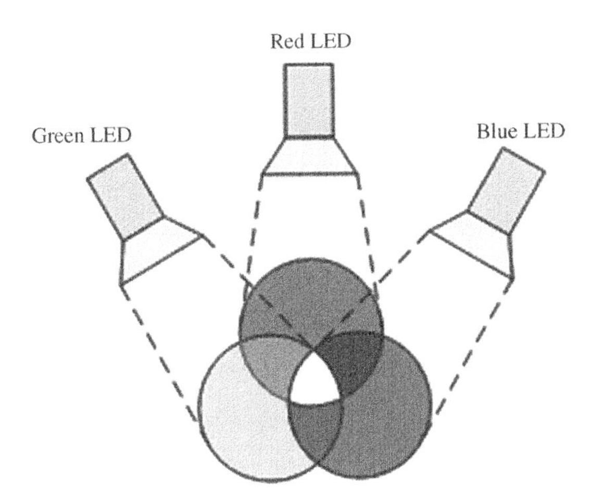

Figure 3.13 Schematic of a white light obtained by mixing several monochromatic lights with different colors. *(Color version of this figure is available online.)*

(i) Mixing Different Lights to Generate White Light

In Section 3.1.6, we saw that white light can be obtained by mixing several monochromatic lights with different colors which is shown in Figure 3.13. In the 1931CIE chromaticity diagram, the color coordinates of the mixed-light are in the polygon forming by the corresponding color coordinates of light sources. By adjusting the radiant flux's ratio of the light source in each part, white light with different color coordinates can be obtained after mixing. R. Mueller-Mach and other researchers have studied two types of trichromatic LED generating white light: 460–530–630 (nm) and 450–550–610 (nm) [4]. The white light adopting these three colors of 460–530–630 (nm) is provided with high lumen efficiency k but low color rendering index Ra; While the white light adopting the three colors of 450–550–610 (nm) is provided with high color rendering index Ra, but it is affected greatly by temperature and is not stable.

The light-mixing method is the simplest one to generate white light by using an LED with different colors. However, the color rendering index of the generated mixed-light Ra, lumen efficiency k and the stability of chromaticity should be taken into consideration comprehensively.

(ii) Using Phosphor Conversion to Generate White Light

At present, the white LED normally uses blue chips and yellow phosphor or ultraviolet chips and RGB phosphor to generate white light. The phosphor is coated on the luminous surface of the chip in LED packaging. When using blue chips and yellow phosphor, part of the blue light generated by the chip is converted into yellow light through phosphor while the other part comes out directly from the packaging module. Then the blue light and yellow light are mixed to generate white light. The working schematic diagram is shown in Figure 3.14.

At present, most white light LEDs use this method. By adjusting the mixing proportion and concentration of phosphor, white light with different color temperatures can be obtained.

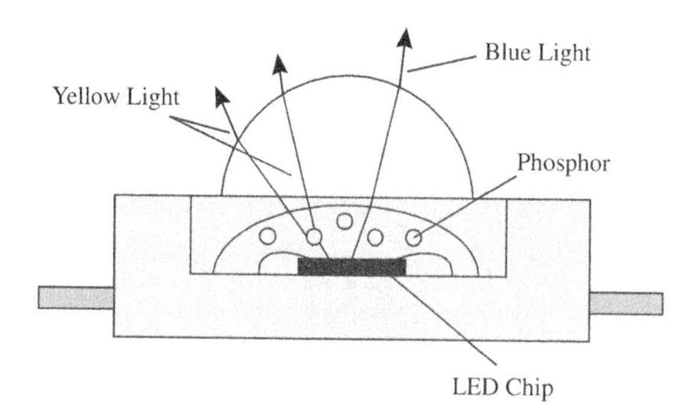

Figure 3.14 Schematic of structure generating white light with the blue light chip and yellow phosphor.

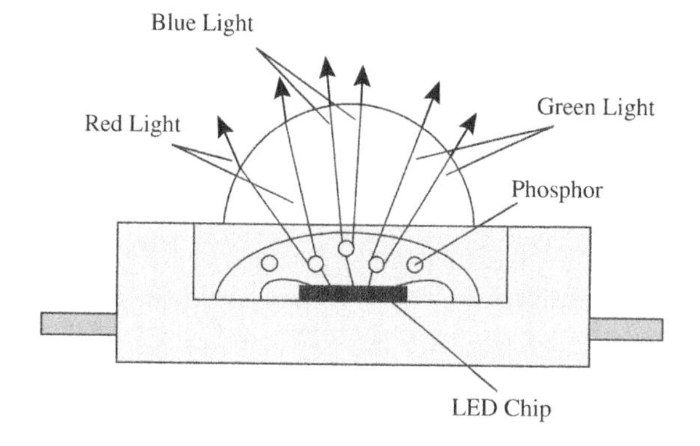

Figure 3.15 Schematic of structure generating white light with ultraviolet light chip and RGB phosphor.

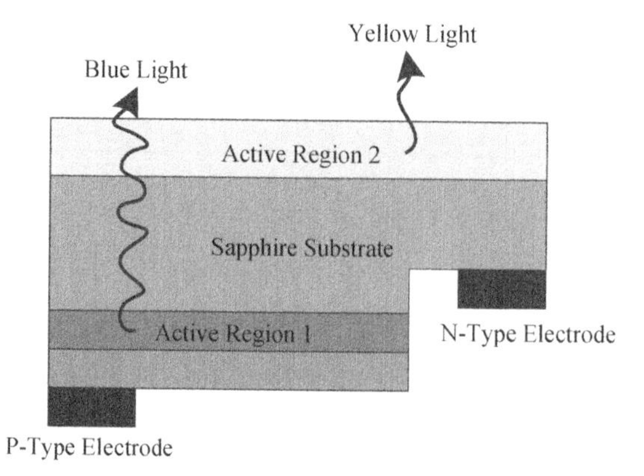

Figure 3.16 Schematic of structure for direct white light chip.

When ultraviolet chips and RGB phosphor, the ultraviolet light generated by the chip is converted into red, green, and yellow light through RGB phosphor. Then these lights are mixed to generate white light [5]. A schematic of the structure is shown in Figure 3.15.

(iii) LED with White Light Chip

White light chip LED means that the LED chip itself can give off white light directly. This requires the LED chip to give off at least two kinds of light which will be mixed to generate white light. X. Guo *et al* [6] made the chip generate blue light and yellow light by photon recycling, and then mixed them to generate white light. A schematic of the chip structure is shown in Figure 3.16.

3.2 Key Components and Packaging Processes for Optical Design

As depicted in Figure 3.17, the basic packaging components of phosphor-converted white LEDs (pcLEDs) include chip, phosphor, lens, reflector and substrate. Reflector is not necessary but it can concentrate the light and change the intensity distribution. The initial optical performance of white LEDs is mainly determined by the optical properties and structures of the packaging components, and the processes. The chip, phosphor, and lens are considered to be the key components in LED packaging. Chip determines the initial optical power, which affects the final maximum luminous efficiency of white LEDs. The emission spectrum of phosphor determines the luminous flux according to visual sensitivity function. Color indices such as correlated color temperature (CCT) and color rendering index (CRI) also depend on the characteristics of phosphor. The lens controls the luminous intensity distribution by changing lens shape and enhances the extraction efficiency of chip by providing encapsulant materials with high refractive index.

3.2.1 Chip Types and Bonding Process

In high power LED packaging, there are mainly three types of LED chips. As shown in Figure 3.18, the first is the conventional chip with horizontal electrodes, the second is the vertical injection chip, and the third is the flip chip. The choice of chip type for LED packaging makes no significant difference in terms of the optical performance. But the general view is that the vertical injection chip and the flip chip can present higher light extraction than that in the

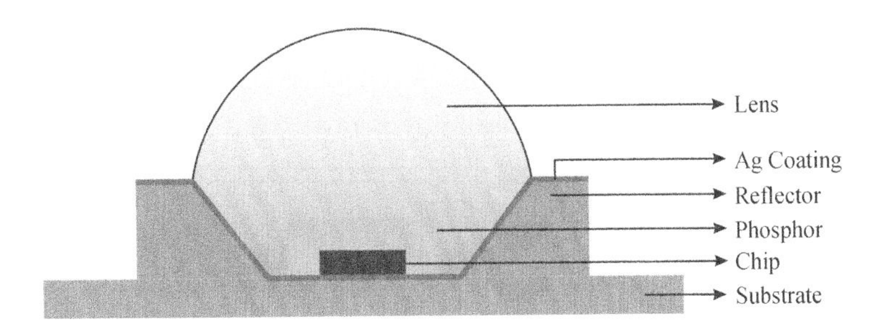

Figure 3.17 Schematic structure of typical phosphor-converted white LEDs. *(Color version of this figure is available online.)*

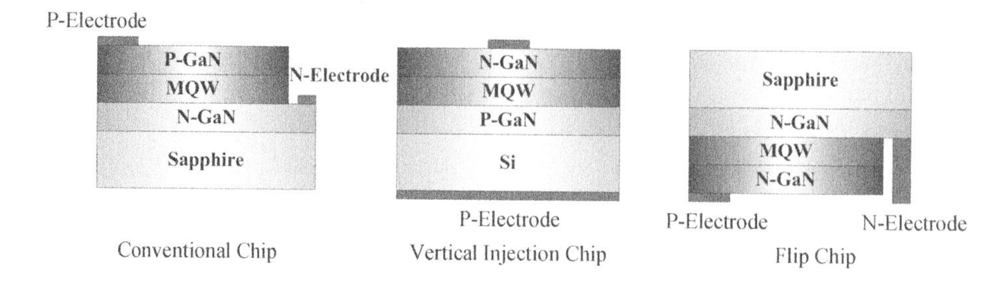

Figure 3.18 Three typical LED chips used for high power LED packaging.

conventional chip. This is because the top area of the chip used for the electrodes in the vertical injection chip is smaller than that in the conventional chip, and the low refractive index of sapphire in the flip chip can give light a larger escaping cone than that in the conventional chip.

Although the conventional chip has a slightly lower light extraction than the other two chip types, this chip is still the major type used for LED packaging. This is due to the fact that the chip manufacturing process of the conventional chip is simpler than the other two chip types.

The bonding process of a chip is critical in LED packaging. Generally, the conventional chip and the vertical injection chip are bonded on a metal substrate by solder or silver paste. The solder or silver paste should be pre-coated on the metal substrate and then the chip is mounted on the solder or pasted with slight pressure. After that, the solder is melted and cooled by reflow soldering, or the silver paste is cured at high temperature. Since the solder and paste tend to be soft and have better flow properties, the chip location on the metal substrate can be changed. In addition, if the pressure used to locate the chip cannot be uniformly applied across the chip surface, the chip will lean. Figure 3.19 shows some typical variations of chip location.

In type I, the chip is slightly deviated from the center. Since light emitted from different parts of the chip propagates with different lengths in phosphor and the light is focused on different locations, this can easily result in non-uniform spatial color distribution and asymmetry luminous intensity distribution. In type II, the solder or paste is too thick. The inconsistency of solder or paste thickness will cause the luminous efficiency and CCT of products to be varied with each other. If the variation of solder or paste thickness is too large, some products may not meet the specifications of users. This will increase the total cost of products. In type III, the chip is leaned. This can also induce non-uniform spatial color distribution. All of these variations of chip location can affect the light extraction due to the scattering of phosphor particles, demonstrating the coupling effect of packaging manufacturing/assembly and the optical quality of LED modules.

In addition, the chip may also rotate around the vertical axis during the manufacturing and handling in these three types. This makes the variation of chip location more complicated.

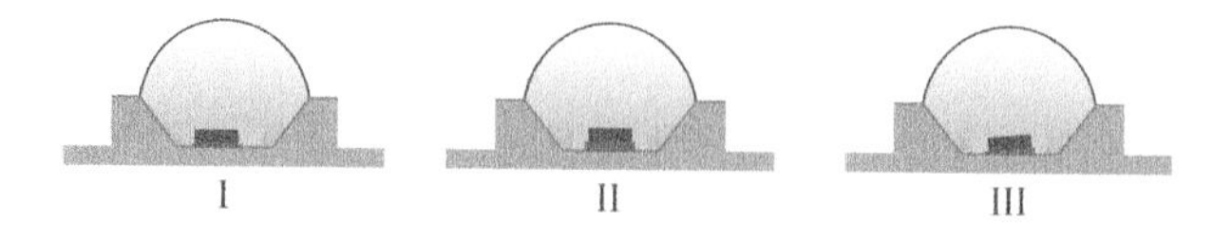

Figure 3.19 Schematic illustration of variation of chip location. *(Color version of this figure is available online.)*

The solder ball array used for flip chip bonding can self-align the chip and therefore reduce the variation of the chip location. To control the chip location as precisely as possible, the thickness of solder and paste should be as thin as possible.

For the conventional chip, special attention should be paid to the issue that the thickness of solder or paste is too thick. The chip can be embedded into the solder due to the weight effect. Some part of the sapphire substrate is therefore covered by solder, which can block the light extraction of the chip. It is also true for the paste, as it may over-flow to the edges of the chip.

3.2.2 *Phosphor Materials and Phosphor Coating Processes*

Generally, phosphor materials are sorted into two categories: one is used for single phosphor-converted (SPC) LEDs, the other is used for multi-phosphor-converted (MPC) LEDs. In SPC LEDs, the optical requirements for phosphor materials include high quantum efficiency pumped by blue light and wide emission spectrum, which are essential for high luminous flux and high CRI. Yellow phosphor, the dominant wavelength of which is in the yellow-green spectrum, is the most widely used phosphor material for SPC LEDs. The materials under development include Ce^{3+} doped garnet materials [7] and Eu^{2+} activated nitride and silicate compounds [8–10]. Among these materials, YAG: Ce phosphor presents the highest quantum efficiency, the best optical-thermal stability, [11] and the widest waveband [8] and is therefore chosen for the generation of natural and cool white LEDs. Figure 3.20 shows the absorption and emission spectrums of one typical YAG: Ce phosphor. The MPC warm white LEDs also prefer YAG: Ce phosphor to be the yellow-green spectrum converter to increase the luminous efficiency.

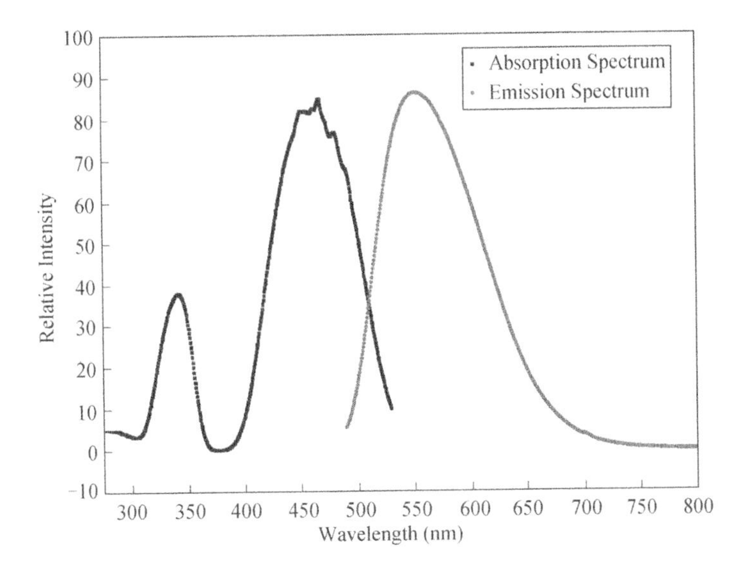

Figure 3.20 Absorption and emission spectrums of one typical YAG: Ce phosphor.

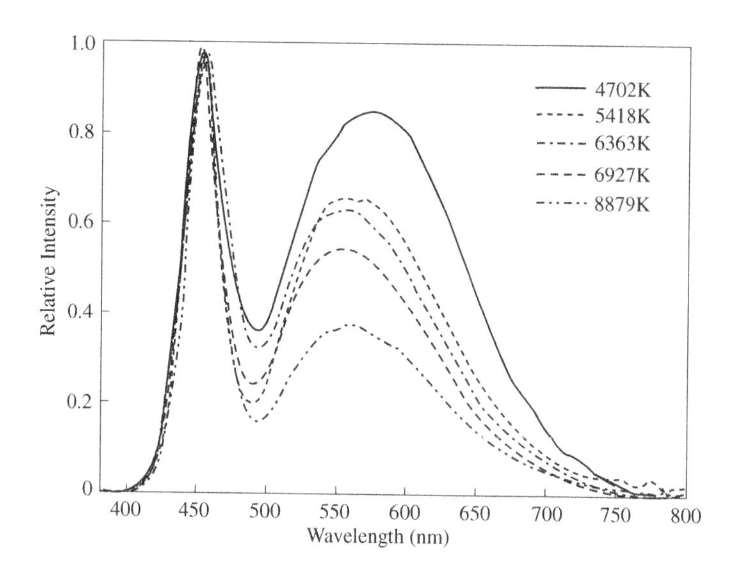

Figure 3.21 Spectrums of SPB LEDs with various CCT. Increasing thickness or concentration can change CCT to be lower.

In SPC LEDs, through adjusting the thickness and concentration of the phosphor layer, the color of light can be changed from cool white (5000–10000 K) to natural white (4000–5000 K) as shown in Figure 3.21. This method is simple and can be easily handled to reduce manufacturing cost. In addition, the excitation spectrum of YAG: Ce phosphor is wide (100–200 nm) [8], which can provide higher CRI than traditional lamps such as the cool, white fluorescent lamp (CRI = 63) and high pressure sodium lamp (CRI = 20) [12]. The CRI of SPB method normally ranges from 70 to 80, which can meet the demands of most applications such as road light and spot lighting.

MPC LEDs normally require two or more types of phosphor materials to generate white light. There are generally two configurations of phosphors for blue chip pumped MPC LEDs. One is adding phosphors with longer wavelength such as red or orange red phosphors into YAG: Ce phosphor [13,14], the other is combining the blend of green and red phosphors with the blue chip, which generates white light by three primary colors [8,15,16]. Figure 3.22 exhibits the typical spectrums of these two configurations. Since the spectrum of MPC LED is much wider than that of SPC LED, especially in the region of red color, CRI of higher than 90 can be easily achieved, and lowering of CCT is realizable. CCT can be controlled by changing the ratio of phosphors in the blend. Increasing the amount of red phosphor will emit more red light and generate warm light.

The rapid development of white LEDs spurs the studies of novel host lattices to obtain more efficient phosphors. Table 3.3 lists the representative phosphors for MPB LEDs. Eu^{2+} activated nitride and silicate compounds are the most concerned materials and a wide class of novel phosphors has been developed [8,9,17]. These phosphors present high quantum efficiencies, good thermal-chemical stability and have the potential to be excellent candidates of high quality white LEDs in solid state lighting applications. Other phosphors such as

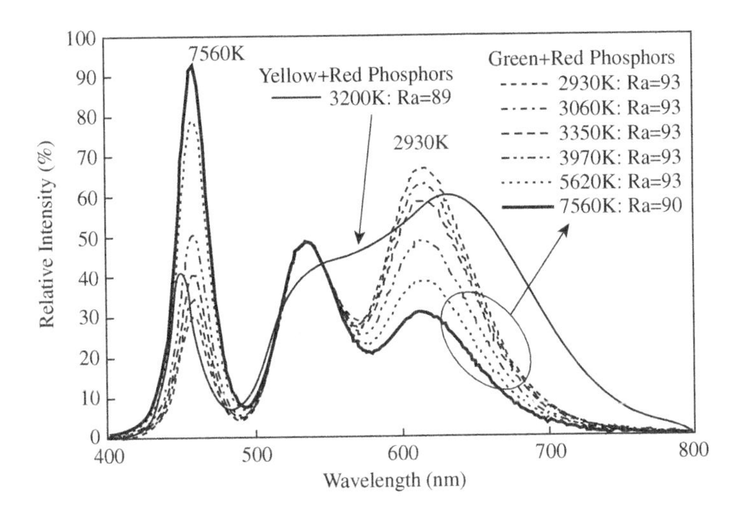

Figure 3.22 Spectrums of MPC LEDs. Green line is the first configuration of MPC. In the second configuration of MPC, it can be found that CCT can be controlled from warm white to cool white while keeping high CRI [15].

sulfides and thiogallates based normally show insufficient stability at high humidity and temperature.

Phosphor coating is one of the most important steps in packaging processes. Figure 3.23 shows the SEM picture of YAG: Ce phosphor particles. Scattering enhancement particles such as SiO_2 are preferred to be added in phosphor to obtain uniform white. The average radius of phosphor particles is 5–8 μm. In one LED module, the number of phosphor particles is generally more than 100,000 per mm^3. This indicates that there may be millions of phosphor particles in a phosphor layer. Therefore, light will encounter many particles and be multi-scattered during the propagation. Each YAG phosphor particle absorbs blue light and emits yellow light. It means that the blue light is weakened gradually whereas the converted yellow light is increased with the increase of scattering. Therefore, the finally emitted white light

Table 3.3 Representative Phosphors for pcLEDs

Phosphor Composition	Emission Color
$SrSiON:Eu^{2+}$	Yellow-Green
$(Ca,Sr,Ba)_5(PO_4)Cl:Eu^{2+},Mn^{2+}$	Yellow-Orange
$Sr_2Ga_2S_4:Eu^{2+}$	Green
$SrAl_2O_4:Eu^{2+}$	Green
$Sr_2P_2O_7:Eu^{2+},Mn^{2+}$	Yellow-Green
$(Y,Gb,Tb)_3(Al,Ga)_5O_{12}:Ce^{3+}$	Yellow-Green-Orange
$(Ba,Sr,Ca)_2Si_5N_8:Eu^{2+}$	Orange-Red
$(Sr,Ca)S:Eu^{2+}$	Red

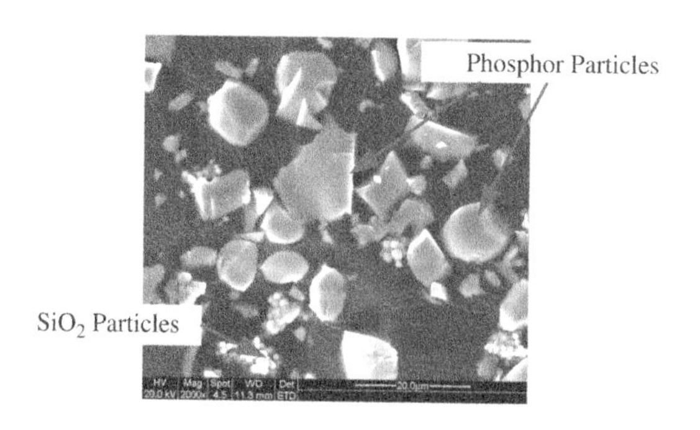

Figure 3.23 SEM picture of phosphor particles. The sizes of phosphor particles are irregular. Study normally applies the average radius to represent the dimension of phosphor particle. SiO_2 particles are used to enhance the scattering.

significantly depends on the scattering chances of light. If the propagation length of one ray is longer or the number of scattering is more than that of other rays, the color of this ray may tend to be yellow. If the opposite case applies, then the color tends to be blue. Since the propagation length and number of scattering are related to the thickness and concentration of the phosphor, when phosphor coating you should try to make the thickness and concentration as uniform as possible.

Normally there are three phosphor coating approaches to LEDs packaging as shown in Figure 3.24. One is freely dispensed coating, the second one is conformal coating, and the third

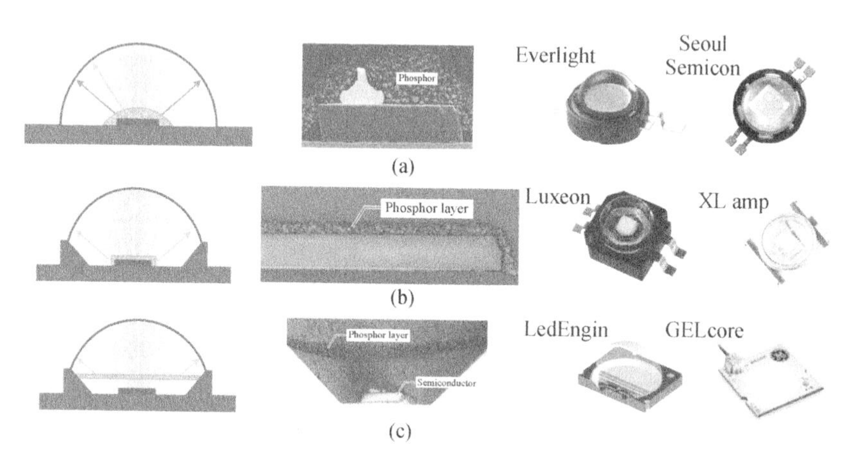

Figure 3.24 Illustrations for three phosphor coating technologies. (a) Freely dispensed coating, light color is varied from yellow to white to blue. Representative corporations are Everlight and Seoul Semicon. (b) Conformal coating, light color is almost white. Representative products are Luxeon from Lumileds and XLamp from Cree. (c) Remote coating, light color is varied from white to yellow. The surfaces of pre-cured encapsulants and phosphor can be found to be concave. Representative corporations are LedEngin and GeLcore. *(Color version of this figure is available online.)*

is remote coating. In the first and third coating approaches, phosphor particles are uniformly mixed with silicone to obtain phosphor silicone. Then phosphor silicone is fabricated to be a film to form a phosphor layer. Phosphor silicone can be conveniently handled by dispensing equipment. However, special attention should be paid to the phosphor settling caused by weight and large particles, which can result in inhomogeneous concentration distribution.

Freely dispensed coating is the oldest method, which was developed from low power white LEDs. This approach dispenses phosphor silicone on the chip without a mold restricting the flow of phosphor silicone until a surface force balance is achieved. The shape of the phosphor layer is normally convex and the thickness of the central zone on a chip is higher than that of the edge zone. Therefore, the color of light emitted from the central zone tends to be warm white, whereas color from the edge zone tends to be cool white. The advantage of this method is that the thickness of the phosphor layer can be easily controlled by the volume of phosphor silicone and the size of dispensing zone. No special techniques are required and therefore the manufacturing time and cost are reduced.

However, this approach cannot fabricate high quality white LEDs. To obtain a balanced white light across the chip, the thickness of phosphor layer is normally in the range of 0.2–0.5 mm to avoid insufficient conversion for side emitting blue light. Therefore, the concentration of phosphor silicone cannot be too high, otherwise the color will tend to be yellow. Increased propagation length induces the reduction of luminous flux. In addition, to enhance the throughout, the viscosity of phosphor silicone is controlled in the range of 3000–5000 mPa·s to enable the phosphor silicone to more easily flow and achieve surface force balance. The dispensing time is longer than 500 ms. The inner radius of the nozzle is normally larger than 0.25 mm. This indicates that controlling the volume of phosphor silicone with micron precision is difficult. Therefore, the repeatability and consistency are relatively low and may reduce the yield of products.

Unlike freely dispensed coating without special manipulation of phosphor, conformal coating is really an advanced packaging process. This approach fabricates the phosphor layer extraordinarily thinly by stacking phosphor particles to obtain high concentration. Phosphor film is uniformly coated on the chip surface to generate uniform white light. Conformal coating was firstly developed by Lumileds, which applied the electrophoretic method [18] to deposit charged phosphor particles on the chip surface. Controlling the voltage and deposition time can adjust the thickness of phosphor film. Therefore, conformal coating can easily realize micron precision. Other developed approaches such as slurry, settling [19], evaporating solvent [20], wafer-level coating [21], and direct white light [22] can also conformally coat the phosphor. Since the thickness is significantly reduced, shortened propagation length will decrease the number of redundant scattering and thereby the lumen output is increased. Conformal coating requires special handling in the following process steps, since the phosphor film is thin and fragile. Small force may damage the film and induce delamination between the film and chip.

The former two technologies both fabricate the phosphor layer directly on the chip surface. The advantage is that the size of LED is minimized and can be used for high density packaging. However, experiments confirmed that there is approximately 50%–60% light back-scattered by the phosphor layer [23,24]. These light rays will be re-absorbed by the chip and part of energy is lost. In addition, the localized heating caused by the high power chip can induce thermal quenching and reduce the quantum efficiency (QE) of phosphor [25]. Temperature of phosphor particles may be as high as 120 °C and QE can be decreased to 70%.

Adjusting the phosphor layer to a remote location can reduce the chip absorption and increase the light extraction [26–28], since only a small part of light rays will reach the chip and be absorbed. Increased distance also improves the color stability by lowering the surface temperature of phosphor [29]. This coating approach normally requires a reflector to fix the phosphor layer. Surface treatment of the reflector is essential to offer high reflectivity. Ag is the most favorable coating material. After a soft encapsulant layer is cured on the chip, phosphor silicone is dispensed in the reflector. Applying phosphor silicone will also confront the limitations as discussed in freely dispensed coating. Techniques from conformal coating such as settling and evaporating solvent can be utilized in remote coating to improve the thickness and concentration control. However, the main disadvantage of remote coating is that the shape of the phosphor layer is not a perfect plane. Affected by the surface tension of liquid, the pre-cured encapsulant materials and phosphor layer normally present concave surfaces. Curvatures of these surfaces are dependent on the dimensions and surface roughness of reflector, viscosity of phosphor silicone, operation temperature, and so on. With the increase of reflector angle, the surfaces will tend to be flat. However, the packaging size will be significantly increased if the angle is larger than 45°. High viscosity or low temperature can also reduce the wetting angle to make the surfaces be more flat. However, these methods cannot fundamentally solve this issue. Since the viscosity of phosphor silicone is generally higher than that of pre-cured encapsulant, the curvature of phosphor silicone will be larger. This will cause the thickness of the phosphor layer to be slightly more in central zones.

3.2.3 Lens and Molding Process

There are two roles for the lens in LED packaging. One is protecting the chip and gold wire from being damaged by harsh environment, and the other is enhancing the light extraction by high refractive encapsulants and controlling the radiation pattern of LEDs by a specific shape. The basic requirements of lens materials include high transparency for visible light, high thermal-optical-mechanical stability, high surface hardness to resist scratch, excellent molding properties, and so on.

The initially developed material used for LED packaging is epoxy resin. The refractive index of epoxy resin is around 1.52, and the surface hardness of epoxy resin after curing is also high. Since the cost of epoxy resin is very low, it has been widely used in low power LED packaging. The disadvantage of epoxy resin is that the thermal stability is poor. In elevated temperature or long-time testing or operation, epoxy resin will turn yellow and the transparency is reduced. The life of LEDs is therefore shortened.

When the high power LED packaging emerges, the thermal challenge prompts the adoption of advanced silicone materials. A more technically correct name of silicone is polysiloxane. By changing the branched species, silicones can provide a variety of different materials that can be chosen according to the specific applications. Silicones have excellent optical-thermal-mechanical properties such as temperature stability (–115 to 260 °C), fuel resistance, optical clarity (>95%), variable refractive index, low moisture absorption, low curing shrinkage (<2%), and low shear stress, all of which make silicones favorable in high power LED packaging.

The generally branched species in silicones include dimethyl and diphenyl. Dimethyl silicones are the most common silicone polymers used in LED packaging due to the cost effective mass production. The refractive index of dimethyl silicones is normally around 1.41–1.43. This value is lower than epoxy resin and reduces the light extraction of the chip. Diphenyl silicones present a higher refractive index and can improve the refractive index to be approximately 1.60. Considering the excellent thermal-mechanical properties, diphenyl silicones are developing fast and begin to be adopted in LED packaging widely. Certain caution should be exercised, as there have been some reports on the passive degradation of white LEDs fabricated by these types of silicones.

The cured modulus of silicones can also be adjusted by changing the crosslink density and the ratio of linear to branched silicon species as shown in Figure 3.25. Hard resins and elastomers are preferred for the molding and fabrication of optical lenses, whereas soft gels are used for encapsulating stress sensitive regions such as wire-bonds.

In conventional high power LED packaging, such as Luxeon, there is a polycarbonate cap in the outer of the lens. This polycarbonate cap is used to inject silicone into the inner space of the cap to protect the LED chip and gold wires. The silicone materials include silicone gel and elastomer. Comparing with elastomer, gel presents better stress relaxation and higher interface adhesion properties, and the low viscosity of gel makes the injection-filling process more efficient. However, the low viscosity of the gel also causes the silicone gel to flow out through the edges of cap during the curing due to the high temperature. Small bubbles can be generated during this process. Elastomers present better flow properties during the curing and are now widely used in high power LED packaging.

There are other methods to fabricate the lens. In the products of Cree, such as XLamp-XR-E, the lens is made by hemispherical silicone resin, which can provide better optical stability for UV light than polycarbonate material. In LedEngin, the lens material is fabricated with glass,

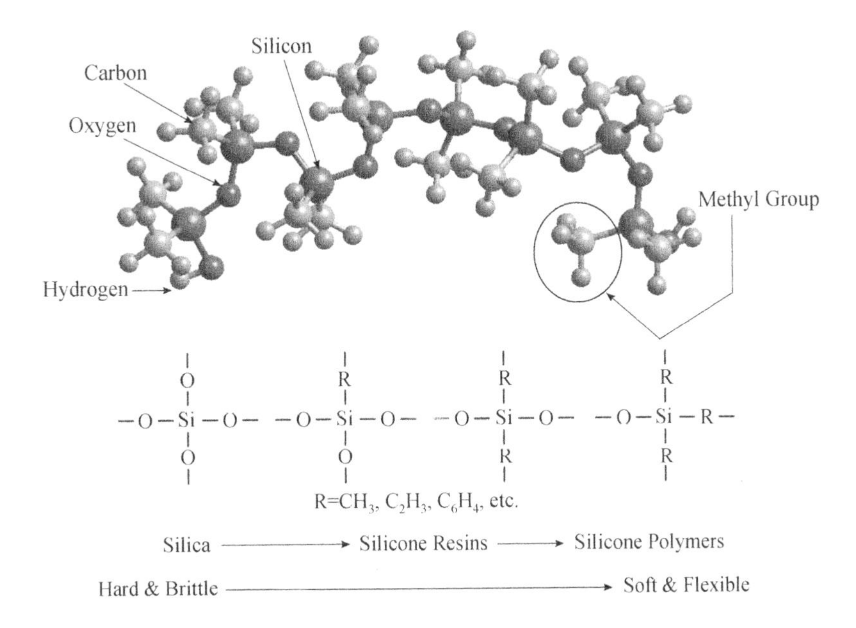

Figure 3.25 Constitution of silicon materials. Increasing the ratio of R branches can decrease the modulus of silicones.

which has higher refractive index and better thermal stability. Recently, the packaging method without the lens cap was developed. This method directly fabricates the lens with soft silicone elastomer. These three methods can avoid the disadvantage of the conventional polycarbonate cap, which will be softened and deformed when the temperature is higher than 100 °C and therefore this means that these types of white LEDs are not usable in the solder reflowing process. By adopting these three lens fabrication methods, white LEDs can be directly bonded on the printed circuit board (PCB) by solder reflowing, therefore surface mountable (that is SMT compatible process), and simplifies the assembly steps in the manufacture of LED luminaires.

The lens molding for packaging method without a lens cap requires more rigorous handling in the manufacturing. Figure 3.26 shows the typical molding approaches in this packaging method. In the first approach, the uncured silicone is pre-filled in the mold, then the board with a chip bonded on it is immersed into the silicone. After curing the silicone, the mold is taken away and the packaged LED is obtained. In the second approach, the mold is assembled with the board. Then the uncured silicone is injected into the mold. Finally, the packaged LED is obtained through the demolding process. In these two approaches, it should be noticed that the alignment of chip and mold is very important. Those other packaging methods with a lens cap can fix the lens one by one and do not have the need to align the lens center and the chip. In the packaging method without a lens cap, LEDs normally are manufactured array by array. Therefore, both the variation of chip location and the alignment error between mold and board array should be considered. Another issue that should be taken care of is the demolding of silicone. Normally, the molds are produced with metal or those materials with

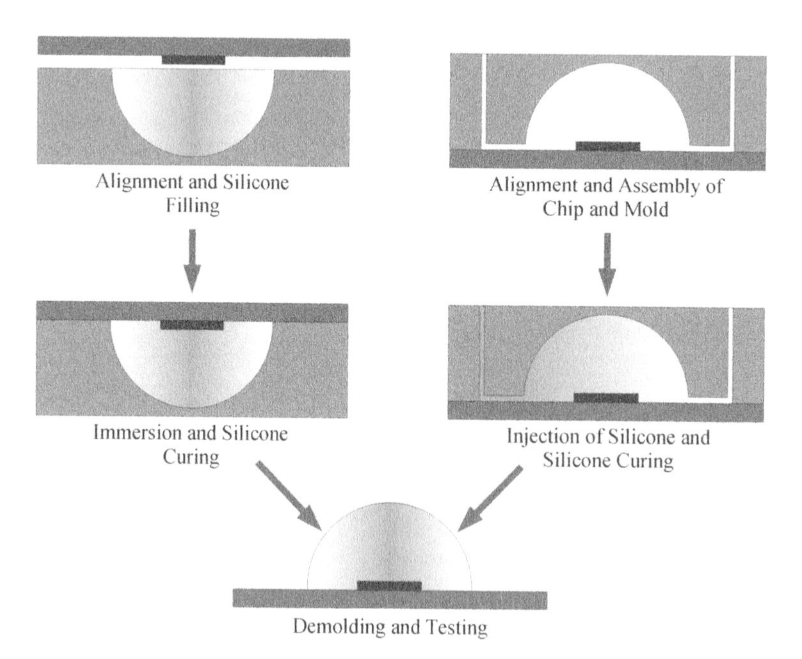

Figure 3.26 Two approaches for the lens molding of packaging method without lens shell.

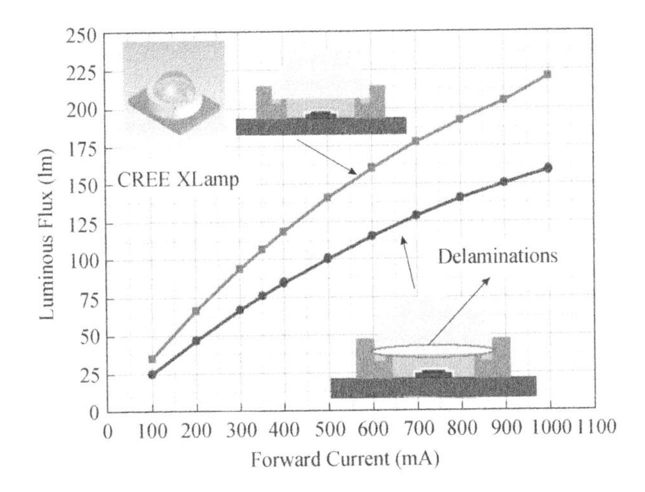

Figure 3.27 Effects of delamination on the luminous flux of a white LED module.

low surface energy. For a metal mold, the mold release agent is required to avoid the adhesion between silicone and mold. The liquid mold release agent should be uniformly coated on the mold surface. As an improvement, mold release film is also used in the demolding. But both these two agents have the risk of producing defective lens due to the adhesion issues.

Although those packaging methods with lens cap have no demolding issues, it does not mean that the lens cap is favorable in LED packaging. Polycarbonate material presents poor adhesion property with silicone. This may induce delaminations which can easily occur along the interface of a polycarbonate cap and the inner silicone. Compared with polycarbonate, silicone resin and glass present better adhesion properties with silicone. However, it should be noted that the delamination still is inevitable. Taking XLamp as an example, if there is delamination in the interface between lens and silicone, the light extraction efficiency can be reduced to be 60% as shown in Figure 3.27.

3.3 Light Extraction

Light extraction is considered to be important in high brightness LED packaging. High luminous efficiency normally requires high light extraction. The refractive index of GaN material is normally around 2.4, meaning that the escaping cone is only 23° when the chip is in air. There are two parts of light to be extracted for phosphor-converted white LEDs. One is blue light from the chip, the other is converted light by phosphor.

The extraction efficiencies of blue light and converted light can be expressed as:

$$\eta_{\mathrm{Blue}} = (1-\eta_{\mathrm{Chip}})(1-\eta_{\mathrm{Phos}})(1-\eta_{\mathrm{Encap}})(1-\eta_{\mathrm{R}}) \tag{3.23}$$

$$\eta_{\mathrm{Conv}} = \eta_{\mathrm{Phos}}\eta_{\mathrm{QE}}\eta_{\mathrm{Stokes}}(1-\eta'_{\mathrm{Chip}})(1-\eta'_{\mathrm{Phos}})(1-\eta'_{\mathrm{Encap}})(1-\eta'_{\mathrm{R}}) \tag{3.24}$$

where η_{Blue} and η_{Conv} are extraction efficiencies for blue light and converted light; η_{Phos}, η_{Chip}, η_{Encap}, and η_R are the efficiencies of phosphor absorption, chip absorption, encapsulant absorption, and light loss of refraction and reflection for blue light, respectively; η'_{Phos}, η'_{Chip}, η'_{Encap}, and η'_{R} are the efficiencies of phosphor absorption, chip absorption, encapsulant absorption, and light loss of refraction and reflection for yellow light, respectively; η_{QE} and η_{Stokes} are quantum efficiency and Stokes efficiency of phosphor materials. The efficiencies and ratios are all calculated by radiometric units.

Luminous flux $L(\lambda)$ of white LEDs is calculated according to the visual sensitivity function $V(\lambda)$:

$$L(\lambda) = P_{\mathrm{Elec}}\eta_{\mathrm{Inj}}\eta_{\mathrm{Int}}\eta_{\mathrm{pc-LEE}}\eta_f(\eta_{\mathrm{Blue}} + \eta_{\mathrm{Conv}})V(\lambda) \tag{3.25}$$

where P_{Elec} is the input electrical power, which is calculated by multiplying the drive current with voltage; η_{Int} is the internal quantum efficiency, which is the ratio of emitted photon numbers to carrier numbers passing through the junction; η_{inj} is the injection efficiency, which is a measure of the efficiency of converting total current to carrier transport in a *p-n* junction; $\eta_{\mathrm{pc\text{-}LEE}}$ is the light extraction efficiency of packaged LED chip; η_f is the feeding efficiency, which is the ratio of photon energy to the total energy of an electron-hole pair.

Another mostly concerned evaluation index for light extraction is packaging efficiency $\eta_{\mathrm{packaging}}$, which is the sum of η_{Blue} and η_{Conv}:

$$\eta_{\mathrm{pack}} = \eta_{\mathrm{Blue}} + \eta_{\mathrm{Conv}} \tag{3.26}$$

Nowadays, η_{pack} is no more than 80% and the expected target is 90% in 2012 [30]. To improve the light extraction, primary solutions are increasing $\eta_{\mathrm{pc\text{-}LEE}}$ and reducing lost energy caused by η_{Chip}, η_{Encap}, η_R, η'_{Chip}, η'_{Encap}, and η'_{R}. Characteristics of phosphor such as η_{QE}, η_{Stokes}, spectrum, and wavebands are also important for high luminous flux.

$\eta_{\mathrm{pc\text{-}LEE}}$ is mainly determined by the chip structure and refractive index of encapsulant materials directly coated on the chip surface. When the top surface of chip is plane, $\eta_{\mathrm{pc\text{-}LEE}}$ is estimated to be approximately 14–16% for cases without encapsulation. Figure 3.28 displays the effect of variation of the refractive index on the light extraction of the chip. It can be found that when the refractive index of encapsulants rises to 1.6, $\eta_{\mathrm{pc\text{-}LEE}}$ is approximately threefold more than that without encapsulant. If the surface of chip is roughened as shown in Figure 3.29, the light extraction efficiency may be as high as 30–50% when the chip is not packaged with encapsulant [31–33]. This is normally twofold higher than conventional LEDs and makes a roughened chip favorable for high brightness LED packaging. However, the improvements of encapsulant on the light extraction will not be as significant as that on conventional LEDs. Estimated enhancement value for encapsulant with refractive index of 1.6 is in the range of 1.1–1.5 as compared with the efficiencies without encapuslant. In addition, the dispensing of encapsulant on a roughened chip should be carefully handled to avoid small bubbles. Since the roughened size is normally in nano-micro scale, air can be easily trapped in small zones in the fast flowing and prototyping process of encapsulants. As shown in Figure 3.29, light may be repeatedly absorbed by the chip, therefore reducing light extraction. In the worst case, $\eta_{\mathrm{pc\text{-}LEE}}$ with encapsulant will be lower than that without encapsulant.

In freely dispensed coating and conformal coating, the first encapsulant layer is mixed with phosphor particles. The refractive index of phosphor is 1.8, whereas refractive indices of embedding materials such as epoxies and silicones are in the range of 1.4–1.6.

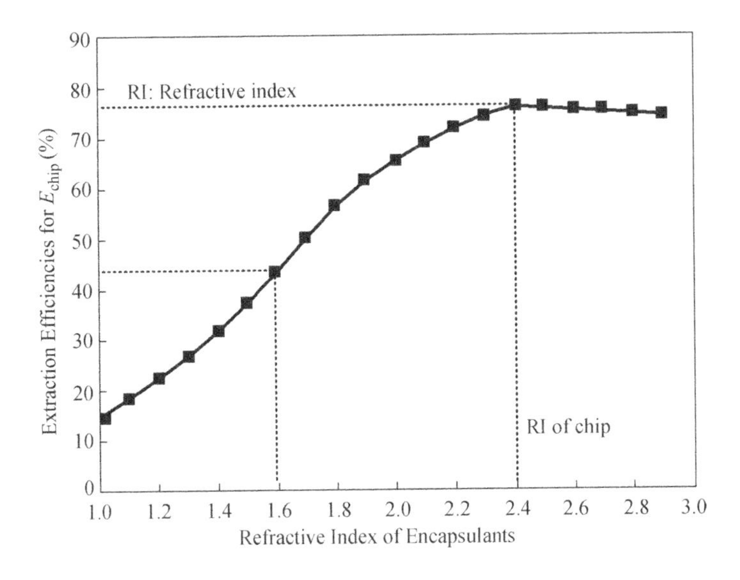

Figure 3.28 Influences of encapsulant refractive index on the light extraction of planar surface chip. Angle is the critical angle in which light can be extracted from chip to encapsulants.

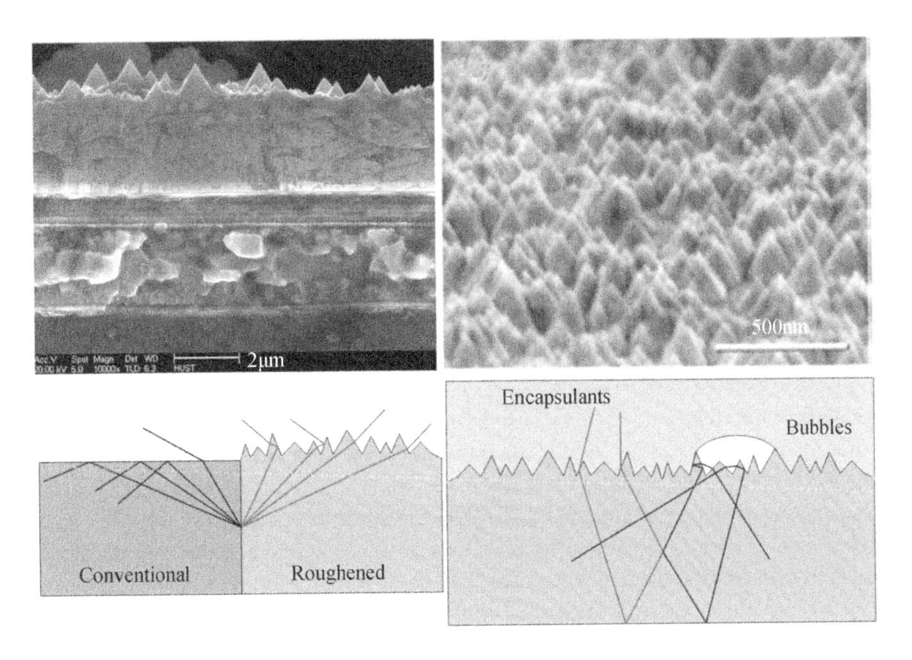

Figure 3.29 Illustrations for chips with roughened surface. It can be found that the size of roughness is nano-micro scale. Roughened surface presents higher opportunity for light extraction. Small bubbles reduce the chances that light can be directly emitted out.

Therefore, mixing silicones with phosphor can provide higher refractive index and $\eta_{\text{pc-LEE}}$ than cases without applying the mixture as first encapsulant layer such as remote coating. However, the refractive index of the mixture cannot exceed 1.8 until refractive indices of encapsulants are further improved. The refractive index of silicone can be controlled by adjusting the ratio of methyl and phenyl units in a silicone system, but the increase will be limited and can induce tremendous research investment. An economic method is adding transparent nano-particles with high refractive index such as TiO_2 to encapsulants [34].

After light is extracted from the chip, high transparency encapsulant and optimized packaging structures are essential to reduce η_{Chip}, η_{Encap}, η_R, η'_{Chip}, η'_{Encap}, and η'_{R}. Silicones are more favorable than epoxies in high power LED packaging. When the power of LED is increased, epoxies show considerable yellowing at higher operating temperature [35]. This yellowing reduces the transparency of epoxies to be less than 70% and presents high absorption for visible light. Higher η_{Encap} and η'_{Encap} will generate significant lumen loss and color variation and consequently shorten the life. Silicones have high optical transmittance in UV-visible region [36,37]. The transparency is normally higher than 95% for thickness of 1 mm. Thermal-opto stability of silicones is also excellent.

To further reduce η_{Encap} and η'_{Encap}, the size of encapsulants should be minimized to decrease the propagation length. However, this will increase η_{Chip} and η'_{Chip}. Reduction of size increases the surface ratio of chip surface to base surface of encapsulant. Back-scattered or back-reflected light will have more chances to enter the chip to be absorbed. Therefore, structures of encapsulant should be designed to decrease back emitted light rays. The first approach is fabricating multi-encapsulants with gradient refractive indices. To avoid totally internal reflection between encapsulants and air, refractive index of the outermost layer should be lower than that of inner layers. Since the refractive index of the inner layer coated on the chip surface is rather high, Fresnel loss in the interface will be high and increase η_R and η'_{R} if there are only two encapsulating layers. Adding intermediate layers to these two layers can guide light rays to the external surface and decrease the chances of total internal reflection. The intermediate layers cannot be too many, otherwise η_R and η'_{R} will be too high and counteract the improvement of η_{Chip} and η'_{Chip}, η_{Encap}, and η'_{Encap}. Process cost also limits the maximum number of layers. Fabricating three to four layers with gradient indices is considered to be acceptable for the requirement of performance and cost.

The second approach is changing the curvatures of local zones and fabricating specific structures in lens and reflectors to decrease the chances of total internal reflection of light when escaping from the external surface of lens [38–42]. Reflector can change the directions of side emitting light rays from chip to central zone. The roughened surface of reflector scatters back reflected light and reduces the chances of light being reflected to the chip. However, it has been found that the improvement is slight if the phosphor layer is directly coated on the chip [43]. The reflector is considered to play a more important role in the remote coating approach and the increase of light extraction can reach 15.4% [27,28]. The len is expected to be more critical in light extraction. Theoretically, the surfaces of lens can be designed to reduce the incident angle of most light and provide more chances for light to be directly emitted out without being re-absorbed by the chip and multi-reflected between lens and reflector. The maximum light extraction could be achieved by a refined lens in principle, which may have many discontinuous surfaces and micro-structures. Actually, the role of today's lens is in controlling spatial intensity distribution and obtaining desired radiation patterns. Figure 3.30 displays three examples of lenses widely used in industry. This design of lens considers the LED chip as a

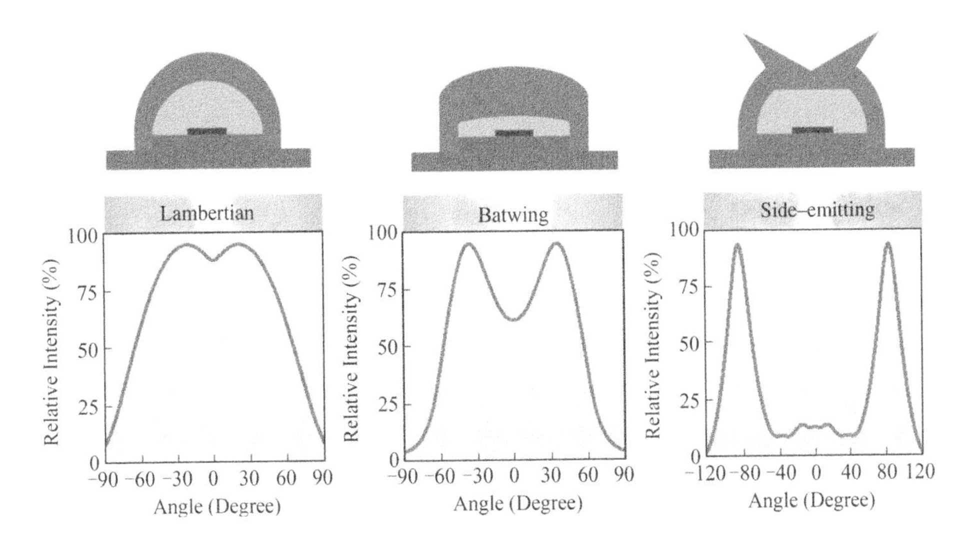

Figure 3.30 Radiation patterns of three lenses. Lambertian lens is the most adopted configuration in LED packaging. Lambertian radiation can be used in applications such as road light, MR16, and so on. Batwing lens and side-emitting lens are suitable for applications such as backlighting and cell phones.

point light source to simplify the optical model. However, the chip essentially is a surface source especially when the size of the LED packaging module is minimized. The lenses designed by the simplified method with point light source presents limited influences on light extraction. However, these lenses can meet requirements of specific applications in which lumen output is not the primary issue but the radiation pattern is the most concerned. Micro-lens array on the chip was believed to be a competitive method to reduce the packaging size and provide high light extraction and various radiation pattern simultaneously [44]. However, the cost will be a serious issue.

The conversion loss caused by phosphor is another important component affecting light extraction. In good LED products such as Luxeon, XLamp, and GoldenDragon, energy loss ratio caused by η_{Chip}, η_{Encap}, η_R, η'_{Chip}, η'_{Encap}, and η'_{R} may not exceed 9% in the extracted light from chip. However, it is estimated that the conversion loss of phosphor may be higher than 12% of $\eta_{\mathrm{pc\text{-}LEE}}$. This is mainly due to the effects of η_{QE} and η_{Stokes}. Nowadays, the highest η_{QE} is achieved by YAG: Ce phosphor, where the η_{QE} is higher than 95% in 75 °C [11]. η_{QE} of other phosphors such as red and green phosphor will be lower than 80% when temperature is higher than 80 °C [45,46]. η_{Stokes} is determined by the excitation wavelength of chip λ_{chip} and emission wavelength of phosphor $\lambda_{\mathrm{phosphor}}$ [47].

$$\eta_{\mathrm{Stokes}} = \frac{\lambda_{\mathrm{chip}}}{\lambda_{\mathrm{phosphor}}} \tag{3.27}$$

Generally, λ_{chip} of InGaN chip is 455–465 nm and $\lambda_{\mathrm{phosphor}}$ of YAG: Ce phosphor is 550–560 nm. Therefore, η_{Stokes} will be lower than 85%. That means the conversion efficiency η_{CE}, which is calculated by multiplying η_{QE} with η_{Stokes}, is lower than 82%. Therefore, the conversion loss is 18%. If η_{CE} is 70%, the energy loss ratio by phosphor will be 12.6%

of $\eta_{\text{pc-LEE}}$. When adding phosphors with wider spectrum or longer wavelength such as red phosphor to compensate for the color of white, η_{Stokes} will be further reduced and thereby increase the conversion loss.

3.4 Optical Modeling and Simulation

To accelerate the penetration of white LEDs into the illumination market especially the general lighting market, a fundamental understanding of the performance fluctuating behaviors of white LEDs during the manufacturing and the operations becomes an urgent task. It is of interest to be able to predict the optical performance of phosphor-converted LEDs (pcLEDs) including light extraction efficiency (LEE), luminous efficiency (LE), correlated color temperature (CCT), and color rendering index (CRI). Since the optical performance of pcLEDs normally coincides with the heat generation from the chip and can be affected by various factors such as material degradations, contaminations, and delaminations simultaneously, experimental investigation of the optical performance behaviors of pcLEDs will be complex. Optical simulation is an effective approach to separately investigating the effects of different factors on the optical performance of pcLEDs by only changing one factor at a time.

The issue is that a well developed simulation method for the optical simulation of pcLEDs is lacking. Monte Carlo is the mostly adopted method to simulate LED packaging. But the accuracy of the results of this method greatly depends on the definitions of material properties of components and the simulation procedure, for which few publications, in any detail, were found. In addition, Monte Carlo is only capable of light ray tracing and cannot directly provide information such as LE, CCT, and CRI. The simulation results of the Monte Carlo method need to be further improved by a suitable method to provide a better prediction of pcLED optical performance.

Before the optical simulation, precise optical modeling of LED packaging is also important. Comparing with the thermal design of high power white LEDs, the optical modeling of white LEDs is more difficult. Although the structure of a single chip LED module is simple, a lack of fundamental understanding of the optical properties of the packaging components makes the optical modeling of white LED packaging rather complicated. There are two barriers affecting the optical modeling: one is the chip modeling, the other is the phosphor modeling.

3.4.1 Chip Modeling

The difficulty of chip modeling is due to the lack of precise optical properties of the chip. Generally, there are three approaches for the definition of the chip, as shown in Figure 3.31. The first approach is treating the chip as a homogeneous material with the same refractive index and absorption coefficient. The blue light is emitted from the top surface of the chip, and in some cases, it is also emitted from the side surfaces [27,48,49]. This approach is very time efficient and suitable for the optical design with the main consideration of the primary/secondary optics [38,42]. This approach cannot reflect the change of radiation pattern of the chip after packaged. This increases the light extraction and changes the emitting pattern of blue light in the package and finally results in remarkable differences between the simulation and experiment [27]. The second approach is distinguishing the differences of refractive indices of AlInGaN alloys but considering the absorption of different AlInGaN alloys to be

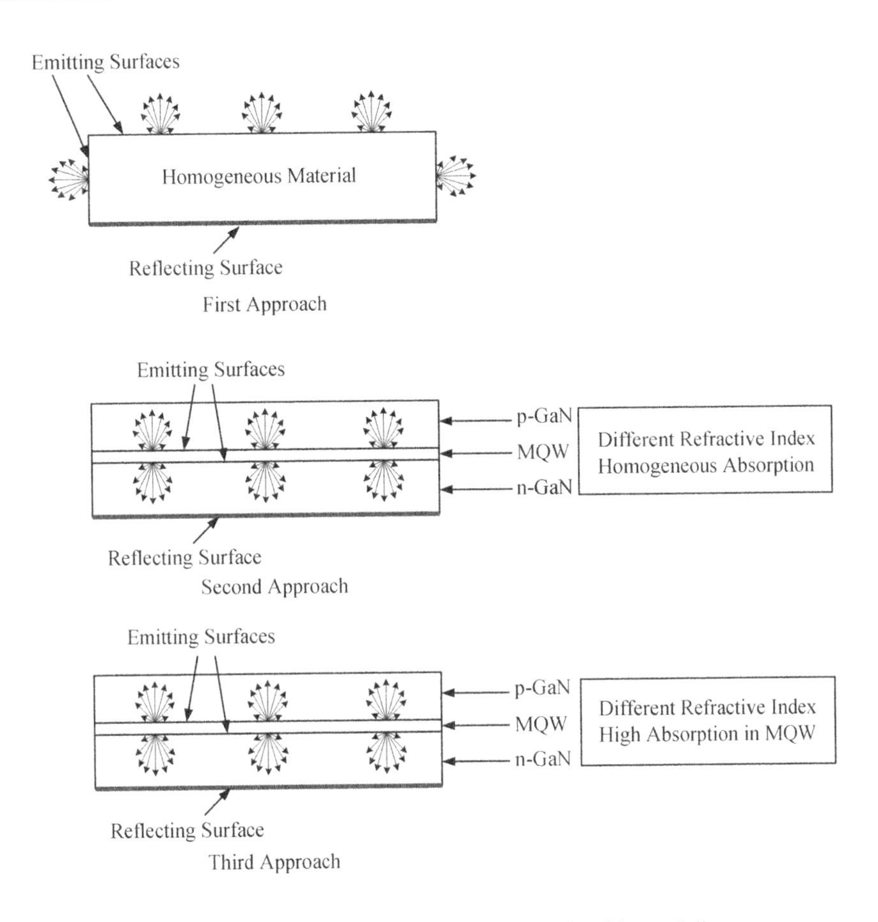

Figure 3.31 Three typical approaches for the chip modeling.

similar [50,51]. The blue light is emitted from the MQW layer with isotropic radiation pattern, so it overcomes the issue of the first approach. The third approach is also giving different refractive index for different AlInGaN alloys but only considering the absorption in MQW layer. Other AlInGaN alloys are defined to be transparent for visible light [52,53]. Normally, the absorption coefficient of MQW in the third approach is significantly higher than that of AlInGaN alloys in the second approach to generate enough chip absorption for blue light. By adjusting the absorption coefficients of AlInGaN alloys to make the light extraction efficiency of the chip (η_{LEE}) match the reported data, both the second and the third approaches can be applied for the optical simulation of LED packaging and provide reasonable results.

It should be noted that both the homogenous absorption of the second approach and the extremely high absorption in the MQW of the third approach are not accurate enough in the description of chip absorption for visible light. When the chip is packaged with encapsulants, those light rays previously confined in the chip can emit and make the light absorption by the chip fluctuate in different chip absorption models. In addition, most of the studies defined that there was no variation of refractive index of AlInGaN alloys for the whole visible spectrum, which may be an issue for optical simulation. Actually, the absorption coefficient and refractive index are both varied with the composites of AlInGaN alloys and the wavelength. But an absolutely precise optical model of the chip is also impossible.

The refractive indices of AlInGaN alloys have been studied fundamentally many times [54–58]. In reference [55], both the refractive index and absorption coefficient of AlGaN epitaxial films were given. Figure 3.32 shows the results [55]. The photon energy E_g has a relationship with wavelength λ as:

$$\lambda(\mathrm{nm}) = \frac{1240}{E_g} \tag{3.28}$$

From Figure 3.32, it can be found that the refractive index of AlGaN is reduced with the increase of the Al fraction in AlGaN alloys. The AlGaN layer is normally used to confine the electrons and holes in the MQW layer to increase the combination efficiency to increase the internal quantum efficiency. Figure 3.32 also gives the refractive index of GaN in visible light, which is varied from 2.4 to 2.55. That means the refractive index of Mg doped and Si doped GaN materials can be approximated by the refractive index of GaN.

Based on the experiments, the theoretical studies of refractive index of AlInGaN alloys have been conducted [54,56–58]. Laws *et al.* [57] summarized the calculation methods of refractive index of AlInGaN alloys and proposed the improved refractive index equations

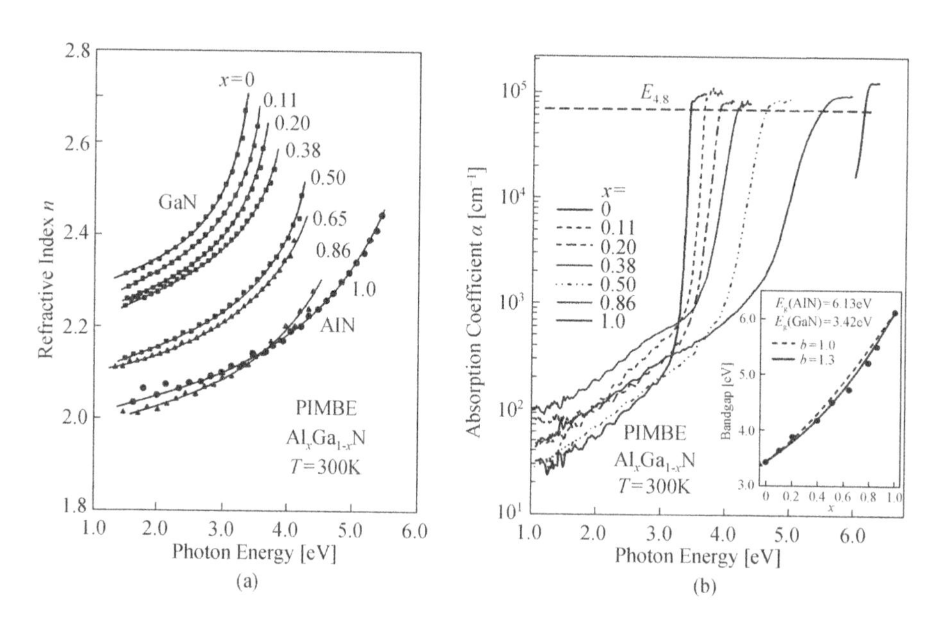

Figure 3.32 Refractive index and absorption coefficient of AlGaN epitaxial films [55]. (Reprinted with permission from D. Brunner, H. Angerer, E. Bustarret *et al.*, "Optical constants of epitaxial AlGaN films and their temperature dependence," *Journal of Applied Physics*, **82**, 5090–5096, 1997. © 1997 American Institute of Physics.)

for AlInGaN alloys. Refractive indices of GaN, $Al_xGa_{1-x}N$, and $In_xGa_{1-x}N$ can be calculated by the band-gap (E_g) of these materials:

$$n_{\mathrm{AlGaN}} = \sqrt{a(x)(hv/E_g)^{-2}[2 - \sqrt{1 + hv/E_g} - \sqrt{1 - hv/E_g})] + b(x)} \tag{3.29}$$

$$n_{\mathrm{InGaN}} = n_{\mathrm{GaN}}(hv - [E_g(x) - E_g(0)]) \tag{3.30}$$

$$E_g(x) = 3.45(1-x) + 6.13x - 1.3x(1-x) \tag{3.31}$$

$$E_g(x) = 3.42(1-x) + 2.65x - 3.9x(1-x) \tag{3.32}$$

where x is a fraction of Al or In in the alloys, h is Planck constant, v is the frequency of light, and $a(x)$ and $b(x)$ are fitting parameters. More precise E_g of $Al_xGa_{1-x}N$ and $In_xGa_{1-x}N$ can be obtained [2]. n of GaN and $Al_xGa_{1-x}N$ can be obtained from Equation 3.29, n of $In_xGa_{1-x}N$ can be obtained from Equation 3.30 by $E_g(0)$, which is the band-gap of GaN. Equation 3.31 is used to calculate the band-gap of $Al_xGa_{1-x}N$ and Equation 3.32 is used to calculate the band-gap of $In_xGa_{1-x}N$. The calculated n of GaN, $Al_xGa_{1-x}N$, and $In_xGa_{1-x}N$ are shown in Figure 3.33.

Absorption coefficients of AlInGaN alloys are more complicated due to the variation of crystal quality. In most of studies [55,59–61], absorption coefficients of GaN and AlGaN for blue light are around 20–50 mm^{-1}, and for yellow light are around 9–15 mm^{-1}. This result may be too high, since η_{LEE} based on this definition is normally lower than 10% by ray tracing. Schad *et al.* proposed that there was a high absorbing layer in the chip due to high dislocation density in the buffer layer and the actual absorption coefficient of GaN materials should be 0.4 mm^{-1} [62]. The testing results of Lelikov *et al.* gave an absorption coefficient of around 2.3 mm^{-1} [63]. We believed that the traditional testing results [55,59–61] ignored the high

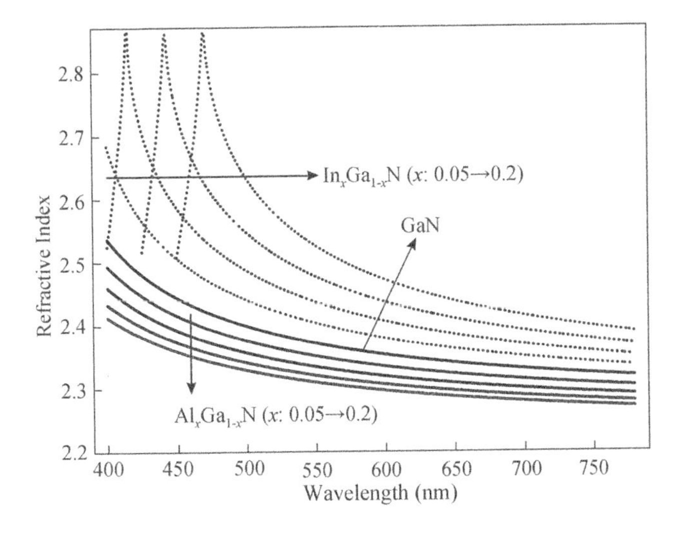

Figure 3.33 Refractive indices of GaN, AlGaN and InGaN in visible light.

absorption of the buffer layer and deduced high absorption coefficient of GaN, but absorption coefficient of GaN cannot be so low as in Schad *et al.* [62]. This is because that E_g of InGaN is narrower than GaN and that means that the MQW layer also presents an extremely high absorption for visible light. Schad *et al* [62] did not consider the absorption of InGaN and believed that most absorption happened in the buffer layer. Some studies [64–66] have reported that the absorption coefficient of InGaN is around 100 mm^{-1}. Therefore, the absorption model of the chip should be carefully treated in defining the absorption coefficients of AlInGaN alloys in the chip.

As an attempt, we studied the absorption model of chip by ray tracing chips with various absorption coefficients. By making η_{LEE} of ray tracing comparable with experiments, we found that if the ray tracing does not consider the roughness on surfaces, the absorption coefficients of P-GaN, N-GaN, and AlGaN should be in the range of 2–7 mm^{-1}, and the absorption coefficient of InGaN is around 20–50 mm^{-1}. It should be noticed that the absorption coefficients of AlInGaN alloys are significantly affected by the crystal quality. This means that the absorption coefficients of AlInGaN alloys can be further reduced with the improvement of epitaxial growth technology.

3.4.2 Phosphor Modeling

The optical modeling of phosphor particles is complicated due to the non-spherical shape of phosphor particles. Figure 3.34a is the SEM photograph of one typical YAG: Ce phosphor material. By improving the milling process, the shape of phosphor particles can be greatly improved and made more spherical. Figure 3.34b is the SEM photograph of YAG: Ce phosphor obtained from Intematix.

The absorption spectrum and emission spectrum of phosphor material have been well known. However, the light absorption and scattering by phosphor particles have not been studied well. Narendran *et al.* [67], Zhu *et al.* [24], and Kang *et al.* [68] tested the optical properties of YAG: Ce phosphor. Narendran *et al.* developed a double-integrating-sphere system to measure the light transmitted and reflected from phosphor [67]. They obtained the light transmittance and reflectance of phosphor as shown in Figure 3.35. For converted light

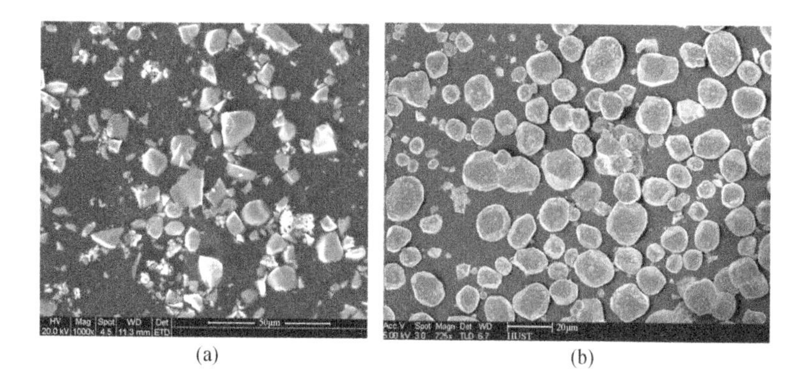

(a) (b)

Figure 3.34 SEM photographs of YAG: Ce phosphor.

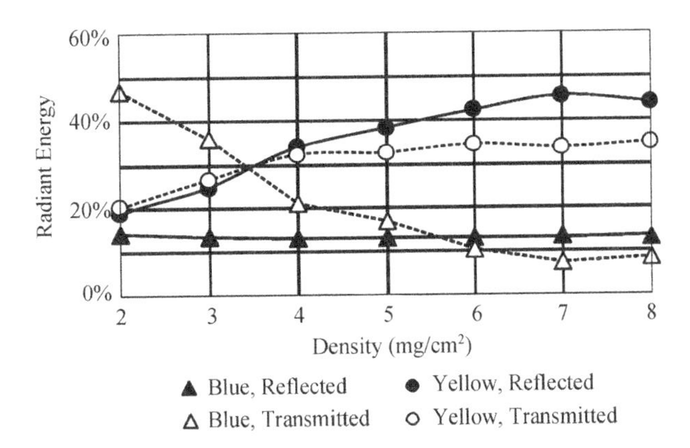

Figure 3.35 Transmitted and reflected radiant energies for blue and yellow light, as a function of phosphor density [67]. (Reproduced with permission from N. Narendran, Y. Gu, J.P. Freyssinier-Nova and Y. Zhu, "Extracting phosphor-scattered photons to improve white LED efficiency," *Physica Status Solidi (a)*, 2005, **202**, 6, R60-R62. © Wiley-VCH Verlag GmbH & Co. KGaA.)

from phosphor, both the transmitted light and reflected light are increased with the increase of phosphor concentration, whereas the transmitted blue light and reflected blue light are reduced. Through the comparison of the transmitted light spectrum and reflected light spectrum, Narendran *et al.* deduced that natural white LEDs can be obtained when the phosphor concentration is around 8 mg/cm^2. Zhu *et al.* obtained the transmittance and reflectance of phosphor for blue light, green light, and red light using the similar double-integrating-sphere system. They gave the conversion efficiency of phosphor to be around 77%. However, Narendran *et al.* and Zhu *et al.* did not give the optical constants of phosphor material, which are very important for the description of phosphor absorption and scattering properties. Kang *et al.* firstly developed the calculation method of optical constants of phosphor by using Lambert-Beer law. By coating phosphor films on the top surface of the chip, they obtained the extinction coefficients of phosphor for blue light and yellow light, respectively. Figure 3.36 presents their testing results. But the extinction coefficients are estimated to be larger than the actual coefficients. This is because that the numerical calculation does not consider the multi-absorption between phosphor and chip due to the high absorption of chip materials.

We improved the double-integrating-sphere system of Narendran *et al.* and Zhu *et al.* by changing the incident light to be parallel [69]. The phosphor slides were fabricated with concentration changed from 0.2 to 0.5 mg/cm^3 and the thickness changed from 0.6 to 0.3 mm. The tested transmittance and reflectance of phosphor films are shown in Figure 3.37. The conversion and emission properties are shown in Figure 3.38.

From Figures 3.37a and 3.37b, it can be found that the YAG: Ce phosphor presents remarkable high absorption for blue light and induces rapid attenuation of η_{BT} and a small reduction of η_{BR} when the phosphor thickness and concentration are increased. η_{YT} is also reduced, but η_{YR} is increased quickly. This is mainly due to the reason that during the light propagation, back scattered blue light will be further absorbed by phosphor particles, whereas back scattered yellow light can be emitted out due to the low absorption of phosphor for long wavelength. From Figure 3.37c, it can be found that η_{CT} and η_{CR} present similar values and

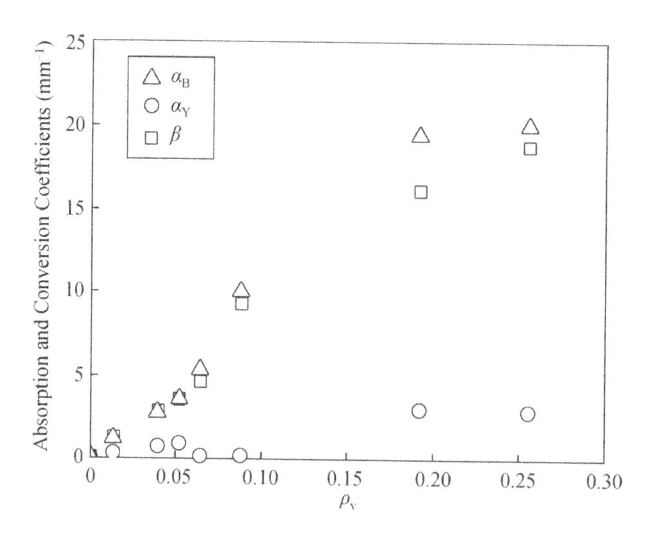

Figure 3.36 Dependence of absorption and conversion coefficients on the volume fraction of phosphor in the phosphor silicone [68]. (Reprinted with permission from D.-Y. Kang, E. Wu and D.-M. Wang, "Modeling white light-emitting diodes with phosphor layers," *Applied Physics Letters*, **89**, 231102, 2006. © 2006 American Institute of Physics.)

tendencies when the phosphor thickness and concentration are increased. This indicates that the emission pattern of a phosphor particle may be isotropic Figure 3.38 shows that η_{CE} of saturated phosphor is slightly higher than 70% and η_{CT}/η_{CR} is approximately 1.05–1.15. Generally, the conversion efficiency is a multiplication of Stokes efficiency and quantum

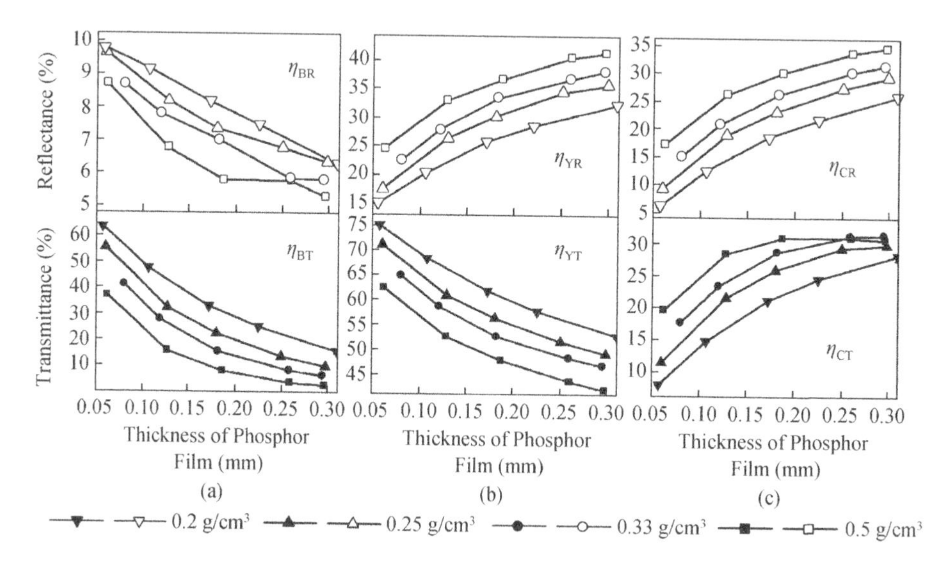

Figure 3.37 Tested reflectance and transmittance of (a) blue light (455 nm), (b) yellow light (595 nm), and (c) converted light for phosphor films with various thickness and concentration.

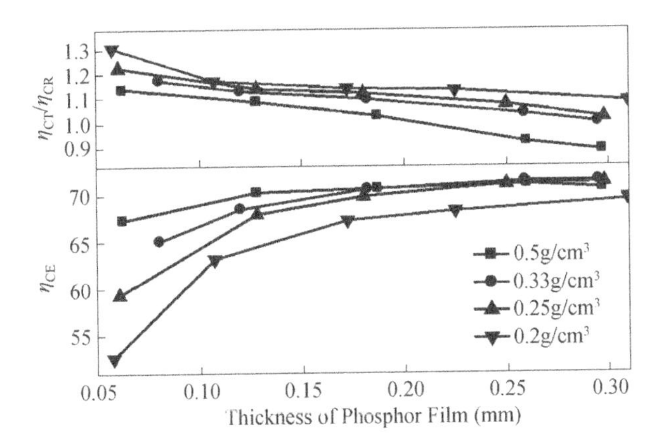

Figure 3.38 Emission and conversion properties of YAG: Ce phosphor with various thicknesses and concentrations.

efficiency of phosphor [47]. The dominant wavelength of the converted light is approximately 562 nm. Therefore, the Stokes efficiency is around 81% and the quantum efficiency of phosphor is around 87%. For cases with low concentration and thin thickness, unsaturated phosphor demonstrates lower η_{CE}. If the thickness is too thick or the concentration is too high, oversaturated phosphor may cause more converted light to be back reflected.

The optical constants of phosphor normally can be calculated by Mie theory [70,71]. We developed the Mie calculation of optical constants of phosphor based on the particle size distribution data. Generally, if the particle size distribution of phosphor and the phosphor concentration are known, the Mie calculation of optical constants can be given as:

$$\mu_{abs}(\lambda) = \int N(r) C_{abs}(\lambda, r) \mathrm{d}r \tag{3.33}$$

$$\mu_{sca}(\lambda) = \int N(r) C_{sca}(\lambda, r) \mathrm{d}r \tag{3.34}$$

$$g(\lambda) = 2\pi \int \int_{-1}^{1} p(\theta, \lambda, r) f(r) \cos\theta \,\mathrm{d} \cos\theta \mathrm{d}r \tag{3.35}$$

where $\mu_{abs}(\lambda)$ is the absorption coefficient, $\mu_{sca}(\lambda)$ is the scattering coefficient, $g(\lambda)$ is the anisotropy factor, $N(r)$ is the number density distribution of particles (mm^{-3}), $C_{abs}(\lambda, r)$ and $C_{sca}(\lambda, r)$ are the absorption and scattering cross-sections (mm^2), $p(\theta, \lambda, r)$ is the phase function of particles, $f(r)$ is the size distribution function of phosphor material, λ is the wavelength of incident light (nm), r is the radius of phosphor particle, and θ is the scattering angle.

$N(r)$ can be calculated by multiplying $f(r)$ with the number density coefficient K_N:

$$N(r) = N_s(r) + N_{phos}(r) = K_N f(r) \tag{3.36}$$

where K_N denotes the number of unit phosphor material in the specific concentration, and can be obtained by dividing the phosphor concentration c (mg/mm^3) with the total mass of unite phosphor material.

$$K_{\mathrm{N}} = \frac{c}{\int M(r)\mathrm{d}r} \tag{3.37}$$

where $M(r)$ (mg) is the mass distribution of unit phosphor material. $M(r)$ can be expressed by:

$$M(r) = \rho V(r) f(r) = \frac{4}{3}\pi r^3 \rho f(r) \tag{3.38}$$

In Mie theory, $C_{\mathrm{abs}}(\lambda, r)$ and $C_{\mathrm{sca}}(\lambda, r)$ are normally calculated by the following relations:

$$C_{\mathrm{sca}} = \frac{2\pi}{k^2}\sum_{0}^{\infty}(2n+1)\left(|a_{\mathrm{n}}|^2 + |b_{\mathrm{n}}|^2\right) \tag{3.39}$$

$$C_{\mathrm{ext}} = \frac{2\pi}{k^2}\sum_{1}^{\infty}(2n+1)\mathrm{Re}(a_{\mathrm{n}} + b_{\mathrm{n}}) \tag{3.40}$$

$$C_{\mathrm{abs}} = C_{\mathrm{ext}} - C_{\mathrm{sca}} \tag{3.41}$$

where C_{ext} is the extinction cross-section, k is the wave number ($=2\pi/\lambda$), and a_{n} and b_{n} are the expansion coefficients with even symmetry and odd symmetry, respectively. a_{n} and b_{n} are calculated by:

$$a_{\mathrm{n}} = \frac{m\psi_{\mathrm{n}}(mx)\psi'_{\mathrm{n}}(x) - \psi'_{\mathrm{n}}(mx)\psi_{\mathrm{n}}(x)}{m\psi_{\mathrm{n}}(mx)\xi'_{\mathrm{n}}(x) - \psi'_{\mathrm{n}}(mx)\xi_{\mathrm{n}}(x)} \tag{3.42}$$

$$b_{\mathrm{n}} = \frac{\psi_{\mathrm{n}}(mx)\psi'_{\mathrm{n}}(x) - m\psi'_{\mathrm{n}}(mx)\psi_{\mathrm{n}}(x)}{\psi_{\mathrm{n}}(mx)\xi'_{\mathrm{n}}(x) - m\psi'_{\mathrm{n}}(mx)\xi_{\mathrm{n}}(x)} \tag{3.43}$$

where x is the size parameter ($=kr$), m is the relative refractive index of particles, and $\psi_n(x)$ and $\xi_n(x)$ are the Riccati-Bessel functions.

For small spheres, the phase function $p(\theta, \lambda, r)$ can be calculated according to:

$$p(\theta, \lambda, r) = \frac{4\pi\beta(\theta, \lambda, r)}{k^2 C_{\mathrm{sca}}(\lambda, r)} \tag{3.44}$$

where $\beta(\theta, \lambda, r)$ is the dimensionless scattering function, which is obtained by the scattering amplitude functions $S_1(\theta)$ and $S_2(\theta)$.

$$\beta(\theta) = (1/2)[|S_1(\theta)|^2 + |S_2(\theta)|^2] \tag{3.45}$$

$$S_1(\theta) = \sum_{n=1}^{\infty}\frac{2n+1}{n(n+1)}\left[a_{\mathrm{n}}\frac{P_{\mathrm{n}}^1(\cos\theta)}{\sin\theta} + b_{\mathrm{n}}\frac{\mathrm{d}P_{\mathrm{n}}^1(\cos\theta)}{\mathrm{d}\theta}\right] \tag{3.46}$$

$$S_2(\theta) = \sum_{n=1}^{\infty} \frac{2n+1}{n(n+1)} \left[b_n \frac{P_n^1(\cos\theta)}{\sin\theta} + a_n \frac{dP_n^1(\cos\theta)}{d\theta} \right] \tag{3.47}$$

where $P_n^1(\cos\theta)$ is the associated Legendre polynomial.

The optical constants of phosphor calculated by Mie theory are not accurate enough in terms of describing the light absorption and scattering properties of phosphor. This is because that Mie theory normally is suitable for those particles with spherical shape, whereas the shape of phosphor particles is non-spherical and varies significantly. Therefore, the Mie's theoretical results should be further modified according to the experimental results. We introduced the reduced scattering coefficient and compared the Mie's theoretical calculation results with the ray tracing results to find reasonable method to modify the Mie's calculation. The reduced scattering coefficient is:

$$\delta_{sca} = \mu_{sca}(1-g) \tag{3.48}$$

Two ray tracing cases are considered. One case assumes $\mu_{sca}(\lambda)$ to be the same as that of Mie's results, and $\mu_{abs}(\lambda)$ and $g(\lambda)$ are changed to make the ray tracing results compatible with the experimental results. The other case assumes $g(\lambda)$ to be the same as that of Mie's results, and $\mu_{abs}(\lambda)$ and $\mu_{sca}(\lambda)$ are changed to make the ray tracing results compatible with the experimental results. The comparisons between ray tracing results and Mie's theoretical results are shown in Figure 3.39. It can be found that δ_{sca} in the two ray tracing cases are almost coincident, implying that δ_{sca} is more precise in describing the light scattering of phosphor with $g(\lambda) > 0.8$. Results show that in Mie's theoretical calculation δ_{sca}(595 nm) is larger than δ_{sca}(455 nm), whereas in ray tracing calculation δ_{sca}(595 nm) is smaller than δ_{sca}(455 nm). This difference indicates that Mie theory is not accurate enough in describing the light scattering properties of phosphor when the incident light is changed from high absorbed spectrum to low absorbed spectrum.

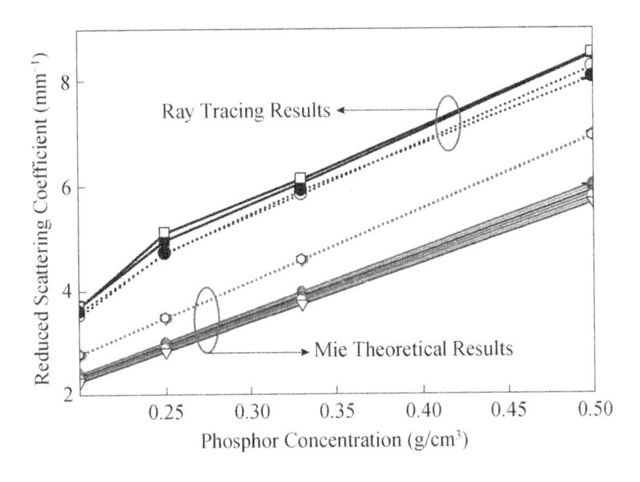

Figure 3.39 Comparisons of reduced scattering coefficients of ray tracing results and Mie's theoretical results. The solid and dashed lines are results for blue light and yellow light, respectively.

The differences between δ_{sca}(455 nm) and δ_{sca}(595 nm) in ray tracing calculation are also smaller than that in Mie theoretical calculation, meaning that the light scattering properties of phosphor in the whole visible spectrum are not varied too much.

The real reason that caused the differences between Mie calculation and ray tracing is that $C_{abs}(\lambda)$ and $C_{sca}(\lambda)$ of Mie calculation are smaller than the actual cross-section. The non-spherical shape of phosphor particles indicates that the size distribution data is difficult to provide the true sizes of phosphor particles. Changing non-spherical phosphor particles to spheres ensures that the true absorption and scattering cross-sections of phosphor particles tend to be smaller. Therefore, a simple method is modifying $C_{abs}(\lambda)$ and $C_{sca}(\lambda)$ by two fitting parameters k_{abs} and k_{sca}:

$$C'_{abs} = k_{abs}C_{abs},\ C'_{sca} = k_{sca}C_{sca} \tag{3.49}$$

According to Equations 3.32–3.34 and 3.49, $\mu_{abs}(\lambda)$, $\mu_{sca}(\lambda)$ and $g(\lambda)$ will be revised to be:

$$\mu'_{abs} = k_{abs}\mu_{abs},\ \mu'_{sca} = k_{sca}\mu_{sca},\ g' = g/k_{sca} \tag{3.50}$$

Therefore, $\delta_{sca}(\lambda)$ will be revised to be:

$$\delta'_{sca} = k_{abs}\mu_{sca}(1 - g/k_{sca}) = \mu_{sca}(k_{sca} - g) \tag{3.51}$$

where C'_{abs}, C'_{sca}, η'_{abs}, η'_{sca}, g and δ'_{sca} are the revised absorption cross-section, scattering cross-section, absorption coefficient, scattering coefficient, anisotropy factor, and reduced scattering coefficient, respectively.

Therefore, if k_{abs} and k_{sca} can be obtained by experiments, the revised Mie calculation will give a precise prediction of phosphor optical properties. In our experiments, the calculated k_{abs} and k_{sca} are around 1.47 and 1.06, respectively. The smaller k_{sca} and the larger k_{abs} indicate that Mie theory has a good description of light scattering but has difficulty in describing the light absorption for non-spherical particles with a high absorption of incident light.

3.5 Phosphor for White LED Packaging

Phosphor is one of the most important components in phosphor converted white LEDs. As a bulk scattering material, phosphor not only absorbs the blue light and converts the blue light to complement light, but also scatters the light to enhance the absorption for blue light and make the spatial color distribution uniform. Many studies have been done to investigate the impacts of phosphor properties on the packaging performance [24,26,27,43,47,51,67,72–76]. The location, thickness, and concentration of phosphor are the most important factors affecting LED optical performance.

3.5.1 Phosphor Location for White LED Packaging

Phosphor location is the primary consideration in LED packaging. alternating the phosphor layer from being close to the chip to being remote, the propagation path and energy of light will be affected in terms of the scattering and absorption of phosphor, the reflection of the reflector,

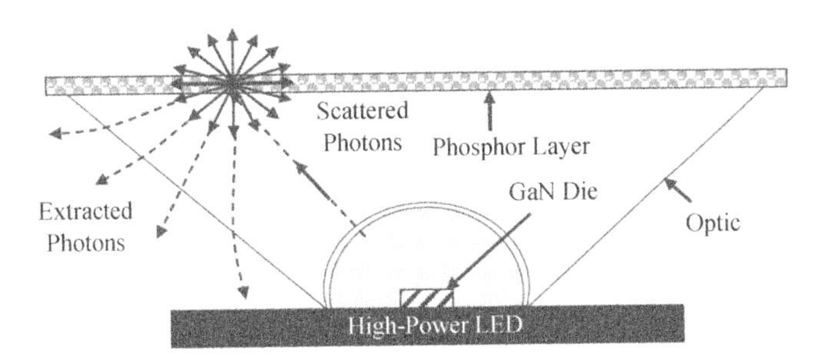

Figure 3.40 Schematic of the scattered photon extraction white LED package [67]. (Reproduced with permission from N. Narendran, Y. Gu, J.P. Freyssinier-Nova and Y. Zhu, "Extracting phosphor-scattered photons to improve white LED efficiency," *Physica Status Solidi (a)*, 2005, **202**, 6, R60-R62. © Wiley-VCH Verlag GmbH & Co. KGaA.)

the absorption of the chip, the refraction of the lens, and so on. The absorption of the phosphor and the chip will influence the output optical power. The scattering of phosphor will disorder the light propagation. The directions of rays could be converged to central zones by the reflection of the reflector and the refraction of the lens or be changed to edge zones. These will induce the variation of light extraction and CCT.

Since almost half of blue light is back scattered and half of converted light is back emitted [24,72], the change of phosphor location is estimated and can effectively reduce the absorption by the chip. Narendran *et al.* [67,72] proposed the scattered photon extraction (SPE) method to improve the light extraction as shown in Figure 3.40. By applying an optic lens with its top surface coated with phosphor, this method can efficiently transfer the light from the GaN die to the phosphor layer and simultaneously make the most of backscattered light from phosphor escape through the lateral surfaces. Experiments revealed that both the light output and luminous efficiency can be enhanced by as much as 61%.

Kim and Luo *et al.* [26,27] compared the light extraction for phosphor directly coated on the chip and remote phosphor location. The surface of the reflector was changed from specular to diffuse to investigate the light enhancement by remote phosphor. The shape of the lens was also changed from flat to hemispherical as comparisons. Their simulation results are shown in Figures 3.41 and 3.42. From Figure 3.41, it can be found that the improvements of light extraction efficiency of remote phosphor arrangement are 36% and 75% for specular reflector cup and diffuse reflector cup, respectively, when comparing with the phosphor in a cup arrangement using a specular reflector cup. Figure 3.42 shows that the phosphor efficiency for phosphor on top configuration with diffuse cup is enhanced 50% compared with that for phosphor in specular cup configuration with flat encapsulation. But their experiments reveal that the actual enhancement of light extraction is not as high as the declared simulations. In experiments, the improvement of light extraction for configuration with specular cup and phosphor on top is 7.8%, and the improvement for configuration with diffuse cup and phosphor on top is 15.4%. This value is lower than that of SPE method.

Allen *et al.* [47,73] change the shape of remote phosphor from flat to be spherical as shown in Figure 3.43. The ELiXIR luminaire developed by them applied the dye as the color converting material to reduce the light loss caused by diffuse scattering of conventional YAG: Ce phosphor.

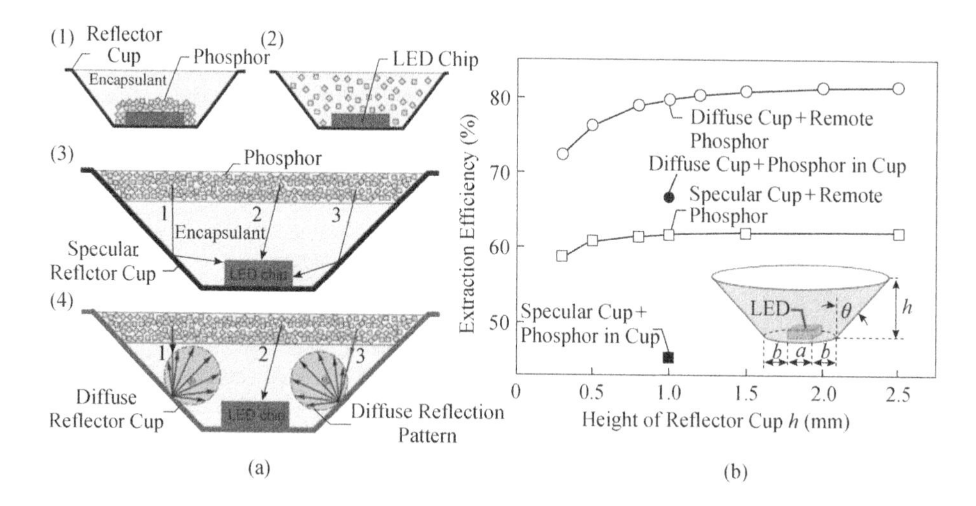

Figure 3.41 (a) Phosphor location in white LED: (1) Conformal distribution directly on LED chip. (2) Uniform distribution in reflector cup (phosphor in cup). (3) Uniform distribution thin layer above LED chip (remote phosphor). (4) Remote phosphor distribution in diffuse reflector cup. (b) Calculated light extraction efficiency as a function the height of reflector cup [26]. (Reprinted with permission from J.K. Kim, H. Luo, E.F. Schubert *et al.*, "Strongly enhanced phosphor efficiency in GaInN white light-emitting diodes using remote phosphor configuration and diffuse reflector cup," *Japanese Journal of Applied Physics*, **44**, L649-L651, 2005. © 2005 The Japan Society of Applied Physics.)

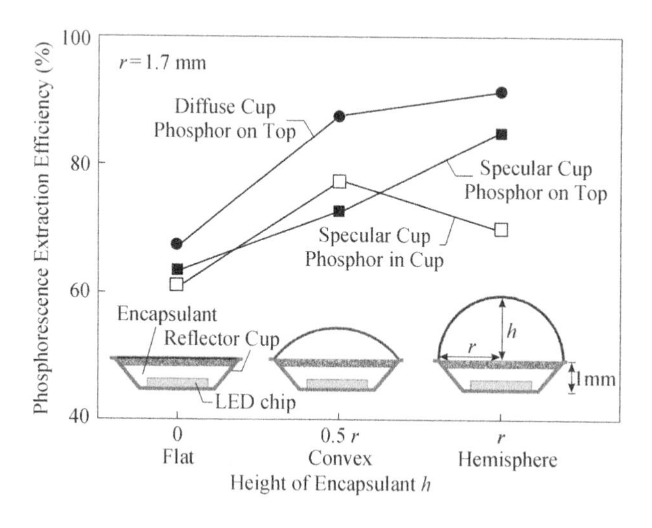

Figure 3.42 Extraction efficiency of phosphorescence calculated by ray tracing for different package configurations [27]. (Reprinted with permission from H. Luo, J.K. Kim, E.F. Schubert *et al.*, "Analysis of high-power packages for phosphor-based white-light-emitting diodes," *Applied Physics Letters*, **86**, 24, 243505, 2005. © 2005 American Institute of Physics.)

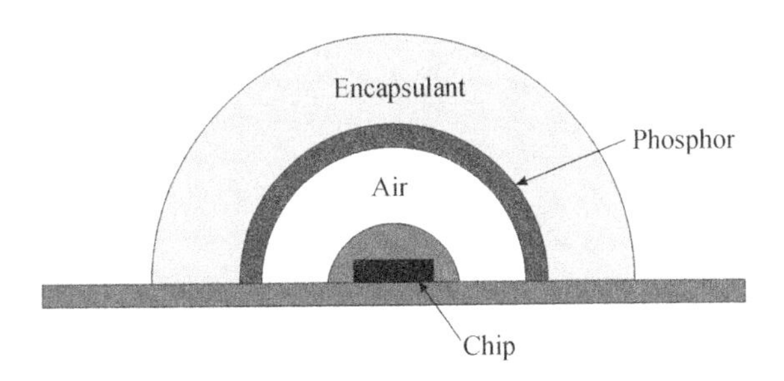

Figure 3.43 Cross-sectional views of ELiXIR remote hemispherical shell semitransparent phosphor package [47]. (Reprinted with permission from S.C. Allen and A.J. Steckl, "ELiXIR-solid-state luminaire with enhanced light extraction by internal reflection," *Journal of Display Technology*, **3**, 2, 155–159, 2007. © 2007 Optical Society of America.)

By using a refractive index matched encapsulation lens to eliminate the total internal reflection, this LED packaging configuration can improve the quantum efficiency of phosphor to be almost 100%, which is significantly higher than those packaging configurations with YAG: Ce phosphor. Since the photostability of dye is not good, Allen *et al.* also developed the composite material that is fabricated by glass and YAG: Ce phosphor. This composite material can decrease the optical scattering and therefore increase the packaging efficiency.

We systematically analyze the effects of phosphor's location on LED packaging performance by ray tracing simulation for five different LED packaging configurations [43]. The five packaging configurations are depicted in Figure 3.44. To evaluate the impact of the reflector on packaging performance, three numerical models in terms of Type III, IV, and V have a reflector on each to compare other two non-reflector models of Type I and II. In Types I and III, the phosphor layer is conformally coated on the chip. The distance between the phosphor layer and the chip is altered from 0 mm to 0.1 mm. The case with 0 mm distance is called the direct-coating case. In Types II and V, the model changes the location by increasing the radius of the phosphor layer. The second model fabricates the phosphor layer with a hemispherical film, and increases the radius from 0.8 mm to 3.9 mm. The radius is increased from 4.25 mm to 10 mm in the fifth model. In Type IV, the height of phosphor layer is increased from 0.2 mm to 1.9 mm. The thickness of phosphor layer is 0.1 mm in all cases.

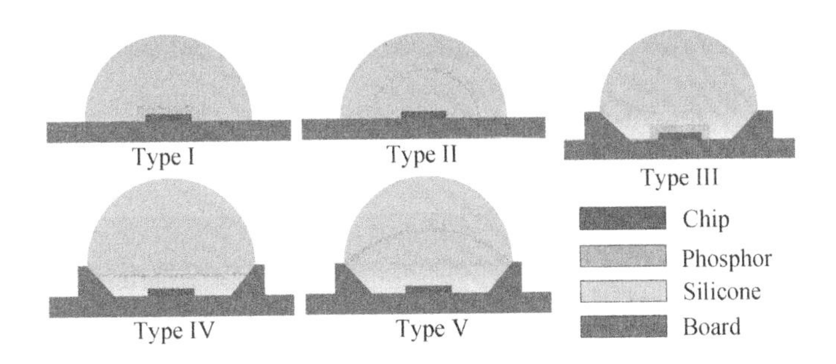

Figure 3.44 Five packaging configurations for the analysis. *(Color version of this figure is available online.)*

We applied the wall plug efficiency (η_{WPE}), nominal packaging efficiency (η_{NPE}), and real packaging efficiency (η_{RPE}) as the evaluation indices of LED performance. They are expressed in the following:

$$\eta_{WPE} = P_{pcLED}/P_{elec} \tag{3.52}$$

$$\eta_{NPE} = \eta_{WPE}/\eta_{LED} \tag{3.53}$$

$$\eta_{RPE} = \eta_{WPE}/(\eta_{inj}\eta_{int}\eta_{f}\eta_{p-LEE}) \tag{3.54}$$

where P_{pcLED} is the extracted optical power from phosphor-converted LEDs including optical power of blue light P_{pcLED} (465 nm) and optical power of yellow light P_{pcLED} (555 nm); $\eta_{p\text{-}LEE}$ is the light extraction efficiency of the packaged LED chip; P_{elec} is the consumed electrical power; η_{inj}, η_{int}, and η_f are injection efficiency, internal quantum efficiency, and feeding efficiency, respectively. The optical power ratio of yellow light and blue light (yellow/blue ratio, YBR) is used here to denote the color of white LEDs.

The numerical results of light extraction efficiency for Type I and III are shown in Figure 3.45. It is clear that real packaging efficiency and wall plug efficiency increases slightly when the phosphor layer is changed to remote place from 0.01 mm to 0.1 mm. However, there is a sudden variation if phosphor is directly dispensed on the chip surface. The variation is strange since the wall plug efficiency is significantly higher than those cases with small distance but the real packaging efficiency is obviously lower.

The variation is mainly due to the relatively lower chip absorption in the direct-coating case. Since the refractive index of the phosphor layer is higher than silicone materials, when phosphor is directly coated on chip, the critical angle is bigger than those cases with silicone

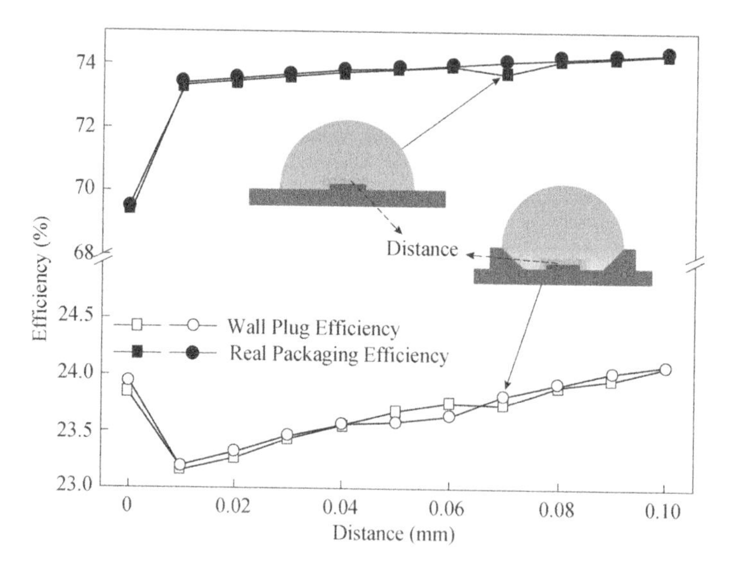

Figure 3.45 Effects of phosphor's location on the light extraction efficiencies of Type I and III. *(Color version of this figure is available online.)*

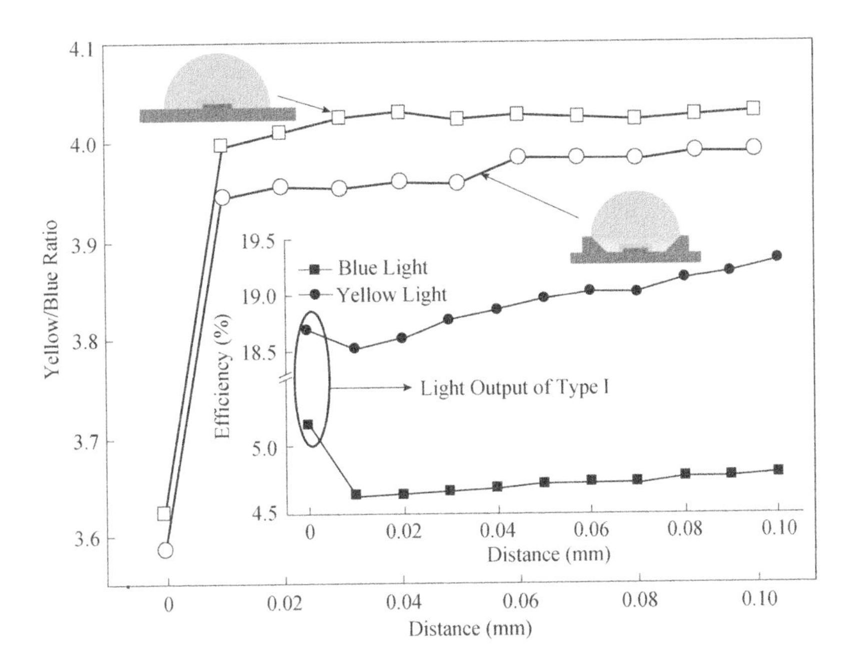

Figure 3.46 Effects of phosphor's location on yellow/blue ratio of Type I and III, blue light output and yellow light output of Type I. *(Color version of this figure is available online.)*

first coated on the chip. Therefore, less blue light will be confined in the chip and emitted out. This induces the lower chip absorption.

As noted in Figure 3.46, the blue light output for a direct-coating case is obviously higher than other cases, which indicates that the effect of refractive index is great. Another important factor is the absorption of phosphor for blue light. Simulation results show that, when phosphor is directly coated on the chip, the totally absorbed blue light by phosphor is 32.49%, but for the distance of 0.01 mm this value is 25.397%. There is at least 8% difference in the absorption of blue light. As a result, the yellow light output could also be higher in the direct-coating case after phosphor's conversion. Consequently, the extracted power from the module is higher but the real packaging efficiency is lower.

It can be found that the increase of yellow light output is not as significant as that of blue light in the direct-coating case, and the yellow light output is increased slowly with the increase of distance in Figure 3.46. This is mainly due to the conversion loss of phosphor and the high absorption of the chip. Since the conversion efficiency is 80%, there will be more energy loss if more blue light is converted. Taking the cases of direct-coating and 0.01mm distance as examples, the difference between absorbed blue light and converted blue light is reduced from 7.09% to 5.67%.

When phosphor is directly coated on the chip, all of the back scattered yellow light must pass through the chip and lose most of the power. However, when there is a gap between the phosphor and the chip, part of the back scattered yellow light could directly emit out after the reflection of board. The further the distance is, the more significant the phenomenon is. This indicates that the remote phosphor location could exhibit a higher light extraction.

As noted in Figure 3.46, the relatively higher blue light output and lower yellow light output also induce the sudden reduction of yellow/blue ratio in direct-coating case. Because of the increased yellow light output for remote location, the color of LEDs should tend to be warm white.

From Figures 3.45 and 3.46, it can be found that the trend and value of light extraction efficiency and CCT in the third numerical model are similar to the first model. This may be caused by the size of reflector. Since the angle of the cone is 102.6°, this may cause most of the lights to be directly emitted out without being reflected. However, if the angle of cone is small enough and the height of the reflector is bigger, the impact on the light propagation may be significant, thus distinguishing the difference between two types. The small difference of YBR indicates that the reflector could change the color to be cooler.

The simulation results for the two structures with a convex phosphor layer are displayed in Figures 3.47 and 3.48. It demonstrates that the influences on light extraction are small when the location is remote enough. The fluctuations for wall plug efficiency and real packaging efficiency are no more than 0.56% and 1.45%, respectively. However, the influences on YBR are significant. This could change the color of light and luminous efficiency.

In the second structure, the efficiency is increased at the beginning but reduced at the end. Inversely, the tendency of the YBR curve is up and then down. The same characteristic is that there exists one balance zone around the radius of 2 mm, which is the half of the lens's radius. This is mainly due to the power ratio between forward scattered light and back scattered light changed with the location. Three factors are considered effective, and they are the size and surface area of phosphor layer, the reflection loss, and the absorption of chip. As shown in Figure 3.49, the blue light output and yellow light output could illustrate the influencing mechanism of these factors.

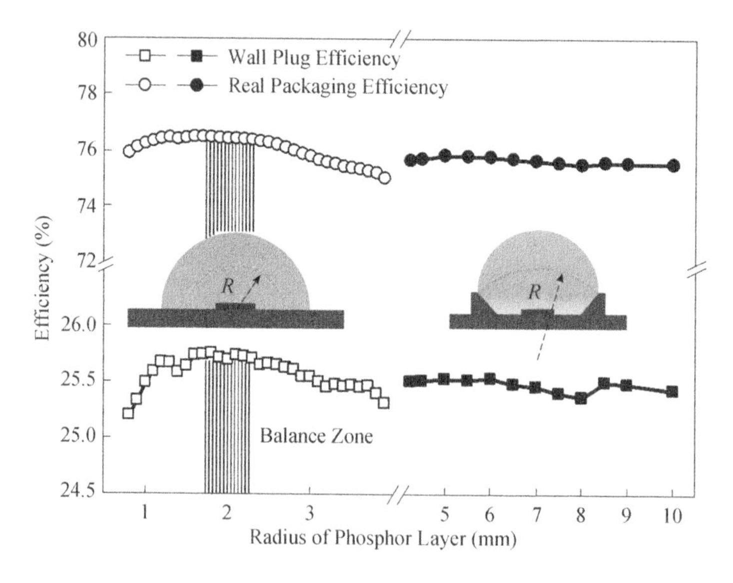

Figure 3.47 Effects of phosphor's location on the light extraction efficiencies of Type II and V. *(Color version of this figure is available online.)*

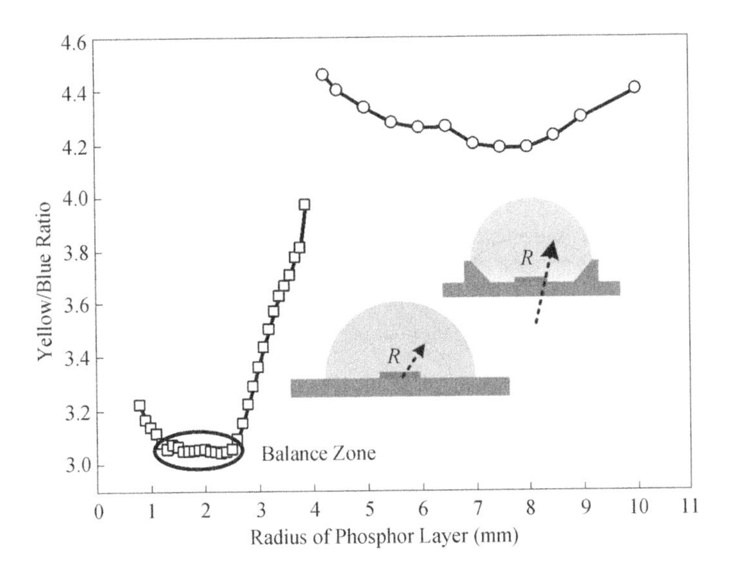

Figure 3.48 Effects of phosphor's location on the yellow blue ratio (CCT) of Type II and V. *(Color version of this figure is available online.)*

In the second structure, the determining factor is the absorption of chip when the location is small. An increased gap between the phosphor layer and the chip could reduce the absorption of the chip, since partial rays could be reflected and emit out. Therefore, there is a slight increase of light output. However, the influence on the blue light is more significant. Blue light is emitted from the chip, which is in the center of the hemispherical phosphor layer. As a comparison, the direction of yellow light is random and disordered. Therefore, the absorbed power of the yellow light is higher than that of the blue light.

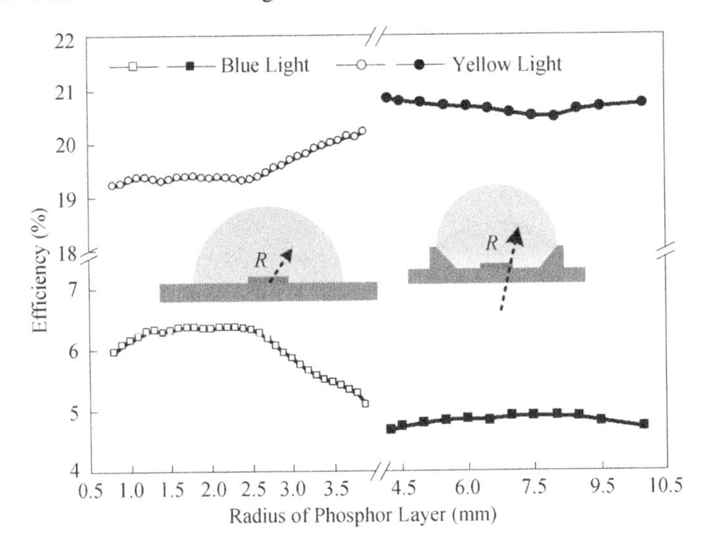

Figure 3.49 Blue light output and yellow light output in Type II and V with the increase of radius. *(Color version of this figure is available online.)*

However, with the increase of the radius, the absorption and scattering of the enlarged phosphor layer, and the reflection loss will be the main factors. The effect of the chip's absorption should be weakened gradually. This results in the phenomenon of the balance zone. Finally, the remoter phosphor layer absorbs more blue light and emits more yellow light. The reduced gap between the lens and the phosphor enhances the recycling of blue light and enables it to be absorbed by the phosphor. The back scattered blue light should also be reflected many times by the board or reflector and finally lose most of its energy in reflection and phosphor layer.

Comparing the obvious change by location in Type II, the variation of Type V is relatively small. This may be due to the location that is too remote. However, the change of curvature still has limited impact on the CCT.

As shown in Figure 3.50, the light extraction and CCT in Type IV has a similar tendency to that of Type II. This indicates that remote phosphor location has a similar influencing mechanism for packaging. However, this case does not present the phenomenon of a balance zone. Therefore, the manufacturing tolerance is relatively lower.

To further discuss the simulation results and obtain more useful conclusions, the light extraction efficiency and YBR for all cases are displayed in one figure. Figure 3.51 presents the tendency of nominal packaging efficiency and real packaging efficiency. Figure 3.52 is about the variation of CCT. Compared to the significant variation of CCT, the impact of phosphor's location on efficiency is small. The fluctuations are no more than 6.5% for η_{NPE} and 4.9% for η_{RPE}. The variations of efficiency tend to be slower and smoother with the increase of location. Results also confirm that remote phosphor location presents higher light extraction than proximate phosphor.

It can be found that the light extraction efficiency of the convex phosphor layer is normally higher than that of the plane phosphor layer, such as Type IV. This may be due to the fact that

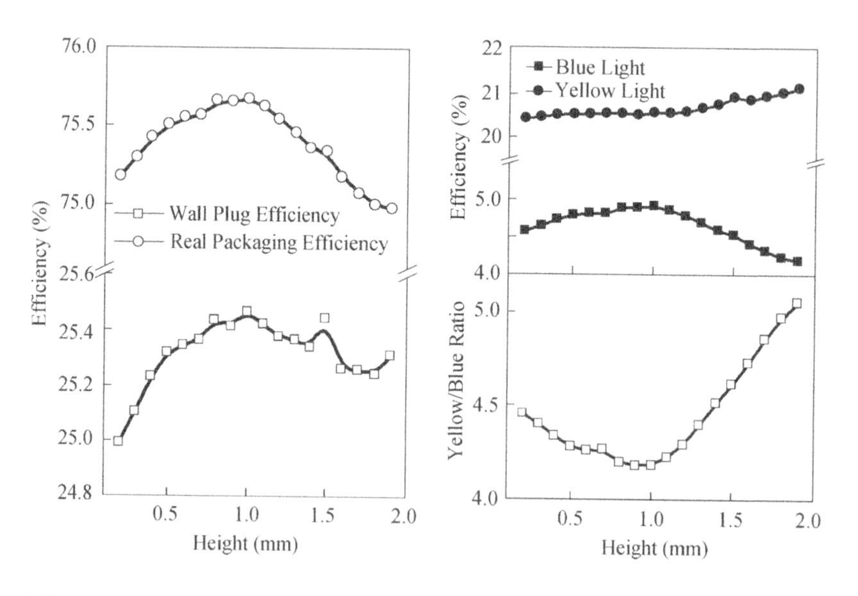

Figure 3.50 Effects of phosphor's location on packaging performance of Type IV. *(Color version of this figure is available online.)*

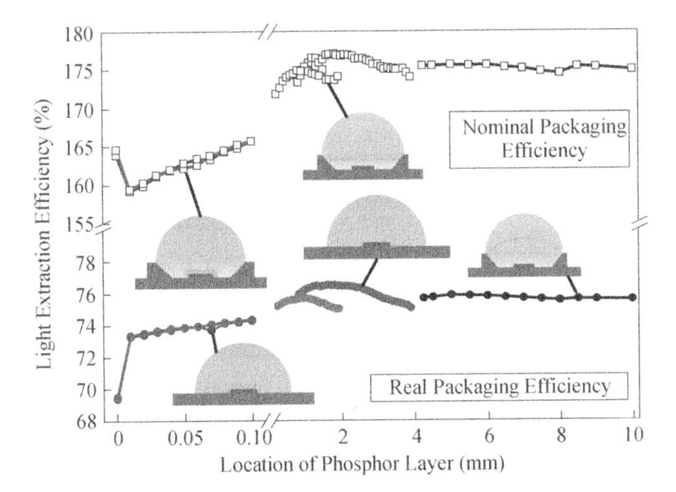

Figure 3.51 Effects of phosphor's location on light extraction efficiencies for all cases. It can be found that the highest η_{NPE} could reach 178%. This is larger than actual test results, which are in the range of 90–110%. *(Color version of this figure is available online.)*

convex surface could improve the critical angle for some random light rays and provide more chances to let these rays be extracted from the surface by reducing the number of multi-scattering. This effect is limited because most of the rays are disordered by phosphor's scattering. This effect plays a special role in structures with light source in the center of convex surface. For example, the blue light emitted from the chip should have relatively higher blue light output in Type II, on the other hand the yellow light emitted from the phosphor layer

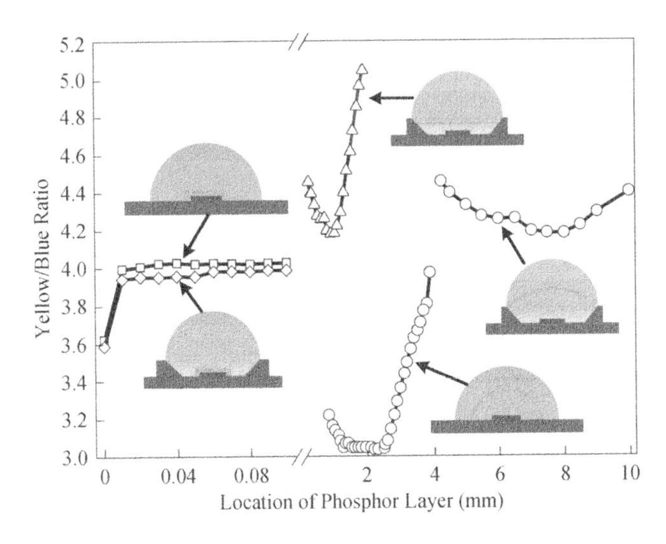

Figure 3.52 Effects of phosphor's location on yellow/blue ratio for all cases. *(Color version of this figure is available online.)*

should present higher yellow light output in Type IV. Therefore, the YBR of Type II is lower than that of Type IV, as shown in Figure 3.52.

Attention should be paid to the calculation of efficiency that is based on the optical power instead of the luminous flux. Luminous flux is related with the visual sensitivity function, which is based on the wavelength. The human eye is more sensitive to yellow light. Therefore, more yellow light output normally generates higher luminous flux. From Figures 3.51 and 3.52, although the variation of optical power output is small, the great change of YBR could influence the luminous flux significantly. Finally, the Type IV with remote and plane phosphor layer is predicted to have the highest luminous flux.

3.5.2 Phosphor Thickness and Concentration for White LED Packaging

Thickness and concentration of phosphor are the second consideration in white LED packaging. This is because the luminous flux and color of LEDs are adjusted mainly through changing the phosphor thickness and concentration after the phosphor converters are chosen. The issue is that the phosphor thickness and concentration can be varied in manufacturing and therefore will affect the optical consistency of white LEDs. The optical consistency, is the ability to control the fluctuation of the produced LED optical performance such as luminous efficiency, correlated color temperature (CCT), and color rendering index (CRI) in a desired range, and is believed to be important for the reduction of LED cost. Poor optical consistency means reduced profit and a higher sale price to the end users, since the material loss for those LED products with poor performance will be included in the price.

Tran *el al.* [74] experimentally studied the effects of phosphor thickness and concentration on LED luminous flux and correlated color temperature. The results are shown in Figures 3.53

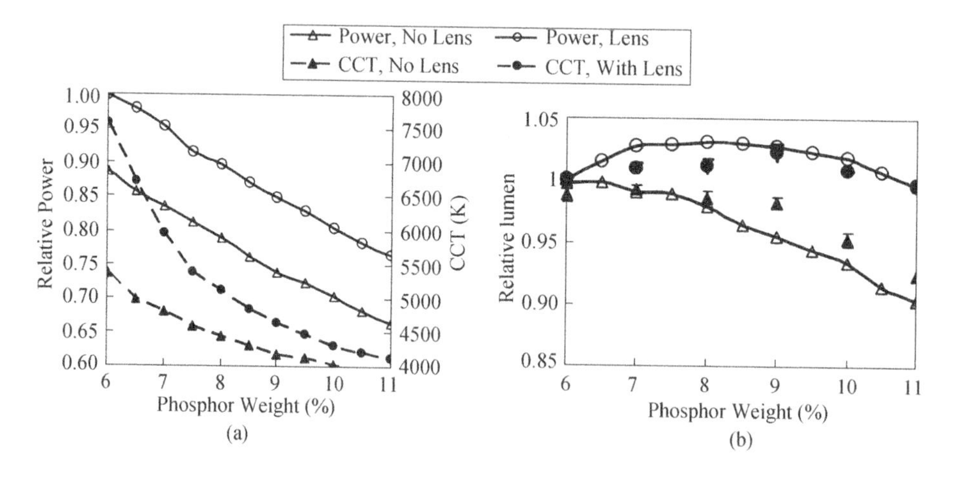

Figure 3.53 Dependence of CCT, radiant power, and lumen on phosphor concentration for the package with a phosphor thickness of 0.8 mm and with two different surface geometries: flat lens and 2.3 mm height lens [74]. (Reprinted with permission from N.T. Tran and F.G. Shi, "Studies of phosphor concentration and thickness for phoshor-based white light-emitting-dioedes," *Journal of Lightwave Technology*, **26**, 21, 3556–3559, 2008. © 2008 Optical Society of America.)

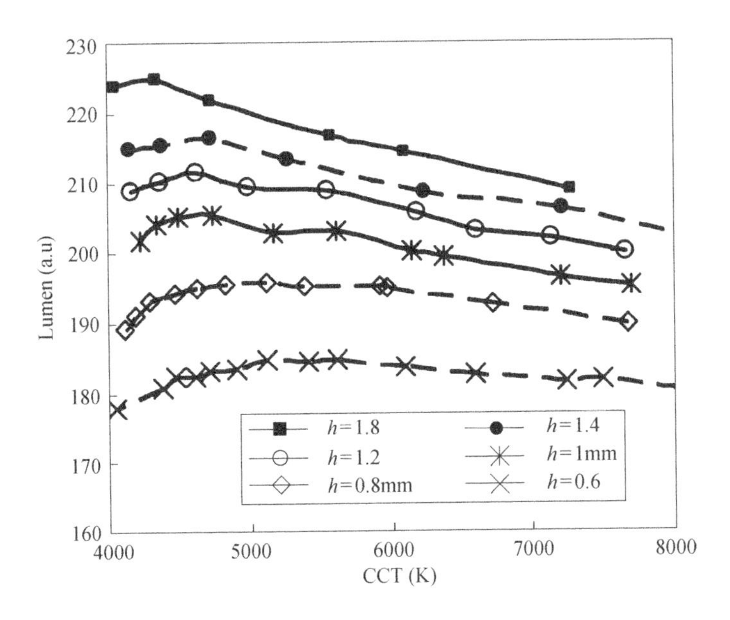

Figure 3.54 Lumen output as a function of CCT and phosphor thickness for a package with a convex lens of 2.3 mm height [74]. (Reprinted with permission from N.T. Tran and F.G. Shi, "Studies of phosphor concentration and thickness for phoshor-based white light-emitting-dioedes," *Journal of Lightwave Technology*, **26**, 21, 3556–3559, 2008. © 2008 Optical Society of America.)

and 3.54. Results show that the package with lower phosphor concentration and higher phosphor thickness has lower trapping efficiency and less backscattering of light, and thus has higher luminous efficacy. When the CCT value is around 4000K, the experimental results show that the lumen output for 1.8 mm thick phosphor package is 23% higher than that for 0.8 mm thick phosphor package.

From Figures 3.53 and 3.54, it also can be seen that the brightness or luminous efficiency of white LED depends largely on the phosphor thickness and concentration. Therefore, a slight variation of phosphor thickness and concentration can change the optical performance significantly. We have systematically analyzed the effects of YAG: Ce phosphor thickness and concentration on the optical performance of phosphor-converted white LEDs including the light extraction, luminous efficiency, and CCT. Five LED packaging methods with different phosphor locations are presented here as a comparison, as shown in Figure 3.55. In Methods I, III, and V we conformally coat the phosphor to replicate the shape of the chip. The difference is that there is a small gap between the phosphor and the chip in Methods III and V whereas, phosphor is directly dispensed on the chip surface in Method I. In Method IV, the phosphor is a planar shape but the location is more remote than those of Methods III and V. Method II fabricates the phosphor with a hemispherical shape, with a patent filed [77]. In all methods, the surfaces of the board and reflector are considered to be coated with Ag to provide high reflection.

The ray tracing simulation results are displayed in Figures 3.56 and 3.57. Our previous paper has reported how the packaging methods affect the initial η_{WPE}, η_{RPE}, and $R_{Y/B}$ [51]. From the

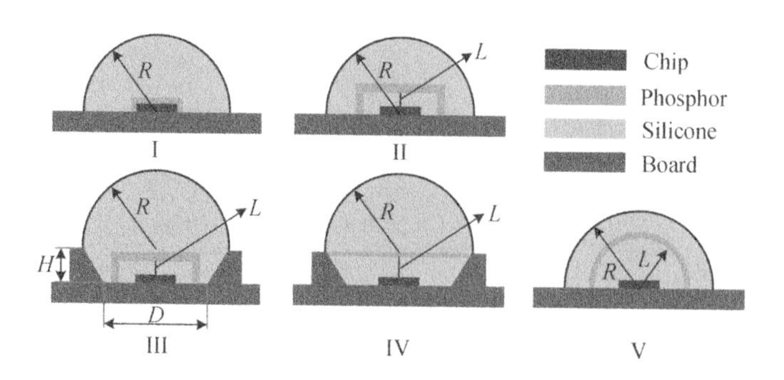

Figure 3.55 Five packaging methods for the analysis. L represents the location of phosphor. In Methods III, IV and V, L is the gap between phosphor and chip. In Method II, L is the radius of phosphor. The radius (R) of lens is 4 mm. The baseline diameter (D) of reflector is 3 mm and the height (H) is 2 mm.

figures, the variations of phosphor thickness and concentration have more significant impacts on η_{WPE}, η_{RPE}, LE, and $R_{Y/B}$. Both the increase of phosphor thickness and concentration can induce rapid reduction of η_{WPE} and η_{RPE} and a significant increase of LE and visible change of color from natural white to yellowish white. However, the different fluctuation behaviors of η_{WPE}, η_{RPE}, LE, and $R_{Y/B}$ for different packaging methods imply that the packaging method is an important factor affecting the optical consistency of white LEDs. Adopting a suitable packaging method can improve the optical consistency when the phosphor thickness and concentration are varied. On the other hand, once the packaging method is chosen, it is important to know the fitting phosphor thickness and concentration. The improvement of optical consistency requires relatively high LE and proper CCT for the packaging method. It can be found that the LE and $R_{Y/B}$ of one packaging method can be better than those of other packaging methods depending on the choice of phosphor thickness and concentration.

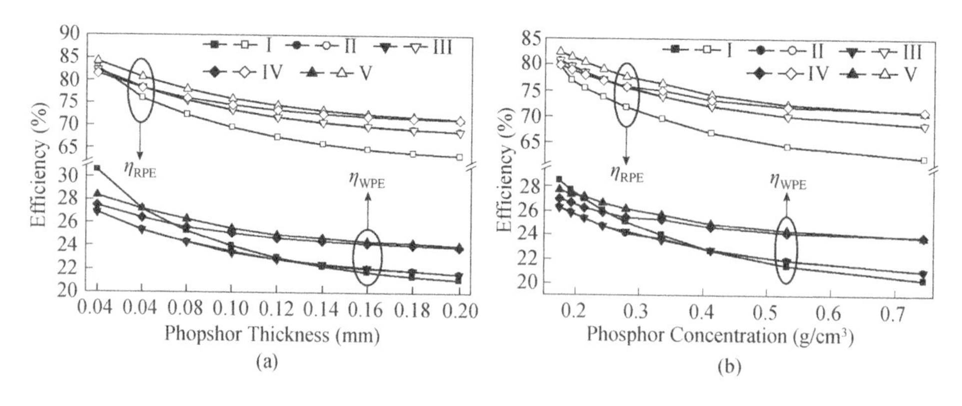

Figure 3.56 (a) Effects of phosphor thickness on η_{WPE} and η_{RPE} (phosphor concentration is 0.33 g/cm^3). (b) Effects of phosphor concentration and η_{WPE} and η_{RPE} (phosphor thickness is 0.1 mm).

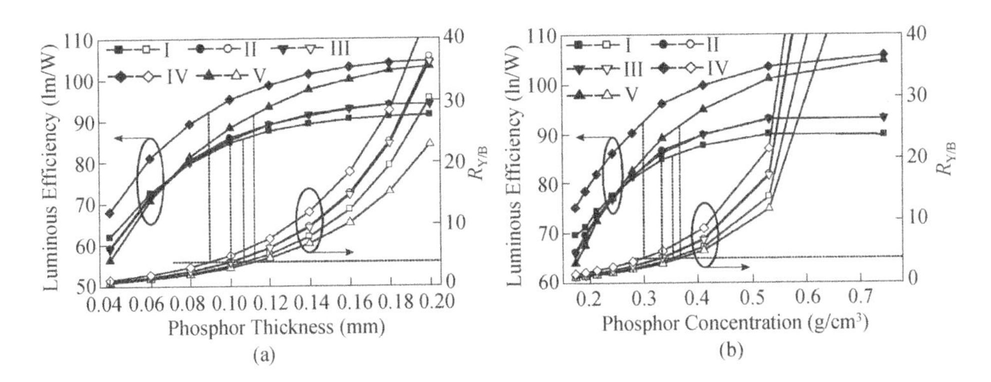

Figure 3.57 (a) Effects of phosphor thickness on luminous efficiency and $R_{Y/B}$ (phosphor concentration is 0.33 g/cm^3). (b) Effects of phosphor concentration on luminous efficiency and $R_{Y/B}$ (phosphor thickness is 0.1 mm). The dashed lines in (a) and (b) refer to the potential maximum luminous efficiency in different methods when $R_{Y/B} = 4$, for which the corresponding CCT and Ra are approximately 5000 K and 70. It can be found that the maximum luminous efficiency of each method are similar in (a) and (b).

Therefore, for the purposes of the discussion here we are proposing suitable packaging methods for white LEDs through the detailed comparison of the optical consistency of the five packaging methods, and providing some suggestions for each packaging method on the choice of phosphor thickness and concentration.

As the most adopted packaging method by manufacturers such as Cree and Lumileds, Method I actually presents the poorest overall performance among all the methods. When the phosphor thickness and concentration are increased, both η_{WPE}, η_{RPE} are reduced more rapidly than those of other methods. The maximum luminous efficiency of Method I is estimated to be 85 lm/W for CCT of 5000 K and CRI of 70, which is the lowest result. However, the color variation of Method I is smaller than that of Methods II, III, and IV. This is due to the higher chip absorptions of blue light and converted yellow light than those in Method I. When the phosphor thickness and concentration are varied, the variations of chip absorption for blue light and converted light will be similar. This feature of high chip absorption is also the fatal defect of Method I, since the luminous efficiency cannot be enhanced as in other methods. But the small variations of luminous efficiency and $R_{Y/B}$ mean that the optical consistency of LED products of Method I can be higher than that of Methods II, III, and IV. Method I also can be an economical packaging method among the five methods, since the process of Method I has been well developed and the chip scale phosphor coating can minimize the size of white LEDs to reduce the total material cost.

Although Method I cannot provide LEDs with the highest luminous efficiency among these methods, the luminous efficiency of Method I can be improved by applying phosphor with a high concentration and thin thickness or a low concentration and thick thickness. From Figures 3.56 and 3.57, it can be seen that when the phosphor thickness <0.06 mm or the concentration <0.2 g/cm^3, η_{WPE} is higher than other methods, and when phosphor thickness <0.07 mm or the concentration <0.25 g/cm^3, luminous efficiency is higher than that of Methods II, III, and V. Results by Tran *et al.* also confirm that low phosphor concentration and thick thickness is beneficial for light extraction [74]. This also explains the reasons why

Lumileds and Cree develop thin film conformal coating of phosphor, whereas the corporations without advanced phosphor coating facilities prefer to control the phosphor silicone mixing ratio to be lower than 0.2 g/cm^3.

For Methods II and III, the tendencies of η_{WPE}, η_{RPE}, luminous efficiency, and $R_{Y/B}$ of these two methods are similar. Therefore, Methods II and III are discussed together. Methods II and III are applied in high reliability LED packaging, in which the silicone layer between the phosphor and the chip is used to protect the electrodes of the chip from the contaminations of phosphor. It can be found that the both methods present poor consistency of η_{WPE}, η_{RPE}, luminous efficiency, and $R_{\mathrm{Y/B}}$ when phosphor thickness and concentration are varied. The variations of η_{WPE} and η_{RPE} are larger than those of Methods IV and V, and the variations of luminous efficiency and $R_{Y/B}$ are larger than those of Method I. The maximum luminous efficiency of the two methods for $R_{\mathrm{Y/B}} = 4$ are estimated to be 86.5 lm/W and 86 lm/W, respectively. These values are only slightly higher than that of Method I. Therefore, although these two methods present higher consistency of η_{WPE} and η_{RPE} than that of Method I, the cost increase caused by the fabrication of the silicone layer meansthat Methods II and III cannot compete with Method I.

Method IV is the common configuration of remote phosphor packaging (RPP), in which the phosphor is planar [26,27]. It can be found that Method IV presents excellent consistency of η_{WPE} and η_{RPE}. The variations of luminous efficiency of Method V are also comparable with those of Methods I, II, and III. The most important feature of Method IV is that the LE of Method IV is the highest among these methods, no matter how the phosphor thickness and concentration are varied. The maximum luminous efficiency of Method IV for $R_{\mathrm{Y/B}} = 4$ is 93.5 lm/W, which is 10% higher than that of Method I. Therefore, Method IV is suitable for those white LEDs with requirements of high luminous flux and moderate consistency of luminous efficiency. But the highest variations of $R_{\mathrm{Y/B}}$ means that Method IV is not suitable for those white LEDs with requirements of high consistency of color. The main reason causing the significant variations of $R_{\mathrm{Y/B}}$ in Method IV is that the chip absorption for converted yellow light of phosphor is reduced effectively. When the phosphor thickness and concentration are increased, the increase of phosphor absorption for blue light and the emission for converted yellow light is more significant than the increase of yellow light loss caused by chip absorption and reflection. Therefore, Method IV is beneficial for yellow light extraction and therefore can enhance luminous efficiency, but the blue light extraction is not good.

Since the color of Method IV is sensitive to the variations of phosphor thickness and concentration, a suggestion for the choice of phosphor thickness and concentration of Method IV is thin thickness and moderate concentration or moderate thickness and low concentration. From Figure 3.57, it can be seen that when the phosphor thickness is <0.09 mm (concentration 0.33 g/cm^3) or the concentration is <0.3 g/cm^3 (thickness 0.1 mm), Method IV can provide proper $R_{Y/B}$ (CCT) for white LEDs, and the luminous efficiency can also be significantly higher than other methods. Therefore, as an example, it is best to fulfill the requirements of consistency control of white LEDs by controlling the phosphor thickness to be thin (that is 0.06–0.08 mm) and the concentration to be moderate (that is 0.3–0.35 g/cm^3), or the phosphor thickness to be moderate (that is 0.09–0.11 mm) and concentration to be low (that is 0.25–0.3 g/cm^3).

Method V is an improvement of Method IV. Changing the phosphor shape to be hemispherical can lead to similar enhancements on the extraction of blue light and yellow light simultaneously [51]. From Figures 3.56 and 3.57, it can be seen that Method V presents the best consistency of η_{WPE}, η_{RPE}, and $R_{Y/B}$ among these methods when the phosphor thickness and concentration are varied. Method V is the only method that can provide better color consistency than Method I among the five methods. This feature means that Method V can be a strong competitor for Method I in those applications demanding more rigorous color consistency of white LEDs, for example, the large size flat panel backlighting. The issue for Method V is that the luminous efficiency consistency of Method V is not good. When the phosphor thickness is thin or the concentration is low, since the extracted blue light of Method V is higher than that of other methods, less converted yellow light makes the luminous efficiency of Method V lower than that of other methods. But a slight increase of phosphor thickness and concentration can enhance the yellow light output more effectively than other methods. The luminous efficiency of Method V is therefore relatively higher among the five packaging methods.

Since the color variation of Method V is the smallest and the luminous efficiency of Method V is lower than that of other methods in the range of thin phosphor thickness and low concentration, the suggestion for the choice of phosphor thickness and concentration of Method V is moderate thickness and moderate concentration. From Figure 3.57, it can be found that when the phosphor thickness is in the range of 0.08–0.113 mm (concentration of 0.33 g/cm^3) or the phosphor concentration is in the range of 0.27–0.37 g/cm^3 (thickness of 0.1 mm), both the luminous efficiency and color consistency can be better than that of Methods I, II, and III. Therefore, taking this simulation as an example, phosphor with moderate thickness (that is 0.09–0.11 mm) and moderate concentration (that is 0.3–0.35 g/cm^3) is believed to be best for the requirements of optical consistency control of white LEDs.

Comparing Figures 3.57a and 3.57b, it can be seen that the variation of phosphor thickness or concentration presents similar effects on η_{WPE}, η_{RPE}, luminous efficiency, and $R_{Y/B}$. Therefore, in the aforementioned phosphor thickness and concentration chosen for each packaging method, both the control of phosphor thickness and concentration are important for the improvement of optical consistency.

As a prospect, Methods I, IV, and V are further discussed and compared here to suggest which method is the most suitable method for future white LED packaging, for which both high luminous efficiency and high optical consistency are required. Method V is predicted to be the best choice. This is because that the moderate thickness and concentration of the phosphor of Method V make the consistency control of Method V easier and the potential maximum luminous efficiency of Method V is only slightly lower than that of Method IV. The thin phosphor thickness of Methods I and IV may cause concern that the mechanical strength of phosphor is poor, and the low phosphor concentration may cause concern that the phosphor settling is more serious. Therefore, the actual consistency of Methods I and IV may be poorer than the simulation, proposing a challenge for precise process control for an optimal optical performance.

3.5.3 Phosphor for Spatial Color Distribution

Spatial color uniformity is important in the efforts to achieve high quality white LEDs. A non-optimized packaging method will induce unexpected phenomenon such as yellow ring. This can be found in some commercial products as shown in Figure 3.58. Color inhomogeneity will influence the actual illumination effects and result in discomfort for the human eye.

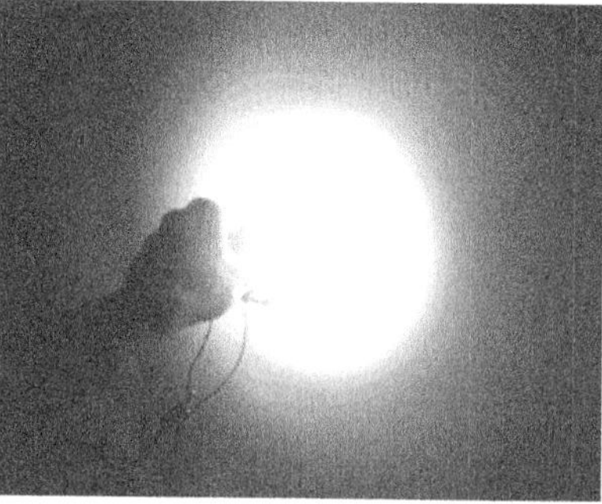

Figure 3.58 Illustration of yellow ring in LED modules. *(Color version of this figure is available online.)*

Nowadays most white LEDs are phosphor converted LEDs. It is perceived that phosphor and packaging structure can influence the spatial color distribution significantly.

Sommer *et al.* firstly studied the effects of phosphor thickness, concentration and size on the spatial color distribution of white LEDs [75]. The results are shown in Figure 3.59. It can be seen that the spatial color uniformity of white LEDs can be realized by changing the phosphor concentration or thickness suitably, and at the same time controlling the phosphor thickness or concentration not to be changed. An increase of the phosphor thickness or concentration can change the color from bluish to yellowish, and the increase in central zone is faster than that in border zone. Therefore, a perfectly uniform white light can be theoretically obtained by adjusting the phosphor thickness and concentration.

We investigated the effects of phosphor location on the spatial color distribution [43,76]. The phosphor location is varied as shown in Figure 3.60. Each packaging method considers four specific cases with different locations. Simulation data are displayed in Figure 3.61 to 3.65. Longitudinal axes for all cases are set with the same length scale to easily compare the variations of curves.

It can be found that YBR at edge zones are normally larger than that at central zones, especially in Methods IV and V. This indicates that there is a yellow ring around the central white zone. The difference cannot be distinguished if the change of YBR from center to side is small and smooth such as in Method I.

Color uniformity is decreased with the increase of location in most methods. However, the influences of packaging methods are more significant than those of location. The variations of color uniformity in each method are 16.67%, 9.8%, 12.5%, 70%, and 31.82% by location, respectively. The average color uniformity is 0.65, 0.67, 0.45, 0.17, and 0.075 in each method. The variation of color uniformity between Methods II and V exceeds 88%. Therefore, the packaging method is most critical to determine the color uniformity, while location plays the secondary role. The tendencies of YBR curves also support this viewpoint. The tendencies are similar with the variation of location, whereas there are obvious differences for YBR curves in various packaging methods.

Methods I and II present higher color uniformity than other methods with a reflector. This is because the reflector can converge the side rays to the central zone especially for blue light. The more blue light is converged to the central zones, then less blue light can emit out at the side zones. Increasing the location of the reflctor can enhance this phenomenon. For example, YBR is reduced from 2.5 to near 1 at central zone in Figure 3.64.

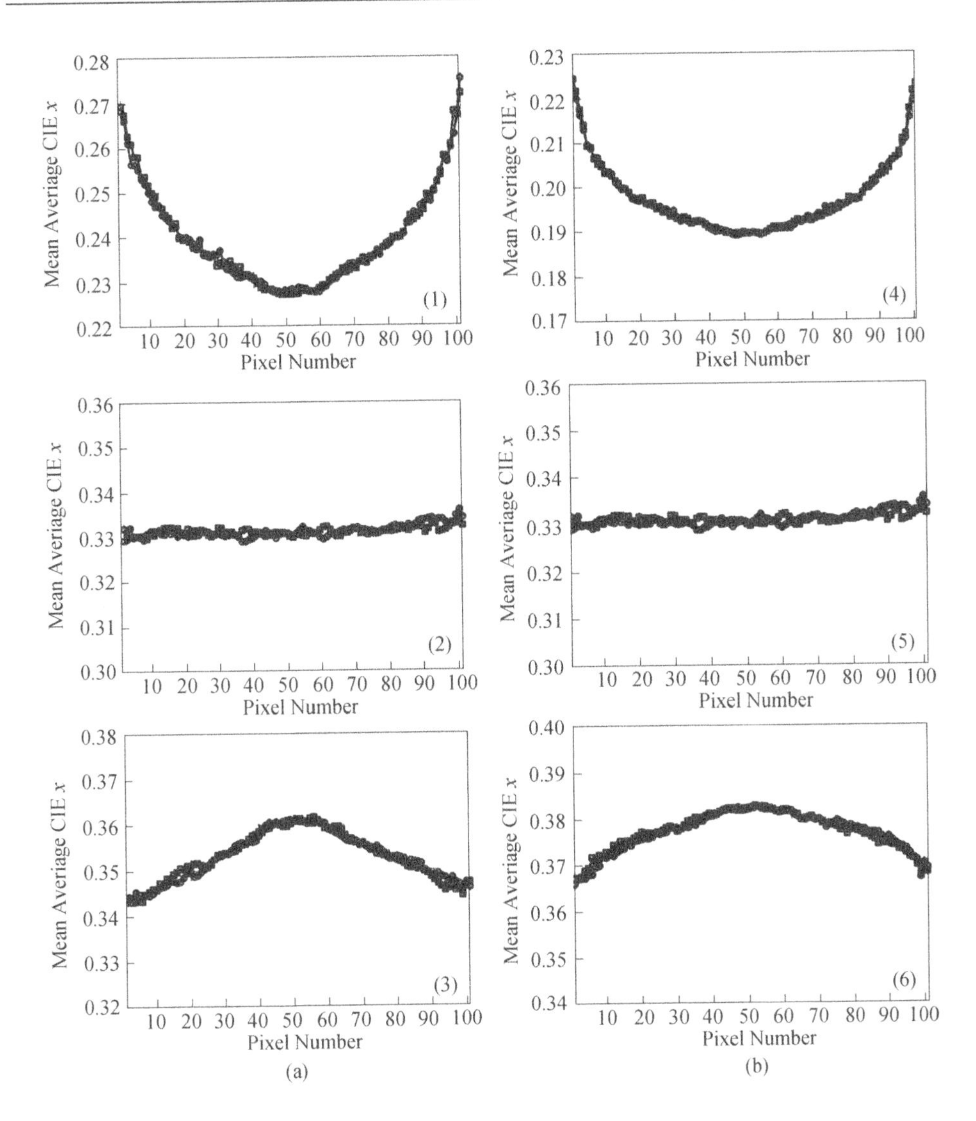

Figure 3.59 (a) Mean average of CIE x chromaticity coordinates for a constant layer broadness of 1040 μm and a constant phosphor concentration of 10% vol. for different heights (1) 100 μm, (2) 400 μm, and (3) 700 μm of phosphor. (b) Mean average of the CIE x chromaticity coordinates for a constant layer broadness of 1040 μm and a constant height of 400 μm for different concentration of yellow phosphor particles in the silicone matrix (4) 2, (5) 10, and (6) 18 vol.% [75]. (Reproduced with permission from C. Sommer, F.-P. Wenzl, P. Hartmann *et al.*, "Tailoring of the color conversion elements in phosphor-converted high-power LEDs by optical simulations," *IEEE Photonics Technology Letters*, **20**, 9, 739–741, 2008. © 2008 IEEE.)

However, the variation of color uniformity in Methods III and V is not as significant as that in Method IV. In Method III, the variation in location is small compared to the size of the reflector, which cannot efficiently affect the light propagation. In Method V, the location is too remote. It can be found that the variation in color uniformity is small when the location is larger than the

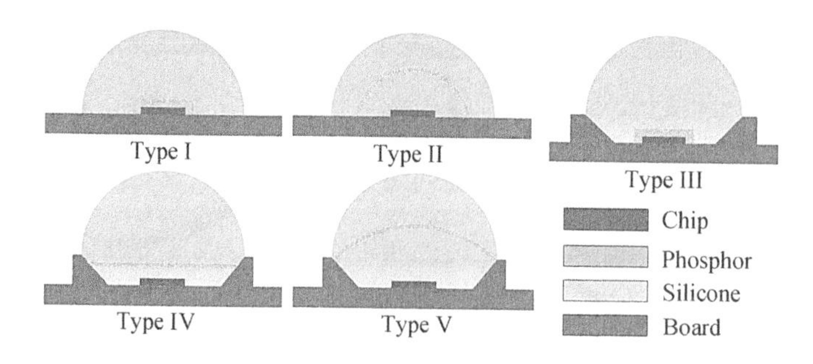

Figure 3.60 Five packaging configurations for the analysis. *(Color version of this figure is available online.)*

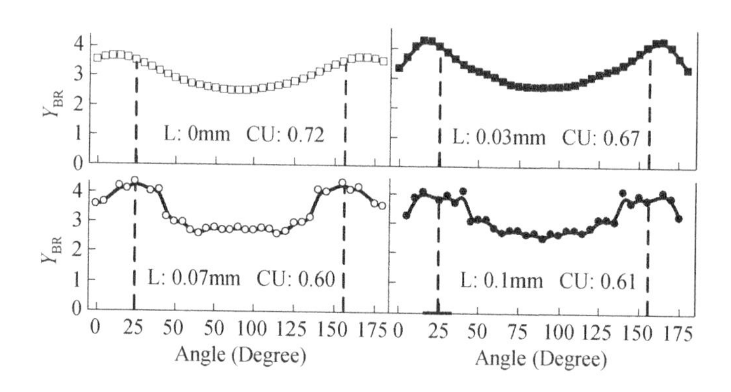

Figure 3.61 YBR curve and color uniformity in Method I. The first specific case directly coats the phosphor on the chip.

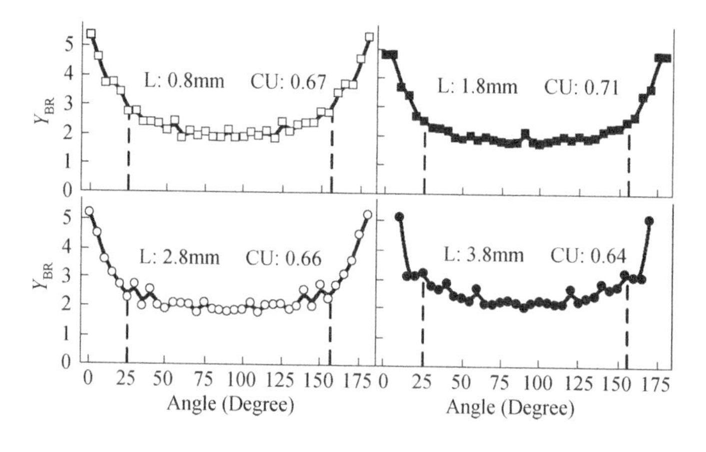

Figure 3.62 YBR curve and color uniformity in Method II.

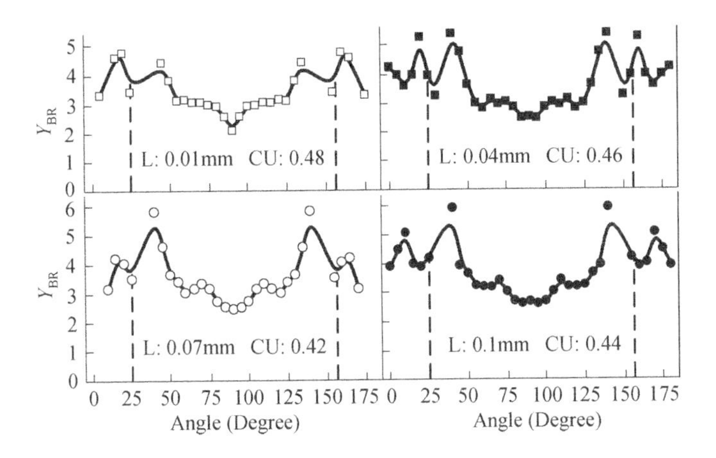

Figure 3.63 YBR curve and color uniformity in Method III.

1.4 mm in Method IV. Therefore, when the location is remote enough, the change of location plays a minor role on color distribution.

The initial color uniformity of Method II is lower than that of Method I, however, the average color uniformity is higher. This is due to the fact that the phosphor layer is hemispherical.

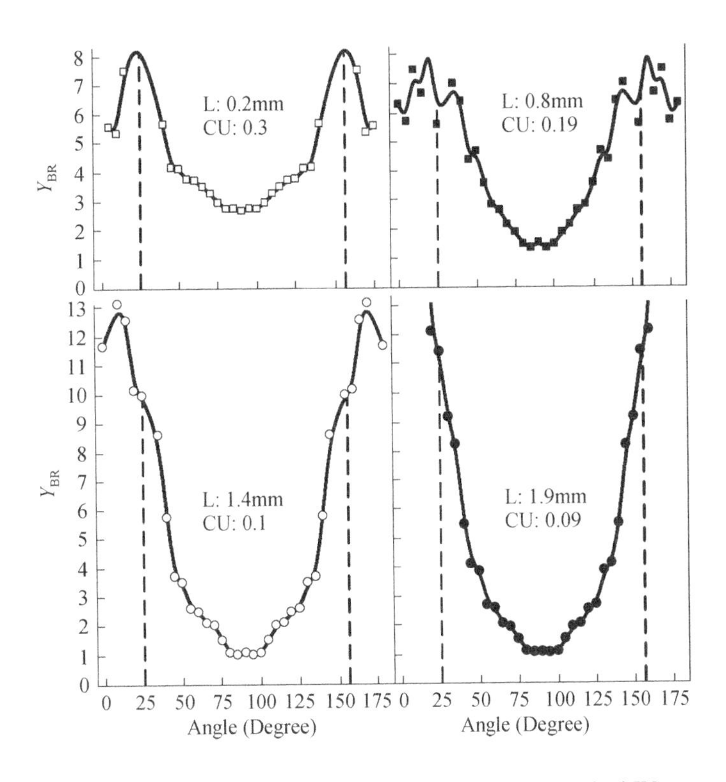

Figure 3.64 YBR curve and color uniformity in Method IV.

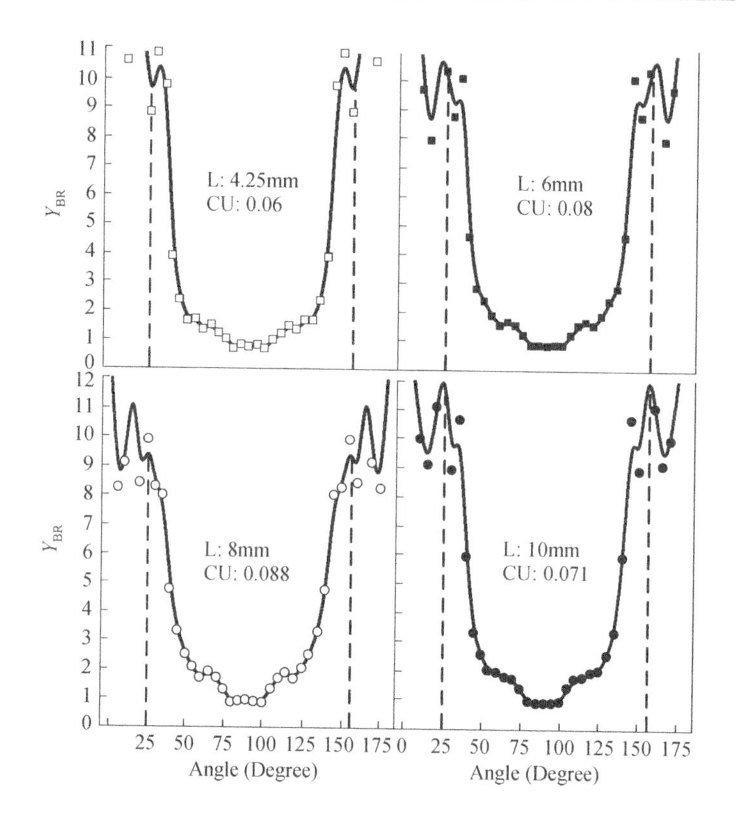

Figure 3.65 YBR curve and color uniformity in Method V.

The radiation pattern of blue light emitted out of chip is similar to Lambertian. Blue light can almost vertically enter the phosphor layer and improve the extraction of side rays. The increase of location cannot affect light propagation fundamentally. Therefore, Method II presents a more stable color distribution.

The reason why the initial color distribution of Method I is more uniform than other methods is that the chip and phosphor are located in the center of the lens. The radiation patterns of blue light out of the chip and yellow light from the phosphor layer are both similar to Lambertian. Since the lens' size is significantly larger than the chip and phosphor layer, they can be treated as small sources. Therefore, most of the rays could directly emit out without being internally reflected. This indicates that the initial color uniformity is high. However, with the increase of location, enlarged dimension and height of the phosphor layer will gradually disorder the directions of blue light and the phosphor layer which cannot be seen as a small source.

Actually, to obtain high color uniformity, packaging elements should make the blue light and yellow light have a similar radiation pattern. That means the packaging elements such as the lens and reflector should affect the propagation of blue light and yellow light simultaneously. This is the fundamental reason why Methods IV and V have so low color uniformity. The reflector affects the propagation of most blue light rays and converges them to center, whereas only part of back scattered yellow light is affected by the reflector. This has the effect that the radiation pattern of blue light and yellow light are obviously different after passing through the phosphor layer. Therefore, the color distribution is significantly non-uniform.

3.6 Collaborative Design

3.6.1 Co-design of Surface Micro-Structures of LED Chips and Packages

At present, high power white LED is realized mainly through GaN based high power blue LED exciting yellow phosphor. However, the highest electro-optical conversion efficiency of GaN based high power LED already discussed does not exceed 60% in the industry [78], which is still far from the ideal 100%. Relatively low light extraction efficiency of LED chips is one of the main factors. In recent years, domestic and foreign researchers have improved its light extraction efficiency mainly through processing various micro-structures on light surface of LED chip, including: surface roughening [31], patterned substrate [79], photonic crystals [80], micro-structure array [81], and so on. Light extraction efficiency can be increased from 20% to 300%. The shapes of various micro-structures are shown in Figure 3.66.

High power LED chips can only be used in various lighting occasions after packaging. The overall light extraction efficiency after LED packaging is its ultimate manifestation of light extraction efficiency. However, most researchers are only concerned about improving the light extraction efficiency of the LED chip itself, often neglecting the effects of follow-up package materials, structure and processes on the light extraction efficiency of the LED, which result in

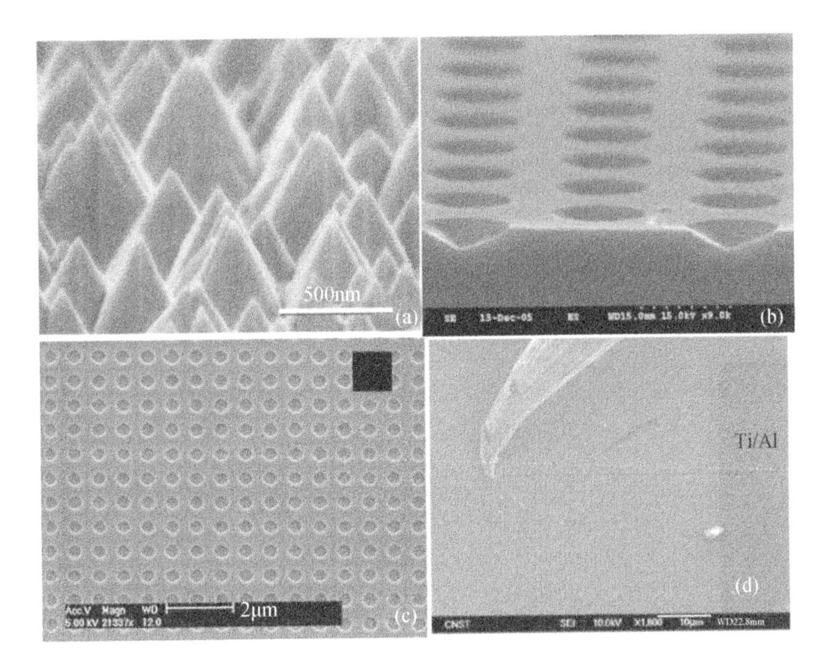

Figure 3.66 Various surface micro-structures on of LED chips: (a) surface roughening; (b) patterned substrate; (c) photonic crystals; and (d) micro-structure array.

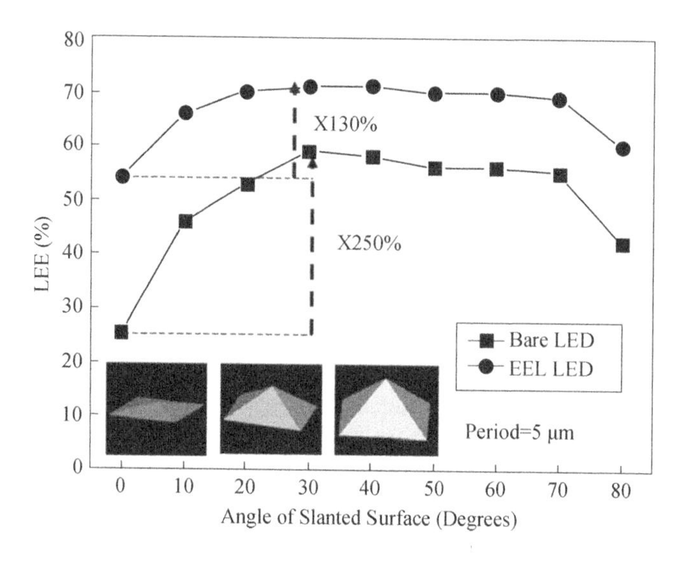

Figure 3.67 Light extraction efficiency (LEE) vs. the angle of the slanted surface in an inversed pyramid structure.

great changes of LED luminous efficiency after packaging. The study of Jiang *et al.* shows that the light extraction efficiency of the LED chip is improved by an average of 50% after surface roughening of the LED chip, but the output optical power is only improved by an average of 15% after packaging [82]. For the photonic crystal LED, air is always adopted as the gap medium when designing its photonic crystal structure, but silicone of a different refractive index in the packaging process will plug the photonic crystal gap and destroy its periodic structure, resulting in defects in photonic crystals, seriously affecting the light extraction efficiency of photonic crystal LED.

At the same time, people currently invest a lot of energy in the careful optimization design of the LED chip structure, with a view to further enhance the light extraction efficiency, but the amount of increase is relatively small, and some even less than 10%. The silicone or phosphor is coated on the light surface of the LED chip in the packaging process, increasing the exiting angle range of chip interface, further improving chip's light extraction efficiency. But for LED chips of different surface micro-structures, the amount of efficiency elevation differs. The simulation studies of Lee T X *et al.* has shown that fabricating a pyramid micro-structure array on the N-type GaN surface can improve the light extraction efficiency of the chip. The simulation results show that (Figure 3.67) [83], the efficiency of 30° angle of inclination improves more than 12% than that of 10 ° angle of inclination, but after the LED is encapsulated with an epoxy lens (simplified EEL–LED) of refractive index of 1.5, the amount of elevation reduces to less than 4%.

Figure 3.68 shows four kinds of different surface micro-structure of LED chips, in which the sample 1 and sample 2 are the vertical electrode LEDs, and the sample 3 and sample 4 are the level electrode LEDs. We have found that the total optical power of four kinds of chips increases considerably after packaging with silicone of refractive index of 1.54, but the rate of light extraction efficiency increasing differs from 31.4% to 59.2% with different surface

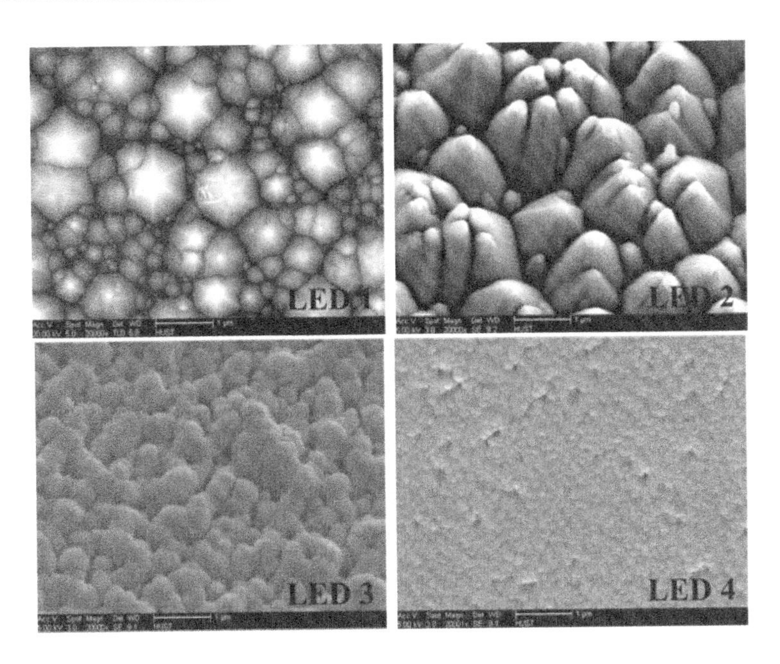

Figure 3.68 SEM diagram of four kinds of surface micro-structures of different LED chips.

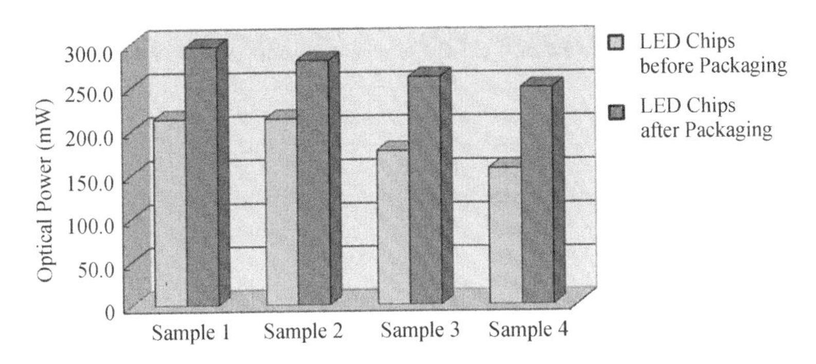

Figure 3.69 Comparison of optical power of four kinds of LED chips with different micro-structures before and after packaging with silicone.

structures (as shown in Figures 3.69 and 3.70). From the figures we can find, the optical power of the sample 2 LED chip is the same as that of the sample 1 LED chip, but after packaging, the total optical power efficiency of the sample 2 LED is lower than that of sample 1 which demonstrates that for LED chips of the same vertical electrode structures, micro-structure on the surface of sample 1 is more advantageous in lighting.

As shown in Figure 3.71, the surface roughening structures of a level electrode LED chip A and a vertical electrode LED chip B are quite different.,We have found from Figure 3.72 to Figure 3.73 that sample B's optical power of bare LED chip is higher than that of sample A, but

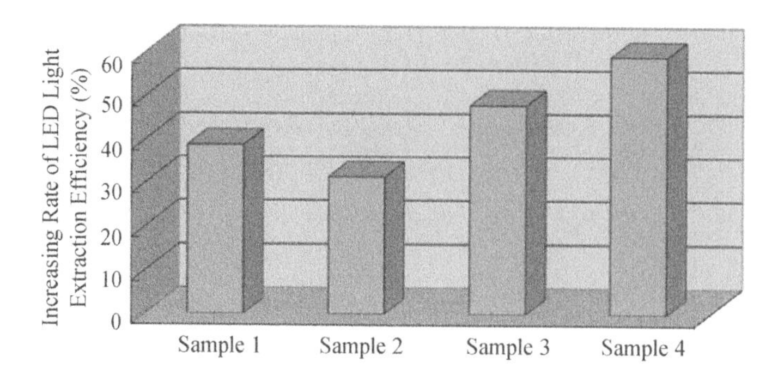

Figure 3.70 Comparison of increasing rate of LED light extraction efficiency after packaging with silicone.

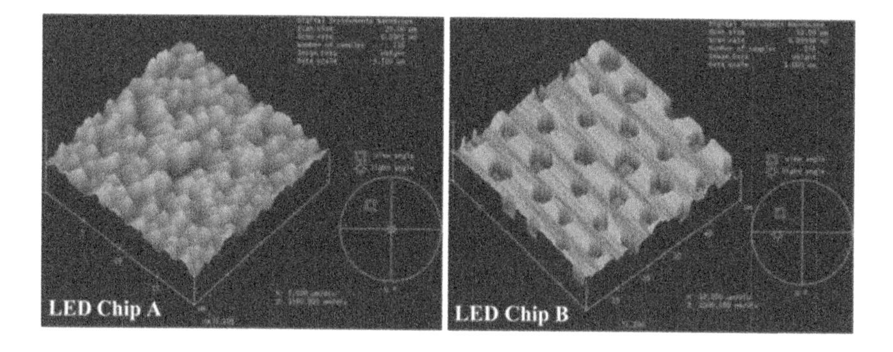

Figure 3.71 AFM diagram of surface micro-structures of sample A and sample B LED chips.

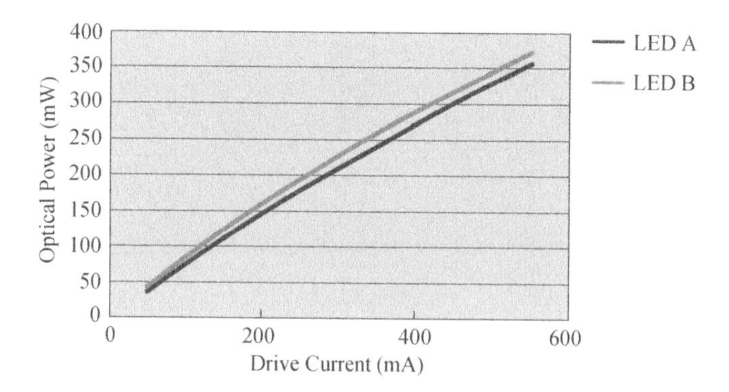

Figure 3.72 Comparison of optical power of sample A and sample B LED chips.

the luminous flux between the two is very close to each other after they are packaged by using the same materials and process, and sample A is even higher than sample B in some occasions. This experiment shows very well that LED chip structure needs co-designing with packages. The output luminous flux is not necessarily high after packaging the LED chip of high bare

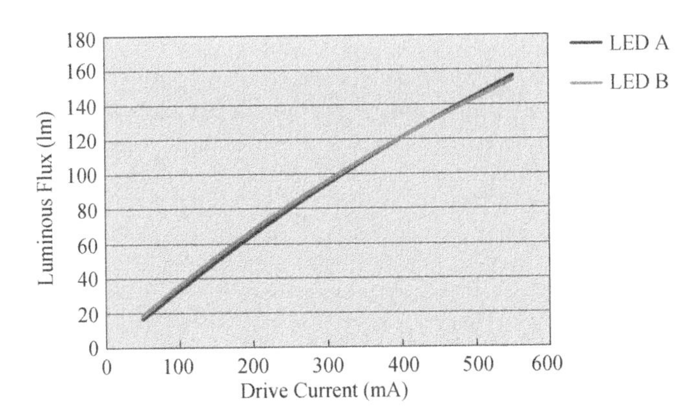

Figure 3.73 Comparison of luminous flux of sample A and sample B LED after packaging.

output optical power. Therefore when designing the chip, the follow-up package situation must be considered and the overall plan must be made to truly improve the cost performance of LED chip.

It can be found from the above analysis that owing to the complexity of LED chip light exiting surface, it is difficult to roughly estimate the impact of packaging on light extraction efficiency increasing of the LED chip with different micro-structures. If considering only from the chips, it is difficult to judge which surface micro-structure is more advantageous and whether there is any further need for careful optimization for certain micro-structures. Therefore, design of the LED chip surface micro-structures must be coordinated with the follow-up package design; quantitatively and systematically analyzing the change between light extraction efficiency and overall packaging efficiency for LED chips of different micro-structure surfaces, guiding the structural optimization design, and processing technology selections of LED chips.

In addition, because light surface micro-structure greatly differs among LED chips, interface behaviors of silicone or phosphor silicone and LED chip light exiting surface are different, and the types, because of defects in the interface and impact of these defects on the LED packaging efficiency and reliability are also different. It is easy for the micro-structure surface LED chip to introduce all kinds of impurities (such as bubbles, dust, and so on.) in the interface of chip and package. The existence of these impurities will not only affect the overall LED light performance, but is also very likely to become the factor inducing crack growth appearing in LED follow-up assembly process (such as reflow), testing and long-term use (temperature, moisture, and so on.), thereby affecting the reliability and durability of the LED. Meanwhile, the interfacial stress situation of the surface micro-surface structure and silicone are different for different chips in the process of silicone curing. There may be relatively large stress concentration points for some micro-structures, easily leading to packaging defects such as delaminations, cracks, and so on in the course of long-term use of LED and a negative impact on the LED light efficiency and reliability. Therefore, co-design of LED chip surface micro-structure and packages will also help guide reliability design of LED chip and packages, as well as the choice of packaging materials and process optimization.

Therefore, co-design of high power LED chip micro-structures and packages must be carried out, integrating all aspects into consideration of the entire course, overall optimizing based on local optimization. The study on co-design of the LED chip micro-structure and packages can guide not only optimization design of LED chips, but also processing technology of the chips and packaging, which is significant to improving the overall LED light efficiency and the reliability after packaging. However, there is a lack of systematic and quantitative research both at home and abroad. LED chip structure design is separated from packaging. Therefore it is important and urgent to carry out co-design of the high power LED chip surface micro-structure and packaging.

3.6.2 Application Specific LED Packages

As shown in Figure 3.74, secondary optics are essential to LED illumination systems because the light patterns of most LEDs are circular symmetrical with non-uniform illuminance distribution, which cannot directly meet the requirements of different illumination applications (for example rectangular light pattern required in road lighting). Freeform lens is an emerging optical technology being developed in recent years with advantages of high design freedom and precise light irradiation control. However, belonging to the category of secondary optics, traditional freeform lenses also have many disadvantages such as being too large for some space confined applications, requiring highly accurate assembly and they can be inconvenient for customers to use. The optical performance of many LED luminaires existing in the market, such as road lamps, MR16 lamps, and so on is poor mainly due to their manufacturers lack of ability of prescribed optical design. Therefore, if the LED package and secondary optics could be integrated within one LED module and could be directly used for some specific applications, this new LED package will be popular in the market.

(i) Application Specific LED Package Single Module

In this section, we will introduce a novel LED package, application-specific LED package (ASLP), which will provide a more cost-effective solution to high performance LED

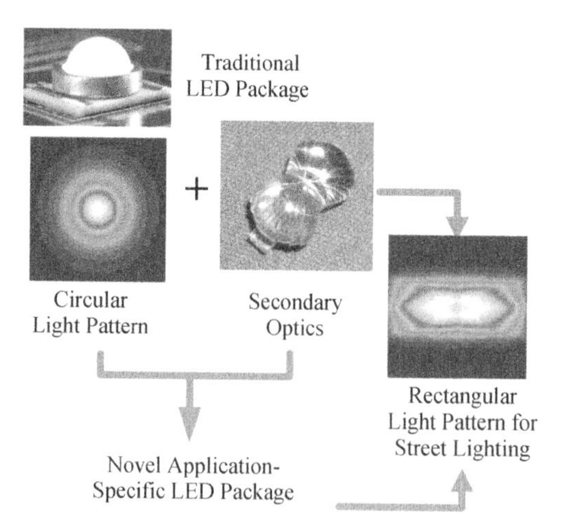

Figure 3.74 Schematic of the concept of an application-specific LED package.

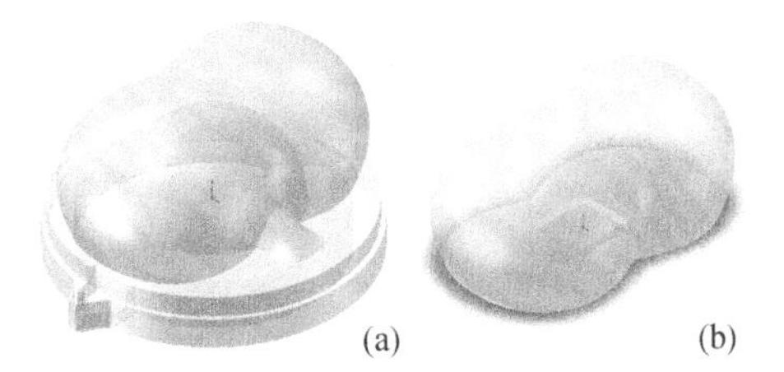

Figure 3.75 (a) Novel compact freeform PC and (b) silicone LED package lenses.

lighting [82–84]. Since road lamps' significance in our LED community signal the open area of general lighting, demonstrated by a recent Chinese government program called 10–city 10,000–lamp and other programs around the world, rectangular light patterns for road lighting are designed for examples. Figure 3.75 shows a novel freeform polycarbonate (PC) LED package lens (LPL) and a silicone LPL with the refractive index of 1.586 and 1.54 respectively for LED packages to form a 32 m long and 12 m wide rectangular illumination area at the height of 8 m. The volume and largest value of length, width and height of the PC lens are 42.9 mm^3, 6.1 mm, 3.8 mm and 2.6 mm respectively, and the values are 44.5 mm^3, 6.3 mm, 3.9 mm and 2.7 mm for the silicone lens. The volume of these two lenses are close to that of the most widely used hemisphere PC LPL.

Figure 3.76a shows one type of the most widely used high power LED package based on leadframe and heat sink. The optical structure of this LED is constructed by LED chip, phosphor, silicone, and PC packaging lens. Since the PC packaging lens is circular symmetrical, the light pattern of this LED is circular with non-uniform illuminance distribution as shown in Fig. 3.76b, which is hard to be directly used in road lighting.

A novel ASLP is obtained only by replacing the circular symmetry PC LPL by the designed compact freeform PC LPL. The optical performance of this ASLP is simulated numerically by the widely used Monte Carlo ray tracing method. The light output efficiency (LOE), defined as

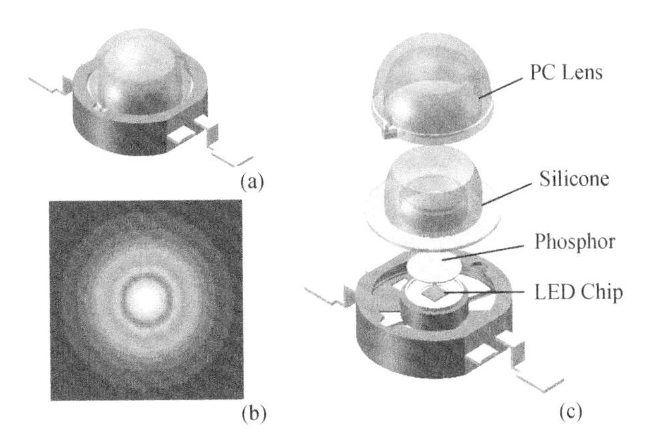

Figure 3.76 (a) Traditional LED package based on leadframe and heat sink; (b) its illumination performance; and (c) its detail optical structure.

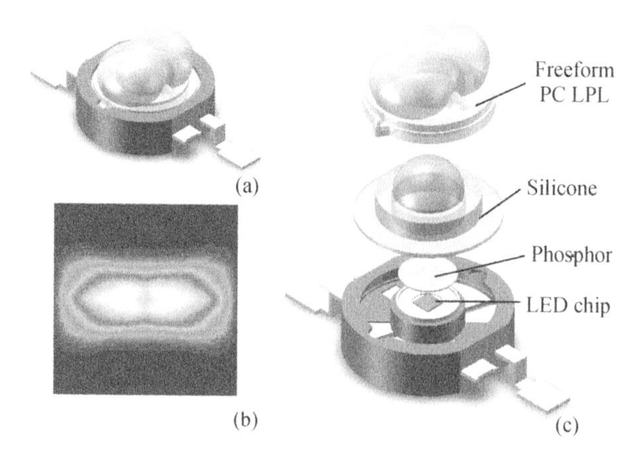

Figure 3.77 (a) Novel application-specific LED package based on leadframe and heat sink; (b) its illumination performance; and (c) its detail optical structure.

the ratio of light energy exits from lens to light energy incidents into lens, of this novel LPL is 94.2% (considering Fresnel loss), which is slightly less than that of traditional hemisphere LPL of 95.0%. As shown in Figure 3.77, more than 95% light energy of the ASLP distributes within an approximately rectangular light pattern with the length of 33 m and width of 14 m at the height of 8 m, which is quite suitable for road lighting. Therefore the ASLP could be directly used for road lighting and no secondary optics is needed, which make it convenient for LED fixtures designers and manufacturers to use and also will furthermore reduce the cost of LED fixtures.

During this novel LED packaging process, the only change we need to make is fixing this novel freeform LPL to the traditional hemisphere PC LPL on the frame, and that is compatible with current LED packaging processes totally, which makes it easier for LED manufacturers to adopt this new technology with little change of existing process.

As shown in Figure 3.78, the freeform PC LPL is manufactured by an injection molding method. The LOE of the freeform PC LPL reaches as high as 94.8%, which is slightly lower than that of traditional hemisphere PC LPL of 95.4%. Considering that the LOEs of secondary optics (for example freeform lenses) are always at a level of about 90%, the system LOE of LED fixture consisting of ASLPs will be about 9% higher than that of traditional LED fixture.

In 2009, Guangdong Real Faith Optoelectronic Co. LTD developed a white light ASLP integrated with the freeform PC LPL for road lighting and its optical efficiency reached as high as 105 lm/W@350 mA (as shown in Figure 3.79). An LED module for road lighting consisting of an LED and a secondary optical element with the kind of freeform lens are shown in Figure 3.80c. From comparisons shown in Figure 3.80, we can find that the height and volume of this novel ASLP are only about a half and an eighth of that of the LED module respectively, which provides an effective way for some size compact LED illumination systems design and more design freedoms for new concept LED lighting fixtures.

Figures 3.81a and 3.81b show the illumination performances of a traditional LED package and the ASLP respectively. The light pattern of the traditional LED package is circular with non-uniform illuminance distribution, while the ASLP redistributes LED's light energy

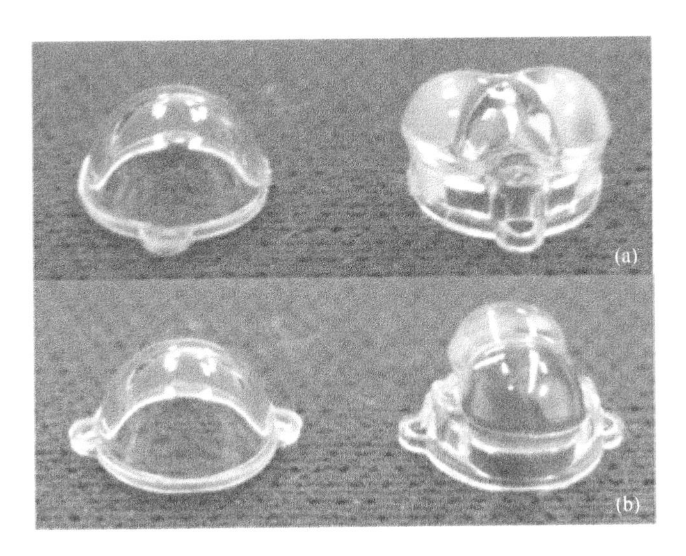

Figure 3.78 (a) Front view of traditional hemisphere LPL (left) and the novel application specific freeform LPL (right) and (b) left view of these two LPLs.

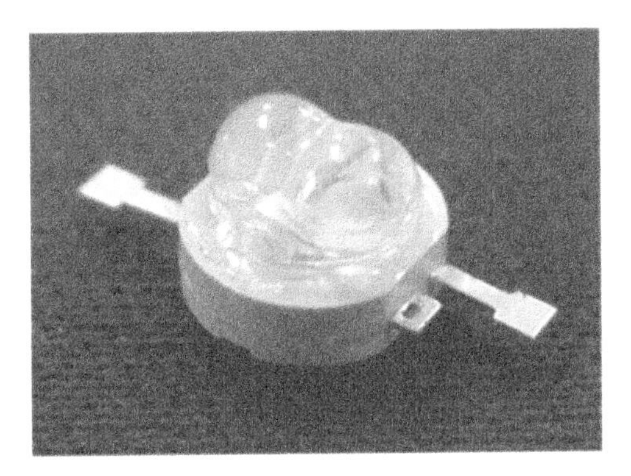

Figure 3.79 Real Faith Optoelectronic white light ASLP.

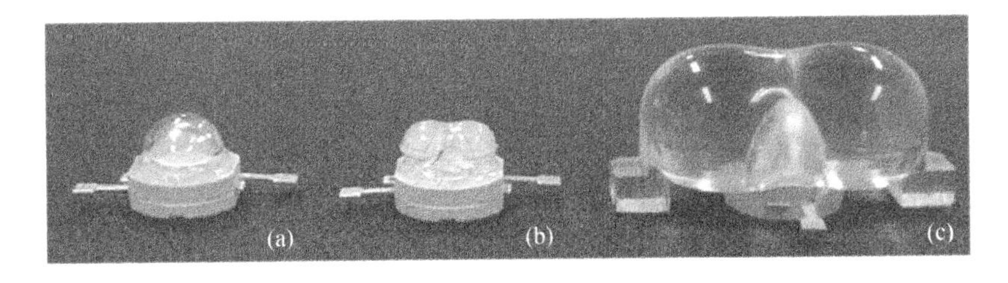

Figure 3.80 (a) Traditional LED package, (b) Real Faith Optoelectronic ASLP and (c) traditional LED road lighting module consisting of a traditional LED and a freeform secondary lens.

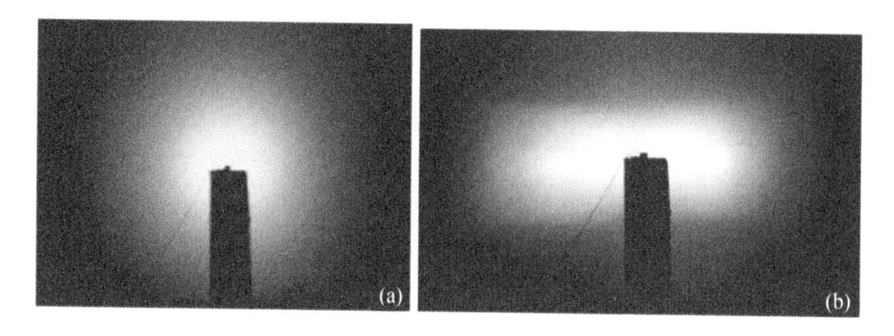

Figure 3.81 Illumination performance of (a) traditional LED package and (b) ASLP. *(Color version of this figure is available online.)*

distribution and forms a rectangular light pattern on the target plane, which is more uniform than the circular light pattern. Therefore, there is no need for this new type of road lighting ASLP which adopts integrated lens to math collocate with the second optical component. Light can be accurately irradiated to the target area, and does not produce any stray light, so saving labour and worry in LED road lighting.

The ceramic board based high power LED package is another kind of traditional LED package with low thermal resistance. The detail optical structures and illumination performances of the traditional LED package and ASLP based on the ceramic board are shown in Figures 3.82 and 3.83. The ASLP is also obtained by replacing the circular symmetry silicone LPL with the designed compact freeform silicone LPL.

A variety of application specific LED package module have also been introduced to the market by OSRAM, such as OSLUX LW F65G and Golden DRAGON oval Plus (http://www.osram.com/). OSLUX LW F65G is as shown in Figure 3.84: the length is 5.3 mm, width 5.2 mm, high 2.8 mm, and the luminous efficiency is 55 lm/W @ 100 mA. Its luminous intensity distribution curve is as shown in Figure 3.85, and compared with Lambert light distribution, it is converging light. The light pattern which is 40 mm away from a sub-rectangle, with its length of 50 mm and width of 45 mm and uniform illumination distribution. This type

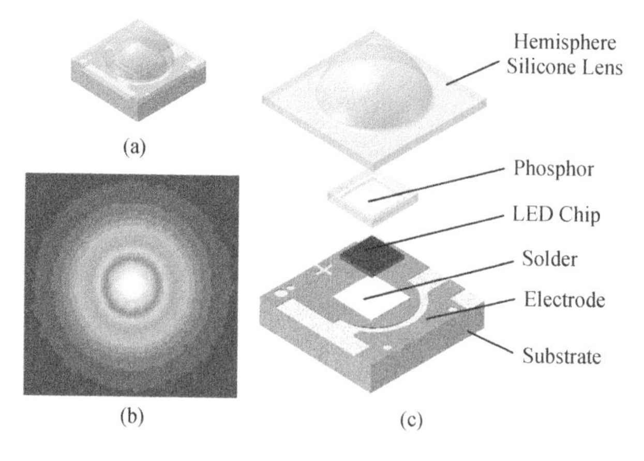

Figure 3.82 (a) Traditional LED package based on ceramic board; (b) its optical performance; and (c) its detail optical structure.

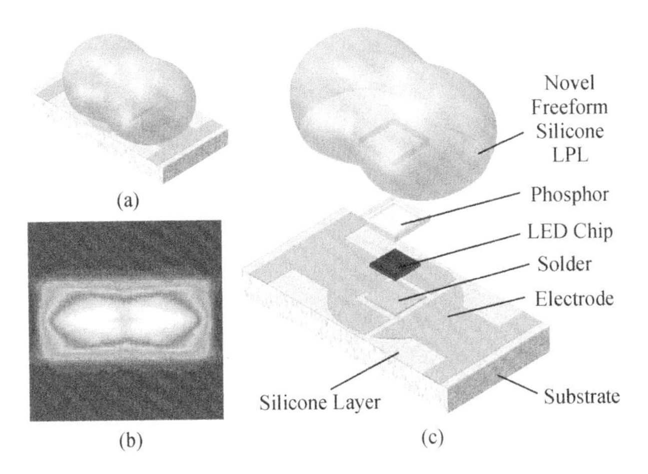

Figure 3.83 (a) Novel ASLP based on ceramic board; (b) its optical performance; and (c) its detail optical structure.

Figure 3.84 OSRAM OSLUX LW F65G.

of LED is mainly applied in photo flash, torch lighting, and high-brightness backlights display lighting, and so on.

Golden DRAGON oval Plus is a type of application specific LED module developed especially for road lighting which was recently introduced by OSRAM in May, 2009, shown in Figure 3.86. Uniform elliptical light pattern (vertical angle 80 °, horizontal angle 120 °) can meet the demands of road lighting and tunnel lighting, and the light extraction efficiency reaches 90 lm/W@350 mA.

(ii) Application Specific LED Package Array Module

ASLP array module is the developing direction of these novel LED packages. ASLP array modules for road lighting, with the type of chip on board (CoB) packaging, could also be

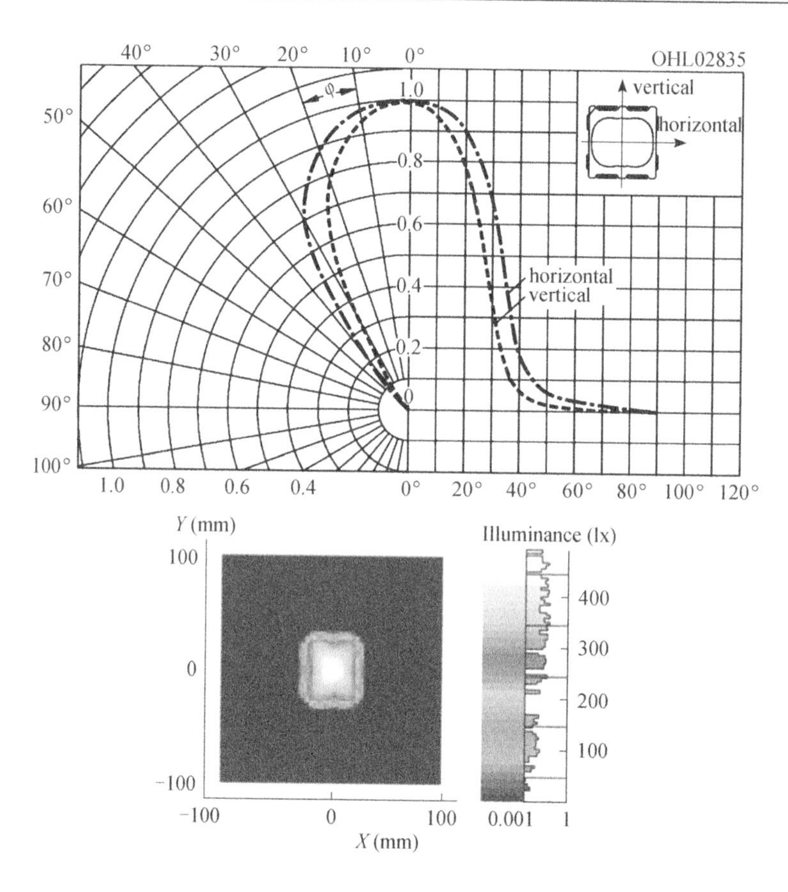

Figure 3.85 Luminous intensity distribution curve and illumination performance of the OSRAM OSLUX LW F65G at 40 mm away.

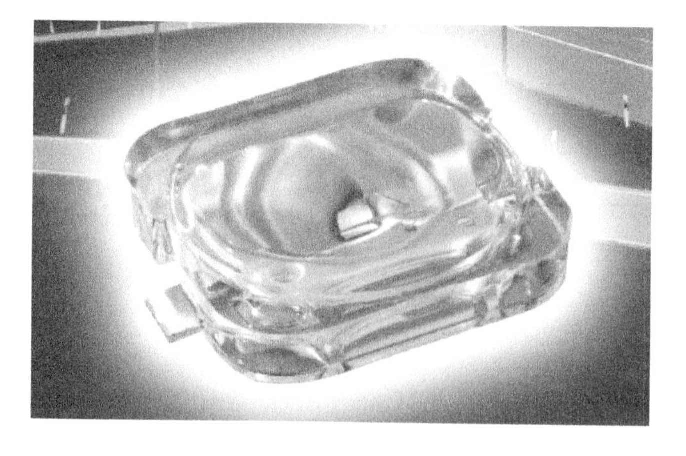

Figure 3.86 OSRAM Golden DRAGON oval Plus.

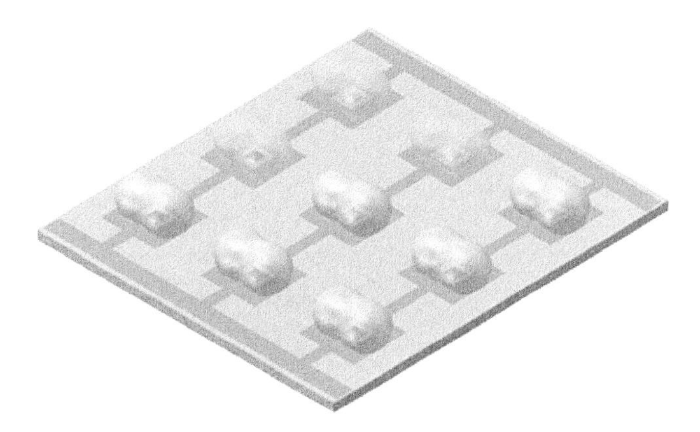

Figure 3.87 A novel 3 × 3 ASLP array module based on ceramic board for road lighting.

obtained by integrating freeform lens arrays with traditional CoB LED modules. Figure 3.87 shows a 3 × 3 LED array module based on ceramic board and it mainly consists of a ceramic board with circuits, LED chips, phosphor, and novel silicone lens array. As shown in Figure 3.88, the simulated light pattern of this LED array module is quite similar to that of the single ASLP and also could be used in road lighting directly. The light source of a 108 watt LED road lamp, which is one of the mostly used types of LED lamps in the market, could be achieved easily by integrating 12 of this type of LED module. Since the length and width of this module are only 34 mm and 30 mm respectively, then the size of the light source of the 108 watt LED road lamp could be less than 120 mm × 105 mm, which will considerably reduce the size and cost of the LED road lamp and throw down a challenge to heat dissipation technologies for LED fixtures in the future.

In most LED applications, since the LED array module with ceramic board will be bonded onto the metal core printed circuit board (MCPCB) before connecting with heat sinks, ASLP array modules directly based on MCPCB will provide a more effective solution with the advantages of low thermal resistance and cost. Figure 3.89 shows a 3 × 3 LED array module based on MCPCB and it mainly consists of a MCPCB with circuits, LED chips, phosphor, and

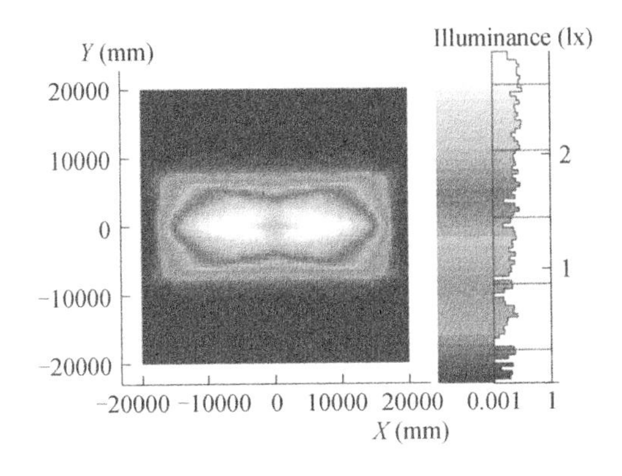

Figure 3.88 Optical performance of the novel 3 × 3 ASLP array module based on ceramic board.

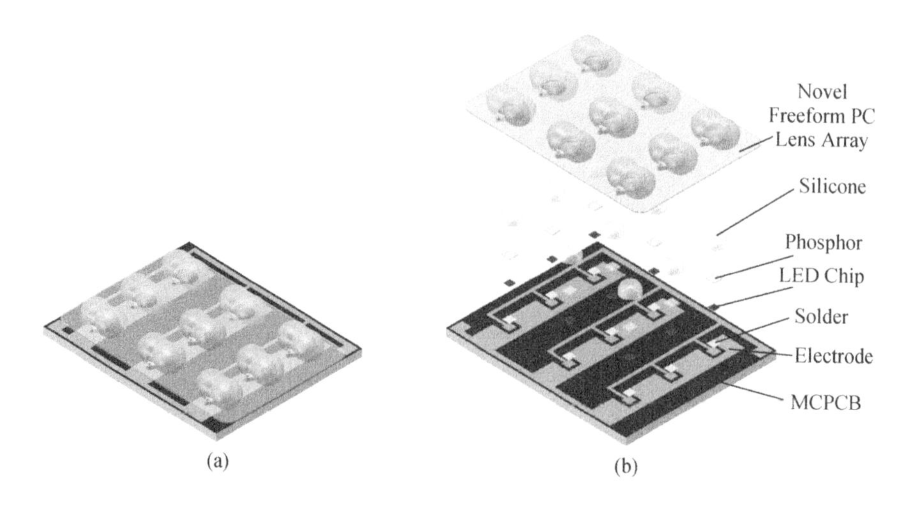

Figure 3.89 (a) A novel 3 × 3 ASLP array module based on MCPCB for road lighting and (b) its detail optical structure.

novel freeform polycarbonate (PC) lens array. This module can be directly bonded onto the heat sink of LED road lamps and reduce one thermal interface between the ceramic board and MCPCB, which will decrease the thermal resistance from LED chips to heat sink. Since the solder mask existing on the surface of the MCPCB will reduce the bonding strength between the silicone lens and the MCPCB, a PC freeform lens array is adopted in this LED module packaging and it could be fixed on the MCPCB by bonding or mechanical fastening. Then silicon will be injected into the cavities of PC lens through the injection holes and fill the cavities. Moreover, the PC lens array will reduce the cost of application-specific LED array modules even more and make it more convenient to assemble.

(iii) Application Specific LED Package System

LED lighting is an integrated system which involves optics, thermal management, electronics, control, mechanical reliability, and other subjects. In addition to good optical design, it also needs appropriate heat dispassion design, driving design, and control design, and so on, and all these are indispensable. Therefore, more and more LED applied commodity producers hope that LED package modules can not only integrate optical systems, but also integrate cooling, driving, controlling systems, and so on. Therefore, the multi-system integrated application that specific LED package modules use will be more convenient for customers to utilize.

Figure 3.90 shows an integrated LED package module with lens and cooling function. Vapor chamber technology is adopted in this LED module to reduce the spreading thermal resistance of the system. Figure 3.91 presents an integrated LED package module with lens, cooling function, powering, and controllers which can be used in lighting applications directly. The LED module can adopt array packaging to increase the total output luminous flux of a single module. Thus, this kind of LED package module can not only work normally, but be even smarter. Controllers comprise micro controller units (MCU), memory, wireless communication modules, and sensors, such as temperature sensor, luminous intensity sensor, audio sensor, accelerometer, and so on. The sensors can monitor the working state of LED

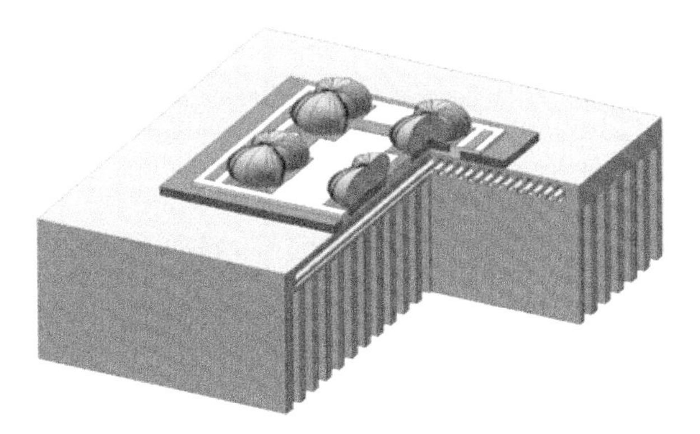

Figure 3.90 Schematic of an integrated LED package module with lens and cooling function.

modules, and by feedback of the control of the driving, make LED modules work in a more healthy state. Figure 3.91 shows an LED light source packaging module which is respectively applied in LED headlamp and rear projection television, introduced by OSRAM (http://www.osram.com/). These modules have integrated optical systems and controlling systems, which are very convenient for customers to utilize.

In summary, by comparing the traditional LED illumination modules consisting of an LED and a secondary optical element, the novel ASLP has the advantages of low profile, small volume, high system light output efficiency, low cost and convenience for customers to use. Moreover, the ASLPs can also be designed to meet other LED lighting applications, such as backlighting for LCD display, automotive lighting, and so on. Therefore ASLPs will provide a more cost-effective solution to high performance LED lighting luminaires and probably become the trend of LED packages.

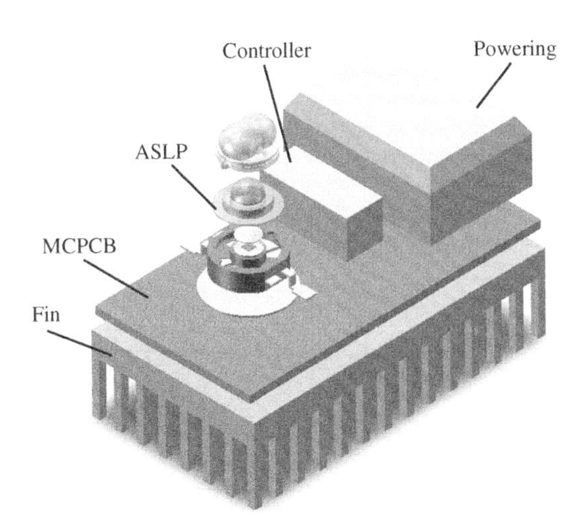

Figure 3.91 Schematic of an integrated LED package module with lens, cooling function, powering and controllers.

3.7 Summary

In this chapter, optical design of high power LED packaging modules was described in detail. First, the properties of LED light were introduced which are important to the module design, with the focus on the definition of those properties such as lumen efficiency, luminous intensity, illuminance and luminance, color temperature, correlated temperature, and color rendering index. Key components and packaging processes for the optical design were also introduced, including chip types, phosphors and their coating processes, and the lens and molding process. Light extraction in the chip and module level was defined. Chip modeling and phosphor modeling were presented, which are essential for white light optical modeling and simulation. Focus was on the phosphors for white LED packaging and detailed discussions were devoted to the various quantities such as location, thickness, and concentration of the phosphor. Spatial color distribution was first proposed as a function of phosphor and packaging structure. A challenging issue of chip surface roughness forced us to propose a co-design issue by considering the interface condition in the optical design model. Finally, a new application specific LED packaging was proposed and discussed, with the objective of providing more functions and chips integrated into one module.

References

[1] Zukauskas, A., Shur, M. S. and Gaska, R. (2002), *Introduction to Solid-State Lighting*, Wiley-Interscience.

[2] Schubert, E.F. (2006), *Light-Emitting Diodes*, New York, Cambridge University Press.

[3] Zhang, Y.M. (2008), *Applied Optics*, Beijing, Publishing House of Electronics Industry. (In Chinese).

[4] Regina, M., Gerd, O.M. and Michael, R.K. (2004), "Phosphors materials and combinations for illumination grade white pcLED," *Third International Conference on Solid State Lighting, SPIE*, 115.

[5] Nishida, T., Ban, T. and Kobayashi, N. (2003), "High-color-rendering light sources consisting of a 350nm ultraviolet lighting-emitting diode and three-basal-colors phosphors," *Applied Physics Letters*, **82**(22): 3817–3819.

[6] Guo, X.Y., Graff, J. and Schubert, E.F. (1999), "Photo recycling semiconductor light emitting diode," *Electron Devices Meeting, IEEE*, 600–603.

[7] Nakamura, S., Pearton, S. and Fasol, G. (1997), *The Blue Laser Diode: GaN Based Light Emitters and Lasers*, Berlin, Germany, Springer.

[8] Mueller-Mach, R., Mueller, G.O., Krames, M.R. and Trottier, T. (2002), "High-power phosphor-converted light-emitting diodes based on III-Nitrides," *IEEE Journal of Selected Topics in Quantum Electronics* **8**(2): 339–345.

[9] Mueller-Mach, R., Mueller, G., Krames, M.R., Hoppe, H.A., Stadler, F., Schnick, W., Juestel, T. and Schmidt, P. (2005), "Highly efficient all-nitride phosphor-converted white light emitting diode," *Physica Status Solidi(a)* **202** (9): 1727–1732.

[10] Xie, R.J., Hirosaki, N., Sakuma, K. and Kimura, N. (2008), "White light-emitting diodes (LEDs) using (oxy) nitride phosphors," *Journal of Physics D: Applied Physics* **41**(14): 144013.

[11] Zachau, M., Becker, D., Berben, D., Fiedler, T., Jermann, F. and Zwaschka, F. (2008), "Phosphors for solid state lighting," *Light-Emitting Diodes: Research, Manufacturing, and Applications XII*, San Jose, CA, USA, SPIE, 691010.

[12] Ohno, Y. (2004), "Color rendering and luminous efficacy of white LED spectra," *Fourth International Conference on Solid State Lighting*, Denver, CO, USA, SPIE, 88–98.

[13] Chou, H.Y., Hsu, T.H. and Yang, T.H. (2005), "Effective method for improving illuminating properties of white-light LEDs," *Light-Emitting Diodes: Research, Manufacturing, and Applications IX*, San Jose, CA, USA, SPIE, 33–41.

[14] Krames, M.R., Shchekin, O.B., Mueller-Mach, R., Mueller, G.O., Zhou, L., Harbers, G. and Craford, M.G. (2007), "Status and future of high-power light-emitting diodes for solid-state lighting," *IEEE Journal of Display Technology* **3**(2): 160–175.

[15] Summers, C.J., Wagner, B.K. and Menkara, H. (2004), "Solid state lighting: diode phosphors," *Third International Conference on Solid State Lighting*, San Diego, CA, USA, SPIE, 123–132.
[16] Hao, W., Hao, W., Xinmin, Z., Chongfeng, G., Jian Xu, A.J.X., Mingmei Wu, A.M.W. and Qiang Su, A.Q.S. (2005), "Three-band white light from InGaN-based blue LED chip precoated with Green/red phosphors," *IEEE Photonics Technology Letters* **17**(6): 1160–1162.
[17] Braune, B., Brunner, H., Strauss, J. and Petersen, K. (2005), "Light conversion in opto semiconductor devices: from the development of luminous materials to products with customized colors," *Optoelectronic Devices: Physics, Fabrication, and Application II*, Boston, MA, USA, SPIE, 60130D-8.
[18] Collins, W.D., Krames, M.R., Verhoeckx, G.J. and Leth, N.J.M. (2001), Using electrophoresis to produce a conformal coated phosphor-converted light emitting semiconductor, *US Patent, Lumileds Light*. 6 576 488.
[19] Yum, J.H., Seo, S.Y., Lee, S. and Sung, Y.E. (2001), "Comparison of Y3Al5O12:Ce0.05 phosphor coating methods for white-light-emitting diode on gallium nitride," *Solid State Lighting and Displays*, San Diego, CA, USA, SPIE, 60–69.
[20] Loh, B.P., JR, N.W.M., Andrews, P., Fu, Y.K., Laughner, M. and Letoquin, R. (2008), Method of uniform phosphor chip coating and LED package fabricated using method, *US Patent, Cree, Inc*. 2 008 007 901 7A1.
[21] Braune, B., Petersen, K., Strauss, J., Kromotis, P. and Kaempf, M. (2007), "A new wafer level coating technique to reduce the color distribution of LEDs," *Light-Emitting Diodes: Research, Manufacturing, and Applications XI*, San Jose, CA, USA, SPIE, 64860X-11.
[22] Liu, S. (2006), Method to fabricate white light-emitting diode, *China Patent application*, 200610029858.6.
[23] Yamada, K., Imai, Y. and Ishii, K. (2003), "Optical simulation of light source devices composed of blue LEDs and YAG phosphor," *Journal of Light and Visual Environment* **27**(2): 70–74.
[24] Zhu, Y., Narendran, N. and Gu, Y. (2006), "Investigation of the optical properties of YAG: Ce phosphor," *Sixth International Conference on Solid State Lighting*, San Diego, CA, USA, SPIE, 63370S.
[25] Arik, M., Setlur, A., Weaver, S., Haitko, D. and Petroski, J. (2007), "Chip to system levels thermal needs and alternative thermal technologies for high brightness LEDs," *Journal of Electronic Packaging* **129**(3): 328–338.
[26] Kim, J.K., Luo, H., Schubert, E.F., Cho, J., Sone, C. and Park, Y. (2005), "Strongly enhanced phosphor efficiency in GaInN white light-emitting diodes using remote phosphor configuration and diffuse reflector cup," *Japanese Journal of Applied Physics* **44**: L649–L651.
[27] Luo, H., Kim, J.K., Schubert, E.F., Cho, J., Sone, C. and Park, Y. (2005), "Analysis of high-power packages for phosphor-based white-light-emitting diodes," *Applied Physics Letters* **86**(24): 243505.
[28] Luo, H., Kim, J. K., Xi, Y., Schubert, E.F., Cho, J., Sone, C. and Park, Y. (2006), "Analysis of high-power packages for white-light-emitting diode lamps with remote phosphor," *Materials Research Socity* **892**: FF09-07.1–FF09-07.6.
[29] Fan, B.F., Wu, H., Zhao, Y., Xian, Y.L. and Wang, G. (2007), "Study of phosphor thermal-isolated packaging technologies for high-power white light-emitting diodes," *IEEE Photonics Technology Letters* **19**(15): 1121–1123.
[30] OIDA. (2002), "Light emitting diodes (LEDs) for general iIllumination, an OIDA technology roadmap update 2002," from *http://lighting.sandia.gov/lightingdocs/OIDA_SSL_LED_Roadmap_Full.pdf*
[31] Fujii, T., Gao, Y., Sharma, R., Hu, E.L., DenBaars, S.P. and Nakamura, S. (2004), "Increase in the extraction efficiency of GaN-based light-emitting diodes via surface roughening," *Applied Physics Letters* **84**(6): 855–857.
[32] Huang, H.W., Kao, C.C., Chu, J.T., Wang, W.C., Lu, T.C., Kuo, H.C., Wang, S.C., Yu, C.C. and Kuo, S.Y. (2007), "Investigation of InGaN/GaN light emitting diodes with nano-roughened surface by excimer laser etching method," *Materials Science and Engineering B* **136**: 182–186.
[33] Gao, H.Y., Yan, F.W., Fan, Z.C., Li, J.M., Zeng, Y.P. and Wang, G.H. (2008), "Improved light extraction of GaN-based LEDs with nano-roughened *p*-GaN surfaces," *Chinese Physics Letters* **25**(9): 3448–3450.
[34] Mont, F.W., Kim, J.K., Schubert, M.F., Luo, H., Schubert, E.F. and Siegel, R.W. (2007), "High refractive index nanoparticle-loaded encapsulants for light-emitting diodes," *Light-Emitting Diodes: Research, Manufacturing, and Applications XI*, San Jose, CA, USA, SPIE, 64861C-8.
[35] Narendran, N., Gu, Y., Freyssinier, J.P., Yu, H. and Deng, L. (2004), "Solid-state lighting: failure analysis of white LEDs," *Journal of Crystal Growth* **268**(3–4): 449–456.
[36] Norris, A.W., Bahadur, M. and Yoshitake, M. (2005), "Novel silicone materials for LED packaging," *Fifth International Conference on Solid State Lighting*, San Diego, CA, USA, SPIE, 594115–7.

[37] Bahadur, M., Norris, A.W., Zarisfi, A., Alger, J.S. and Windiate, C. C. (2006), "Silicone materials for LED packaging," *Sixth International Conference on Solid State Lighting*, San Diego, CA, USA, SPIE, 63370F-7.

[38] Lee, S. J. (2001), "Analysis of Light-Emitting Diodes by Monte Carlo Photon Simulation," *Applied Optics* **40**(9): 1427–1437.

[39] West, R.S. (2002), "Side-emitting high-power LEDs and their application in illumination," *Solid State Lighting II*, Seattle, WA, USA, SPIE, 171–175.

[40] Chang, J.G., Liao, C.L.D. and Hwang, C.C. (2006), "Enhancement of the optical performances for the LED backlight systems with a novel lens cap," *Novel Optical Systems Design and Optimization IX*, San Diego, CA, USA, SPIE, 62890X-1–6.

[41] Chao, P.C.P., Liao, L.D. and Chiu, C.W. (2006), "Design of a novel LED lens cap and optimization of LED placement in a large area direct backlight for LCD-TVs," *Photonics in Multimedia*, Strasbourg, France, SPIE, 61960N-1–9.

[42] Chi, W. and George, N. (2006), "Light-emitting diode illumination design with a condensing sphere," *Journal of Optical Society of America A* **23**(9): 2295–2298.

[43] Liu, Z.Y., Liu, S., Wang, K. and Luo, X.B. (2008), "Optical analysis of color distribution in white LEDs with various packaging methods," *IEEE Photonics Technology Letters* **20**(20): 2027–2029.

[44] Wang, D.M., Chen, S.H., Wang, M. and Xiang, S., Hua (2006), "The design and fabrication of microlens and LED integrated packaging," *2006 7th International Conference on Electronic Packaging Technology*, Shanghai, China, IEEE, 359803-1–3.

[45] Braune, B., Bogner, G., Brunner, H., Kraeuter, G. and Hoehn, K. (2003), "New developments in LED lighting by novel phosphors," *Light-Emitting Diodes: Research, Manufacturing, and Applications VII*, San Jose, CA, USA, SPIE, 87–94.

[46] Mesli, T. (2007), "Improvement of ultra-high-brightness white LEDs," Manufacturing LEDs for Lighting and Displays, Berlin, Germany, SPIE, 67970N-1–9.

[47] Allen, S.C. and Steckl, A.J. (2007), "ELiXIR-solid-state luminaire with enhanced light extraction by internal reflection," *Journal of Display Technology* **3**(2): 155–159.

[48] Lee, S.J. (2001), "Light-emitting diode lamp design by Monte Carlo photon simulation," *Light-Emitting Diodes: Research, Manufacturing, and Applications V*, San Jose, CA, USA, SPIE, 99–108.

[49] Sun, C.C., Lee, T.X., Ma, S.H., Lee, Y.L. and Huang, S.M. (2006), "Precise optical modeling for LED lighting verified by cross correlation in the midfield region," *Optics Letters* **31**(14): 2193–2195.

[50] Borbely, A. and Johnson, S.G. (2004), "Performance of phosphor-coated LED optics in ray trace simulations," *Fourth International Conference on Solid State Lighting*, Denver, CO, USA, SPIE, 266–273.

[51] Liu, Z.Y., Liu, S., Wang, K. and Luo, X.B. (2009), "Optical analysis of phosphor's location for high-power light-emitting diodes," *IEEE Transactions on Device and Materials Reliability* **9**(1): 65–73.

[52] Lee, T.X., Lin, C.Y., Ma, S.H. and Sun, C.C. (2005), "Analysis of position-dependent light extraction of GaN-based LEDs," *Optics Express* **13**(11): 4175–4179.

[53] Nagarajan, S., Oh, T.S., Kumar, M.S., Hong, C.H., and Suh, E.K. (2007), "Structural and Optical Properties of In-Rich InAlGaN/InGaN Heterostructures for White Light Emission," *Japan Journal of Applied Physics* **47**: 4413–4416.

[54] Peng, T. and Piprek, J. (1996), "Refractive index of AlGaInN alloys," *Electronics Letters* **32**(24): 2285–2286.

[55] Brunner, D., Angerer, H., Bustarret, E., Freudenberg, F., Hopler, R., Dimitrov, R., Ambacher, O. and Stutzmann, M. (1997), "Optical constants of epitaxial AlGaN films and their temperature dependence," *Journal of Applied Physics* **82**(10): 5090–5096.

[56] Bergmann, M.J. and Casey, J.H.C. (1998), "Optical-field calculations for lossy multiple-layer $Al_xGa_{1-x}N/In_xGa_{1-x}N$ laser diodes," *Journal of Applied Physics* **84**(3): 1196–1203.

[57] Laws, G.M., Larkins, E.C., Harrison, I., Molloy, C. and Somerford, D. (2001), "Improved refractive index formulas for the $Al_xGa_{1-x}N$ and $In_yGa_{1-y}N$ alloys," *Journal of Applied Physics* **89**(2): 1108–1115.

[58] Anani, M., Abid, H., Chama, Z., Mathieu, C., Sayede, A. and Khelifa, B. (2007), "InxGa1-xN refractive index calculations," *Microelectronics Journal*: 262–266.

[59] Ambacher, O., Rieger, W., Ansmann, P., Angerer, H., Moustakas, T.D. and Stutzmann, M. (1996), "Sub-bandgap absorption of gallium nitride determined by photothermal deflection spectroscopy," *Solid State Communications* **97**(5): 365–370.

[60] Bentoumi, G., Deneuville, A., Beaumont, B. and Gibart, P. (1997), "Influence of Si doping level on the Raman and IR reflectivity spectra and optical absorption spectrum of GaN," *Materials Science and Engineering B* **50** (1–3): 142–147.

[61] Yu, G., Wang, G., Ishikawa, H., Umeno, M., Soga, T., Egawa, T., Watanabe, J. and Jimbo, T. (1997), "Optical properties of wurtzite structure GaN on sapphire around fundamental absorption edge (0.78–4.77 eV) by spectroscopic ellipsometry and the optical transmission method," *Applied Physics Letters* **70**(24): 3209–3211.

[62] Schad, S.S., Neubert, B., Eichler, C., Scherer, M., Habel, F., Seyboth, M., Scholz, F., Hofstetter, D., Unger, P., Schmid, W., Karnutsch, C. and Streubel, K. (2004), "Absorption and Light Scattering in InGaN-on-Sapphire- and AlGaInP-Based Light-Emitting Diodes," *Journal of Lightwave Technology* **22**(10): 2323.

[63] Lelikov, Y., Bochkareva, N., Gorbunov, R., Martynov, I., Rebane, Y., Tarkin, D. and Shreter, Y. (2008), "Measurement of the absorption coefficient for light laterally propagating in light-emitting diode structures with In0.2Ga0.8N/GaN quantum wells," *Semiconductors* **42**(11): 1342–1345.

[64] Martin, R.W., Middleton, P. G., O'Donnell, K. P. and Van der Stricht, W. (1999), "Exciton localization and the Stokes' shift in InGaN epilayers," *Applied Physics Letters* **74**(2): 263–265.

[65] Damilano, B., Grandjean, N., Massies, J., Siozade, L. and Leymarie, J. (2000), "InGaN/GaN quantum wells grown by molecular-beam epitaxy emitting from blue to red at 300 K," *Applied Physics Letters* **77**(9): 1268–1270.

[66] Kvietkova, J., Siozade, L., Disseix, P., Vasson, A., Leymarie, J., Damilano, B., Grandjean, N. and Massies, J. (2002), "Optical Investigations and Absorption Coefficient Determination of InGaN/GaN Quantum Wells," *Physica Status Solidi* **190**(1): 135–140.

[67] Narendran, N., Gu, F., Freyssinier-Nova, J.P. and Zhu, Y. (2005), "Extracting phosphor-scattered photons to improve white LED efficiency," *Physica Status Solidi* **202**(6): R60–R62.

[68] Kang, D.Y., Wu, E. and Wang, D.M. (2006), "Modeling white light-emitting diodes with phosphor layers," *Applied Physics Letters* **89**(23): 231102.

[69] Liu, Z.Y., Liu, S., Wang, K., and Luo, X.B. (2009), "Measurement and numerical studies of optical properties of YAG: Ce phosphor for white LED packaging," Applied Optics, Submitted.

[70] Bohren, C.F. and Huffman, D.R. (1983), *Absorption and scattering of light by small particles*, Wiley.

[71] Jonasz, M. and Fournier, G.R. (2007), *Light scattering by particles in water*, Elsevier.

[72] Narendran, N. (2005), "Improved performance white LED," *Fifth International Conference on Solid State Lighting*, San Diego, CA, USA, SPIE, 594108-1–594108-6.

[73] Allen, S.C. and Steckl, A.J. (2008), "A nearly ideal phosphor-converted white light-emitting diode," *Applied Physics Letters* **92**(14): 143309.

[74] Tran, N.T. and Shi, F.G. (2008), "Studies of phosphor concentration and thickness for phosphor-based white light-emitting-diodes," *Journal of Lightwave Technology* **26**(21): 3556–3559.

[75] Sommer, C., Wenzl, F.P., Hartmann, P., Pachler, P., Schweighart, M. and Leising, G. (2008), "Tailoring of the color conversion elements in phosphor-converted high-power LEDs by optical simulations," *IEEE Photonics Technology Letters* **20**(9): 739–741.

[76] Liu, Z.Y., Liu, S., Wang, K. and Luo, X.B. (2008), "Effects of phosphor's location on LED packaging performance," *2008 International Conference on Electronic Packaging Technology & High Density Packaging*, Shanghai, China, IEEE, 4606982-1–7.

[77] Muller, G., Muller, R., Basin, G., West, R.S., Martin, P.S., Lim, T.S. and Eberle, S. (2008), LED with phosphor tile and overmolded phosphor in lens, *U. S. Patent, Philips Lumileds Light Company,* LLC. 20 080 048 200.

[78] Chen, G., Craven, M., Kim, A., Munkholm, A., Watanabe, S., Camras, M., Götz, W. and Steranka, F. (2008), "Performance of high-power III-nitride light emitting diodes," *Physica Status Solidi (a)* **205**(5): 1086–1092.

[79] Lee, J.H., Oh, J.T., Choi, S.B., Woo, J.G., Lee, S.Y. and Lee, M.B. (2007), "Extraction-efficiency enhancement of InGaN-based vertical LEDs on hemispherically patterned sapphire," *Physica Status Solidi (c)* **4**(7): 2806–2809.

[80] Erchak, A.A., Ripin, D.J., Fan, S., Rakich, P., Joannopoulos, J.D., Ippen, E.P., Petrich, G.S. and Kolodziejski, L.A. (2001), "Enhanced coupling to vertical radiation using a two-dimensional photonic crystal in a semiconductor light-emitting diode," *Applied Physics Letters* **78**(5): 563–565.

[81] Hwang, J. M., Lee, K. F. and Hwang, H. L. (2008), "Optical and electrical properties of GaN micron-scale light-emitting diode," *Journal of Physics and Chemistry of Solids* **69**(2–3): 752–758.

[82] China Solid State Lighting Alliance. (2006), China Solid-State Lighting Industry Development Almanac, *Beijing, Science Press*. (In Chinese).

[83] Lee, T.X., Gao, K.F., Chien, W.T. and Sun, C.C. (2007), "Light extraction analysis of GaN-based light-emitting diodes with surface texture and/or patterned substrate," *Optics Express* **15**(11): 6670–6676.

[84] Wang, K., Liu, S., Chen, F., Liu, Z.Y. and Luo, X.B. (2009a), "Novel application-specific LED packaging with compact freeform lens," *IEEE 59th Electronic Components & Technology Conference*, San Diego, CA, USA.

4

Thermal Management of High Power LED Packaging Module

Thermal dissipation has been a serious issue with the invention of high power LEDs. Constrained by the internal and external quantum efficiencies, non-radiative process converts a significant part of electrical power to heat. There is a power density of almost 70 W/cm^2 for an one watt (lW) LED with an 1 mm^2 area, which is higher than a conventional microprocessor chip. Generated heat will increase the junction temperature significantly, which may damage the PN junction, lower luminous efficiency, increase forward voltage, cause wavelength shift, reduce lifetime, and affect the quantum efficiency of phosphor [1–4]. The degradation of materials may occur when subjected to high temperature. Elevated temperature can induce thermal stresses in packaging components due to the mismatch of the coefficient of thermal expansion (CTE). The active layer of the chip is very sensitive to thermal stresses. For most commercial LEDs, the junction temperature cannot exceed 120 °C. High thermal stress may lead to cracks, delaminations and other failures. Therefore, it is crucial to rapidly remove the heat and keep the junction temperature below a certain limit for the maintenance of performance.

Since a low operation temperature of the LED chip is very important, thermal management is strongly required for LED packaging and application products. In this chapter, we will discuss the thermal issues on LED packaging from the perspective of heat transfer theory.

4.1 Basic Concepts of Heat Transfer

Heat transfer is a science that seeks to predict the energy transfer which may take place in a body or between bodies as a result of temperature difference [5]. Heat transfer is a natural phenomenon, which appears in nearly all the engineering fields. Thermal management design based on heat transfer theory is the bottleneck of many technology applications, and the LED is one of them.

There are three modes of heat transfer: conduction, convection, and radiation as shown in Figure 4.1. In the following parts, we will briefly explain the mechanisms of these modes.

LED Packaging for Lighting Applications: Design, Manufacturing and Testing, First Edition. Sheng Liu and Xiaobing Luo.
© 2011 Chemical Industry Press. All rights reserved. Published 2011 by John Wiley & Sons (Asia) Pte Ltd.

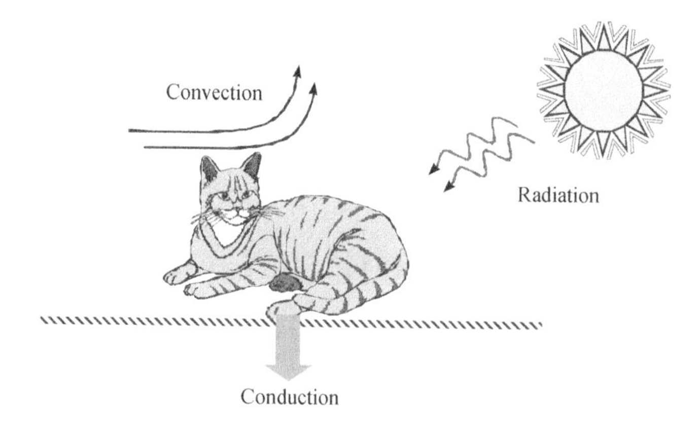

Figure 4.1 Three modes of heat transfer.

4.1.1 Conduction Heat Transfer

There will be an energy transfer from the high-temperature region to the low-temperature region when a temperature gradient exists in a body. We say that the energy is transferred by conduction. In microscopic scale, thermal motion of micro-particles such as molecules, atoms, free electrons, and so on. in a body will induce heat transfer. The induced heat transfer process is called heat conduction [6]. In 1804, French physicist Biot obtained the earliest expression of a heat conduction law based on the experimental result of heat conduction through a plane wall. Fourier in France used mathematical methods to derive a differential form expression named the Fourier Law according to Biot's law. Its expression is given by:

$$q = \frac{\phi}{A} = -\lambda \frac{\partial T}{\partial x} \tag{4.1}$$

where Φ is the heat transfer rate, q is the heat flux density, A is the cross-sectional area which is perpendicular to the direction of heat conduction, λ is the thermal conductivity of the material, and $\frac{\partial T}{\partial x}$ is the temperature gradient in the direction. The minus sign indicates that heat transfers in a direction opposite to that of the temperature rise as shown in Figure 4.2.

4.1.2 Convection Heat Transfer

Heat convection is a heat transfer process caused by the mixing of hot and cold fluids because of the macroscopic motion of fluids which gives rise to the relative displacement within parts of

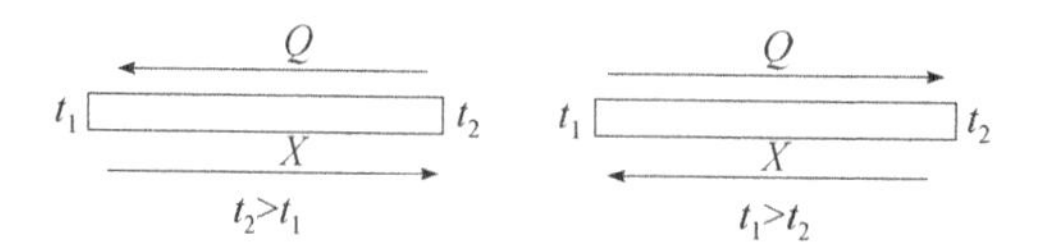

Figure 4.2 Directional relationship between temperature gradient and heat transfer.

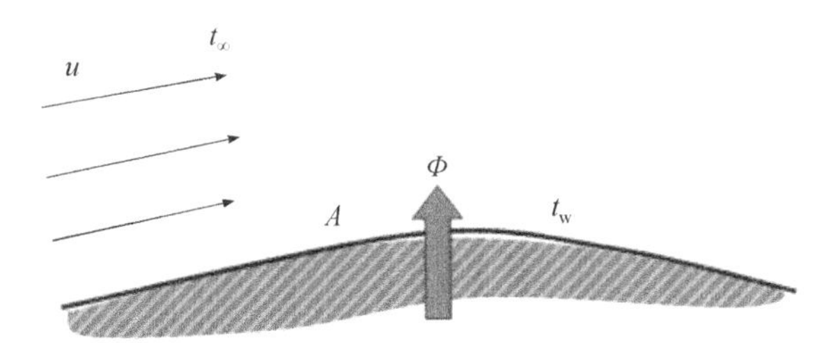

Figure 4.3 Model of convective heat exchange.

the fluids [5]. Convective heat exchange, usually mentioned in engineering, denotes a process of heat exchange between fluids and the solid surface exposed in fluids. It results from the co-effect of heat conduction and convection. According to the possibility whether phase changes occur or not, convective heat exchange can be divided into convections with phase changes and without phase changes. According to the causes of flow, there are free convection and forced convection. Free convection is caused by the circulation of fluids due to buoyancy from the density changes induced by heating itself. Forced convection is due to movement in the fluid which results from many other forces, such as a fan or pump. The heat transfers which occur when fluid boils on a hot surface or when vapor condenses on a cold surface are called the boiling heat transfer and condensation heat transfer (convective boiling of phase changes), respectively. In 1701, British scientist Newton put forward a mathematical expression which was later called Newton's Law of Cooling for calculating the temperature of a glowed iron bar.

Law of Cooling is a commonly-used calculation method for convective heat exchange. Figure 4.3 shows the model of convective heat exchange, with the equation as follows:

$$\phi = hA(t_w - t_\infty) \tag{4.2}$$

$$q = \frac{\phi}{A} h(t_w - t_\infty) \tag{4.3}$$

where Φ is the heat flux, q is the heat flux density, A is the cross-sectional area which is perpendicular to the direction of heat transfer, t_w is the surface temperature of the solid, t_∞ is the fluid temperature, and h is the convective heat transfer coefficient. Convective heat transfer coefficient is a physical quantity denoting the intensity of convective heat transfer. There are many factors to affect h, such as physical properties of fluids (thermal conductivity, viscosity, density, specific heat capacity, and so on), form of flow (laminar flow, eddy flow), the cause of flow (free convection or forced convection), the shapes and sizes of the object's surface, the occurrence of phase change in heat exchange (boiling or condensation), and so on. Different heat transfer coefficient magnitudes with different fluids under various conditions are shown in Figure 4.4.

4.1.3 Thermal Radiation

Thermal radiation is the process by which the surface of an object radiates its thermal energy in the form of electromagnetic waves. It is a kind of electromagnetic radiation whose wavelength

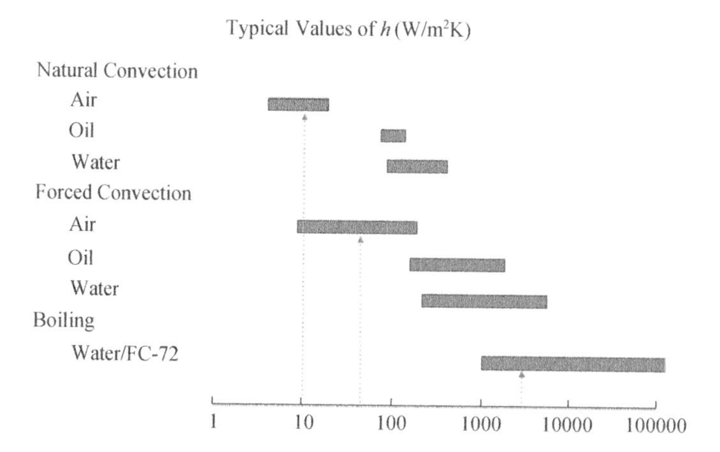

Figure 4.4 Magnitude order of heat transfer coefficient of different fluids under various conditions [7].

is from 0.1 mm to 100 mm. Therefore, heat can directly transfer in a vacuum without any media by radiation. Every object owns a radiating power which is in proportion to the biquadrate of its Kelvin temperature. An object can also absorb the thermal radiation around it. The co-effect of radiation and absorption is called radiation heat transfer [8]. Radiation heat transfer is a dynamic process. When the substance reaches thermal equilibrium with the environment, the rate of radiant heat exchange is zero. However, radiation and absorption are still in process, with a heat balance between radiation and absorption.

All objects with a temperature higher than absolute zero can radiate heat. The higher the temperature, the greater the amount of radiated energy is. The spectrum of thermal radiation is continuous with a wavelength theoretically covering from 0 to ∞. General radiation depends mainly on long-wavelength visible light and infrared ray. Heat radiates mainly by invisible infrared light at low temperature. At temperature of 300 °C, the wavelength of the strongest thermal radiation is in the infrared region. In the range of 500 °C to 800 °C, the wavelength of the strongest thermal radiation is in the visible region. In 1878, Stefan's experiment found that radiance is directly proportional to the biquadrate of absolute temperature. The result was proven theoretically by Boltzmann in 1884 and expressed as Stefan–Boltzman law, commonly known as fourth-power law and is given by:

$$\phi = A\sigma T^4 \tag{4.4}$$

$$\sigma = 5.67 \times 10^{-8}\ \mathrm{W/(m^2 \cdot K^4)} \tag{4.5}$$

where ϕ is the total radiate energy, A is the radiant superficial area, σ the is Stefan–Boltzman constant.

Usually, the actual heat transfer does not occur in a unique mode. It is a process of co-effect of radiation, convection and conduction. Different modes of heat transfer follow different laws. To select the correct mode, three heat transfer modes are studied respectively in research and then integrated together for actual heat transfer process analysis.

4.1.4 Thermal Resistance

There are driving force and resistance in every kind of transfer process, of which the transfer amount is closely related with those two, that is:

$$m = \frac{F}{R} \tag{4.6}$$

where m is the transfer amount in the process, F is driving force, R is resistance of the process. Ohm's Law that we are familiar with in electrics is a typical example:

$$I = \frac{U}{R} \tag{4.7}$$

where I is the current of some critical component, U is the voltage between two terminals of the component, and R is the electrical resistance of the component.

Heat transfer is a kind of transfer process in nature. Similarly, heat transfer process follows the same law as Equation 4.6. Temperature difference is the driving force of heat transfer. Heat flux is similar to m in Equation 4.6. As a result, resistance of heat transfer called thermal resistance is defined as:

$$R_{th} = \frac{\Delta T}{Q} \tag{4.8}$$

where ΔT is the temperature difference of some critical component, Q is the heat flux in the component at steady station.

According to Equations 4.1 and 4.8, the thermal resistance of conduction is given by:

$$R_c = \frac{\delta}{\lambda A} \tag{4.9}$$

Similarly, the thermal resistance of convection is defined as follows based on Equations 4.2, 4.3 and 4.8:

$$R_{\mathrm{h}} = \frac{1}{hA} \tag{4.10}$$

Being similar to electric resistance, the method for calculating electric resistances in series or in parallel is suitable for thermal resistance calculation in series or in parallel. After a system thermal resistance is calculated, the temperature difference between the system and the environment can be calculated by Equation 4.8. The junction temperature of the die can be calculated as the ambient temperature is known.

In electronic or LED packaging, heat generated by chips is dissipated through thermal interface materials and a small heat sink inside the chips to the outside of the packaging structure in the way of conduction. It is a thermal conduction process of multi-layer structure.

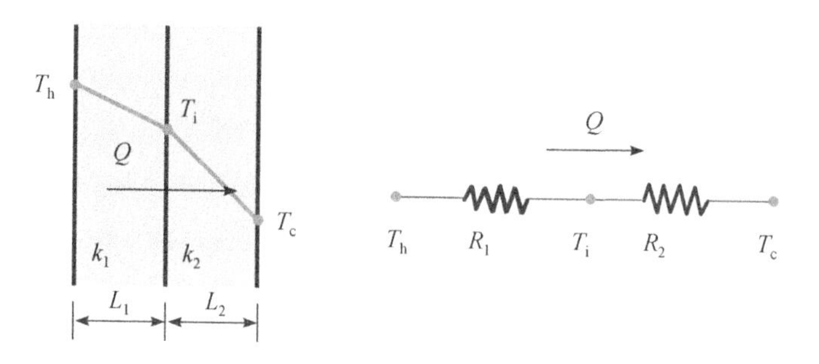

Figure 4.5 Heat transfer of double-layer materials.

On the basis of Fourier's law, heat flux density is proportional to the temperature gradient.

$$q = \frac{Q}{A} = k\frac{T_1 - T_2}{L} = k\frac{\Delta T}{\Delta x} \tag{4.11}$$

where k is the thermal conductivity, A is the cross-sectional area, Δx is the thickness of the heat conduction material, and q is the heat flux density or the dissipation power per unit area. For multi-layer composite materials, entire thermal resistance can be simplified as:

$$\frac{\Delta T}{Q} \sim \sum_i \frac{\Delta x_i}{k_i} \tag{4.12}$$

Using the case shown in Figure 4.5 as an example, there is:

$$R = R_1 + R_2 = \frac{L_1}{k_1 A} + \frac{L_2}{k_2 A} \tag{4.13}$$

When $k_1 > k_2$, $R_1 > R_2$, the above equation can be simplified as $R \approx R_2$.

From the above equations, reducing thicknesses of materials and using materials with high thermal conductivity will minimize thermal resistances. The following formula is usually used to describe the thermal resistance in packaging:

$$R_{\text{ja}} = \frac{T_{\text{j}} - T_{\text{a}}}{Q} \tag{4.14}$$

where T_j is the juncture temperature of chip, T_a is the ambient temperature, Q is the heating power input of the chip. Under the condition of the same heating power and ambient temperature, the bigger the thermal resistance, the lower the reliability is.

4.2 Thermal Resistance Analysis of Typical LED Packaging

The LED chip's operation temperature generally should be maintained below 120 °C. Therefore, in heat transfer analysis of LED packaging, only a small amount of heat is radiated

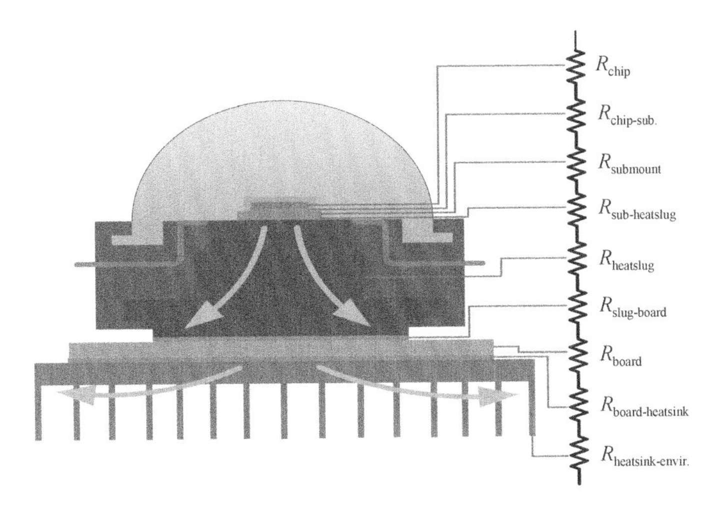

Figure 4.6 Thermal resistance network of Luxeon LED. *(Color version of this figure is available online.)*

out from the LED as compared with other heat transfer modes because of relative low temperature. It means that conduction and convection should be the main heat transfer modes in LED packaging and application products. It should be cautioned that radiation heat transfer can be important in those cases such as LED bulbs, which are going to be discussed in Chapter 7.

There are two paths for heat dissipation in LED packaging. One is through the encapsulant component, the other is through the chip and leadframe. Since the encapsulants are polymer and heat insulated, all of the heat must be conducted through the leadframe. Considering the high heat flux density of the chip, the leadframe should not only present high thermal conduction, but also rapidly spread the heat to circumstance with effective configuration.

In 1998, Lumileds developed the first high power LED packaging-Luxeon, which was embedded with a metal slug of large volume for heat dissipation. This leadframe based plastic package has become the main packaging type adopted by many corporations. This packaging method has reduced the thermal resistance to 4–10 K/W and can dissipate chip power up to 5 W [9]. Figure 4.6 shows the thermal resistance network for a complete Luxeon LED module. Thermal resistance network is generally utilized to evaluate the performance of heat dissipation. Low system thermal resistance implies that heat can be conducted to the environment rapidly and therefore the temperature difference between junction and environment is reduced. In 2006, with the improvement of base materials and attachment technologies, Luxeon K2 package can allow the junction temperature to rise to 150 °C for the white LED with a drive current up to 1.5 A [10].

Instead of a separate heatsink/leadframe assembly packaging as shown in Figure 4.6, another approach for high power LED solutions is the Chip-on-Board (CoB) technology. The chip is directly mounted on the board with a circuit. Therefore, the size of the CoB can be more compact. Figure 4.7 is the thermal resistance network of CoB. Two thermal interfaces between chip and heatsink are reduced. Therefore, heat can be more efficiently conducted to the heatsink. Another advantage of CoB is that the packaging density can be significantly higher with hundreds or thousands of chips packaged on one board. Increased lumen output in the unit packaging area can decrease the manufacturing cost due to less packaging material usage.

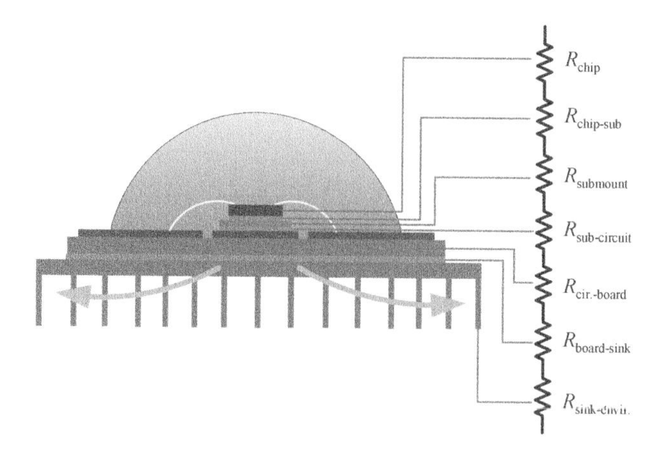

Figure 4.7 Thermal resistance network of CoB packaging. *(Color version of this figure is available online.)*

The main constraints are the light intensity control and thermal dissipation because of the high heat flux density.

4.3 Various LED Packages for Decreasing Thermal Resistance

4.3.1 Development of LED Packaging

LED packaging aims to guarantee electrical and mechanical connection between LED chip and the circuit and to protect the chip from external shocks such as mechanical force, heat, moisture, and so on. In the packaging development, to implement the optical functionality of the LED, design of the packaging should firstly meet the thermal requirements, while implementing the optical functionality of the LED [3].

After 40 years' development, the LED industry has undergone four stages: Lead LED, SMD LED, Power LED, and High Power LED [11]. Figure 4.8 shows the development of the

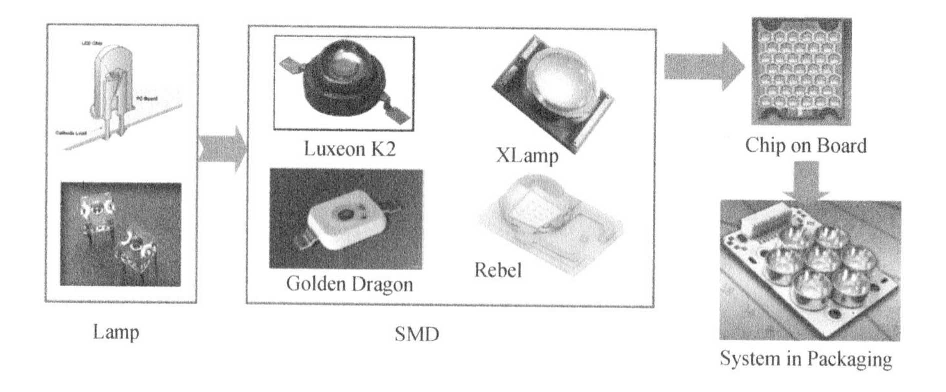

Figure 4.8 Development of LED packaging technology and structure.

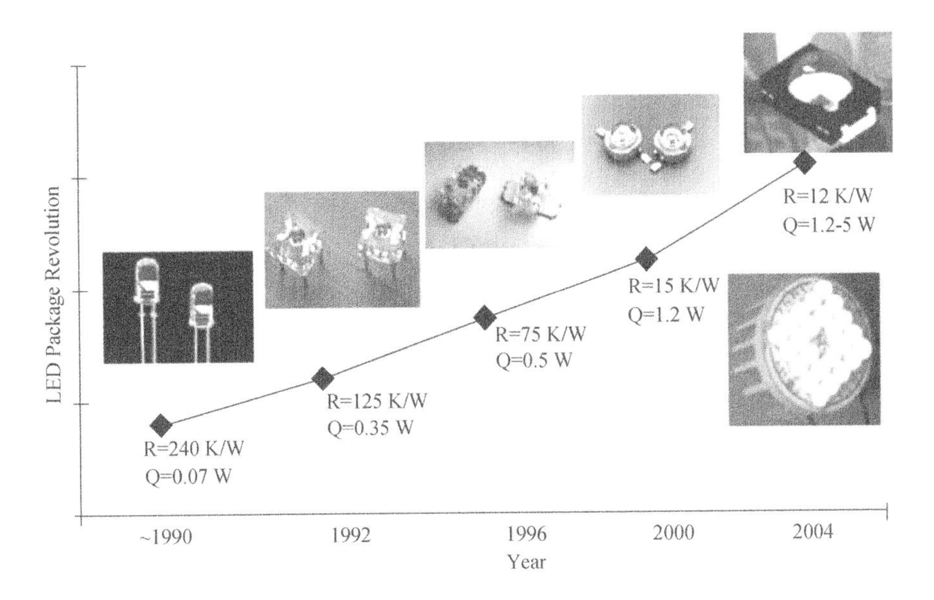

Figure 4.9 Thermal resistance changing history of LED module [2–4].

packaging structures of the LED module. Figure 4.9 shows the thermal resistance evolution history. From Figure 4.9, we can see that the value of thermal resistance of the packaging is becoming smaller and smaller.

LED lead packaging, using leadframe as the lead of various packages, was the earliest and most successfully developed packaging structure in the market, which is packaged with highly-matured technology. LED standardization is recently regarded by most customers as the most convenient and economical solution. The traditional LED is placed in the package which can only bear 0.1 W input power. Ninety percent of generated heat is dissipated from the cathode pin frame to the PCB (printed circuit board), and then into the air. How to decrease the temperature rising of PN junction during work is an issue that must be considered in packaging and applications. In 2002, the SMD LED was gradually accepted by the market, and gained a large share of the market. The packaging change from Lead to SMD followed the development trend of electronic industry. Many manufacturers launched such kind of products. To make sure that luminous flux which is 10–20 times bigger than that ϕ5 mm LED can be generated at the high current, an effective thermal dissipation structure should be designed and stable packaging materials should be used [12] to solve the issue of optical attenuation. The key techniques also cover the shell-tube design and packaging. High power LED packagings such as 5 W–series LED in white, green, cyan, and blue have been supplied to the market since 2003.

From the perspective of applications, simple, high-power, and high-brightness LED devices in smaller volume will replace the traditional lower-power LED devices in most lighting applications. Lighting fixtures using lower-power LEDs have achieved high brightness. However, their disadvantages are obvious: the leads are extremely complex and the cooling effect is very poor. A complex power-supply circuit is designed to balance the discrepancy of voltage current relation among each LED. In comparison, a single-chip high-power LED whose power is equal to the total power of tens of low-power LEDs has simpler power-supply circuit and thermal dissipation structure, and the physical features of high-power LED are stable.

For the high-power LED devices, traditional packaging methods and materials as used in lower-power LED devices cannot be easily used. Simply enlarging the light-emitting area cannot radically solve the issues of thermal management and lighting. As a result, expected luminous flux in an application cannot be achieved. The thermal characterization of the high-power LED directly affects the operation temperature of the LED, luminous efficiency, luminescence spectrum and service life, and so on. Therefore, it is very important to develop an effective packaging structure of high-power LEDs.

Flip chip bonding has been recently applied in the manufacturing of the high-power LED. In flipchip bonding, the removal of the gold wire bonding pad on the top of the chip surface improves the brightness. Because the distance that the current flows is shortened and electric resistance decreases, the generated heat is relatively smaller. Moreover, this kind of bonding transfers heat effectively to the substrate lower and outer layers. By the application of the technology to SMD, the LED will not only increase light output but also reduce the volume of the products. There are two main proposals in the development of flip chip LED technology. One is lead-tin ball bonding. The other is ultrasonic thermal bonding. Lead-tinball bonding has been applied in IC packaging for a long time. The technology is highly matured. For the target with low cost and small leads, the ultrasonic thermal flip-chip technology is applicable to the bonding of the high-power LED. Gold is used at the interface for welding, for which its melting-point temperature is higher than silver paste. Therefore, the design of manufacturing after die bonding can be more flexible. Besides, there are some other advantages, such as lead-free processing, simple working procedure, and an accurate joining place. After years of research and experience, the ultrasonic thermal bonding based flip-chip technology has mastered the most optimized process parameters. A large number of LED manufacturers have successfully put it into mass production.

4.3.2 Thermal Resistance Decrease for LED Packaging

According to Figures 4.6 and 4.7, the thermal resistance control of LED packaging needs a system process. Every thermal resistance in the network should be effectively reduced. In this section, the reduction method for every thermal resistance will be discussed one by one.

A chip is normally mounted on the metal heatslug or substrate with copper circuit because of the high thermal conductivity of metals. However, there is a large CTE mismatch between metal slug/circuit and GaN chip, which may generate seriously thermal stresses. Therefore, it is essential to add a submount between the chip and slug/circuit to relieve the stress by compensating for the CTE mismatch. Silicon is considered to be a suitable choice due to the fact that the silicon has a similar CTE to the chip and can integrate electronics. However, if a GaN chip can be separated by laser lift-off or other means from sapphire and bonded onto the Si, a submount is not necessary for die attach.

Die attach materials between the chip and submount should be carefully selected. Expected materials should present excellent adhesion between the bonded surfaces, good stress relaxation at the interface, and effective heat dissipation. Eutectic solder is considered to be a suitable choice. $Au_{20}Sn_{80}$ eutectic alloy is one of the most frequently used lead-free solders with superior thermal properties (60 W/mK). However, this alloy is sensitive to the compo-

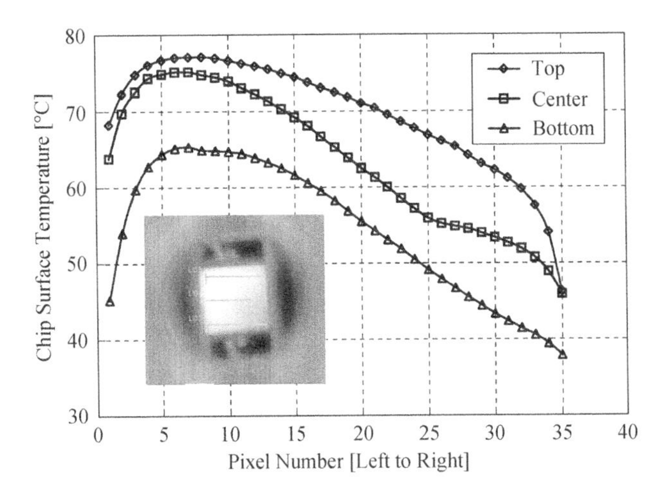

Figure 4.10 Temperature distribution for a defected LED chip [13]. (Reprinted with permission from B. F. Fan, Y. Zhao, Y.L. Xian and G. Wang, "Thermal simulation studies of high-power light-emitting diodes," Advanced LEDs for Solid State Lighting, *Proceedings of SPIE Vol. 6355*, 63550D, 2006. © 2006 SPIE.)

sition and shows poor reflow behavior. A perfect die bonding is important for the heat conduction. It has been found that the thermal resistance of die solder constitutes a large portion of system resistance. Defects such as pores and intermetallic compounds (IMC) in the bonding layer will increase the junction temperature dramatically, leading to premature failure of LEDs. Figure 4.10 shows the temperature variation across the surface of a defected chip.

In leadframe packaging, the choices for heatslug materials are limited for industrial applications in terms of cost and reliability. Copper is the most frequently used material. Increasing the volume of heatslug is the generally adopted approach to enhance the heat dissipation, which will induce a higher cost. CoB can provide more flexible and powerful methods for heat removal by keeping the size compact. Figure 4.11 displays six technologies based on CoB, in which the chips are all vertical electrode structures.

The board in the first method is called the Metal Core Printed Circuit Board (MCPCB). Aluminum, is a low cost material, and is normally utilized as the metal core with a typical thermal conductivity of 160 W/mK. However, constrained by the low thermal conductivity (2–10 W/mK) of dielectric layer, it is estimated that the total thermal resistance may as high as 50 K/W. Substituting the dielectric layer with a thin alumina film can lower the thermal resistance to be approximately 35 K/W. Another alternative approach for the metal core board is removing the dielectric layer and soldering the chip directly onto the substrate. The benefit of Method II is that the heat can be more efficiently dissipated through metal substrate. However, for electrical design considerations, an electrically active metal base may generate reliability issues. Method II is expected to be suitable for parallel electrode chip. But the mismatch of CTE between the chip and substrate may induce high thermal stresses.

Compared to the overmolded leadframes and metal core based board, ceramic materials are considered to be more suitable for high power LED packaging. Table 4.1 lists the thermal properties of ceramics, metals, composites, and other materials as a comparison.

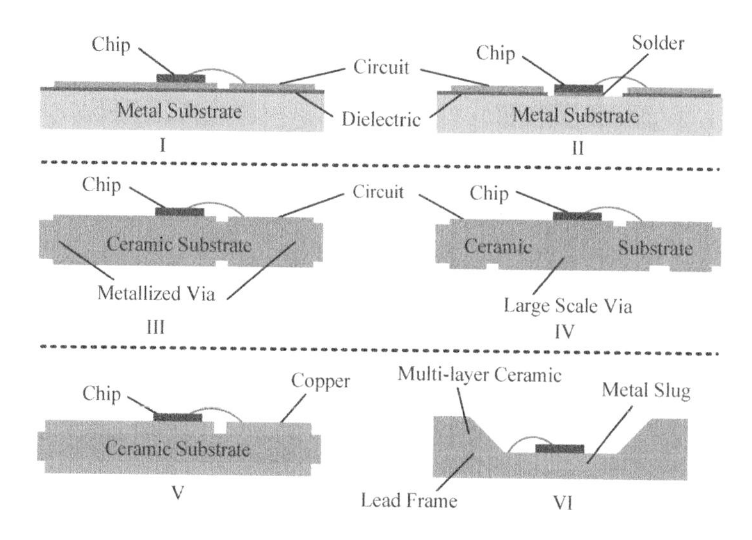

Figure 4.11 Schematic diagram for six CoB technologies [14].

Ceramic substrate has advantages in terms of compact size, surface mountability, endurance at high temperature and UV radiation, long-term thermal-mechanical stability, and so on. The reflecting cavity, signal via, interconnections and various integrated electronics can be fabricated onto theceramic substrate soforming a multi-layer circuit. Due to the low CTE, thermal stress between the chip and ceramic board is also reduced and thereby the reliability is enhanced.

The board in Method III is a thick film ceramic substrate. Anode and cathode are routed to the back of the substrate by metallized via. Ceramic substrate itself is utilized for thermal dissipation. Al_2O_3 ceramics is considered to be preferable for low cost industrial requirements, the typical conductivity of which is 27 W/mK. Other ceramics such as aluminum nitride, and

Table 4.1 Thermal properties of materials

Materials	Thermal Conductivity (W/mK)	CTE ($\times 10^{-6}$/K)
Cu	398	16.5
AlN	175	4.5
AlSiC	200	7.4
Si	148	4.0
Al	160	23.6
Al_2O_3	27	6.9
LTCC	3	5.8
Au	318	14.1
Ag	429	19.1
CuMo	165	6.6
CuW	175	6.8
Cu/Diamond	600	5.8

berallium oxide (BeO) with high thermal conductivities are not suitable for LED packaging due to their high cost. Aluminum silicon carbide (AlSiC) is evaluated as a potential ceramic substrate for LED, because it can provide higher thermal conductivity, more compatible CTE with Si, and higher strength and stiffness than Cu and Al. The cost of AlSiC is also in an acceptable range.

However, compared to metal substrate, the thermal conductivity of ceramic is obviously lower, which restricts the development of ceramic based CoB. A modified method is fabricating a large scale via a metal slug beneath the LED chip. The thermal conductivity can be improved to more than 200 W/mK and thermal resistance can be reduced to 8.5 K/W [15–17].

Direct Bonded Copper (DBC) provides another competitive approach to improve the thermal dissipation of ceramic based CoB. In Methods III and IV, there actually exists a compound interface between the circuit and ceramics. This interface increases the total thermal resistance of the board. DBC bonds the copper circuit to ceramics by high temperature eutectic soldering. Therefore, there is no identifiable interface between the copper and the ceramic. The eutectic bonding layer constrains the expansion of copper and therefore lowers the CTE of the board. In addition, DBC presents a higher thermal conductivity since the copper is purer than that in Methods III and IV. The controllable thickness of the copper layer also provides competitive thermal management solution. The disadvantage of DBC is that it cannot fabricate metallized via through ceramics, which indicates that the DBC board cannot be applied to a surface mounting process.

Since ceramics is capable of forming a multi-layer substrate, embedding leadframe with a large metal slug in ceramic is becoming the most potential solution with low cost for high power LEDs. Both the thermal dissipation and electrical insulation can be solved. This method is totally compatible with the IC process and only one issue should be improved on the molding of the reflector. In addition, electronic devices such as the driver, sensor, and controller can be integrated into the package to develop the System in Packaging (SiP) technology.

To further improve the thermal performance of substrate, other composite materials such as monolithic carbonaceous materials (MCMs), metal matrix composites (MMCs), carbon/carbon composites (CCCs), and ceramic matrix composites (CMCs) are developed. These materials present controllable CTE and remarkable thermal performance.

Another alternative approach to lowering the thermal resistance of substrate is thinning the substrate. Kim *et al.* [18] developed a competitive aluminum-based packaging platform by selectively anodizing Al. The total thickness of the package is only 500 μm and the base for chip attachment is 180 μm thick. Thermal resistance was reduced to approximately 2 K/W, which is a significant advancement for today's LED packaging. However, thinner substrate may be more fragile due to thermal stress and operation. Therefore, the mechanical reliability must be evaluated.

The final factor affecting the system thermal resistance is the interface among submount, heatslug/substrate, and heatsink. Since these components may be warped and the surfaces are roughened, there exist small gaps between two components. Thermal interface materials (TIMs) are therefore essential to fill these gaps to increase thermal conduction. TIMs should be soft and wettable to avoid residual air in the interfaces. SnAgCu alloys are normally utilized for soldering between the submount and heatslug/substrate, whereas thermally conductive epoxy is chosen as the attachment of heatslug/substrate and heatsink. TIMs should be as thin as possible since the thermal conductivities of these materials are low. Precise dispensing and optimized curing processes are important to enable a thin film for TIMs. However, compared to the bulk thermal resistance of TIMs, it has been found that the contact resistance presents more significant impacts on the total interface resistance.

Adding aligned carbon nanotubes (CNT) to thermal interface materials may improve the thermal conduction in the long run. However, compared to the ultra high thermal conductivity of CNT itself, the effects are not as significant as expected. This is due to the fact that the small size of CNT leads to many interfaces and thereby blocks the heat conduction. There needs to be further improvements on alignment, straight morphology, and bonding strength for CNT when the density is increased.

For single-chip packaging and low power multi-chip packaging, thermal dissipation must rely primarily on natural conversion cooling to achieve low cost and high reliability. However, there may be tens or hundreds of chips integrated in one package in some specific illumination applications such as airport illumination, head lamp of tank, and so on. This demands a semi-active cooling or active cooling system integrated to the package, with some examples to be presented in Chapter 6.

4.3.3 SiP/COB LED Chip Packaging Process

A typical laboratory example for LED packaging by using CoB packaging technology is given below.

(1) As shown in Figure 4.12, the electrical connection and copper board were designed; the distances between the chips in the module were optimized based on thermal dissipation solutions.
(2) The LED chips were bonded with the board through a series of processes such as solder coating and reflow, the wires were connected by wire-bonder; the final prototype is shown in Figure 4.13.
(3) The LED chips were cured in an oven for 1–2 hours after phosphor was dispersed on the chips and the temperature was maintained between 120–150 °C. Figure 4.14 shows the LED chips after phosphor coating.

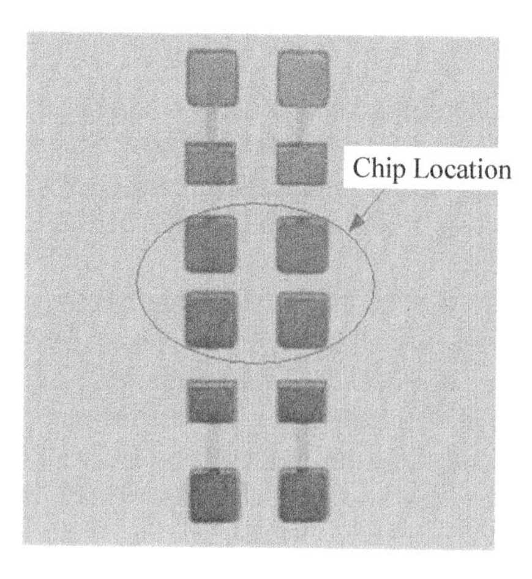

Figure 4.12 Two-by-two chip array on copper board.

Figure 4.13 Two-by-two chip array after wire bonding.

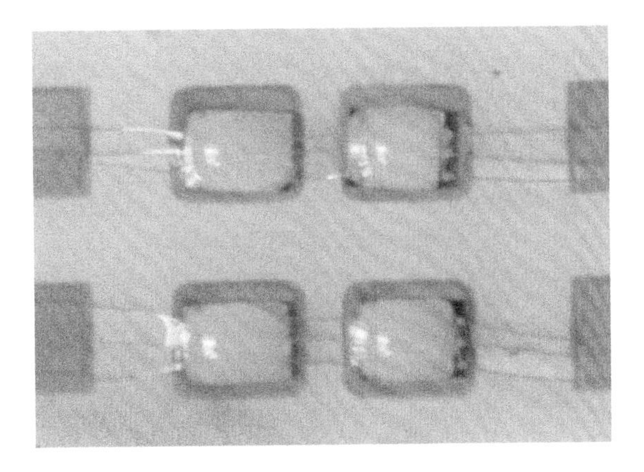

Figure 4.14 LED array after phosphor coating.

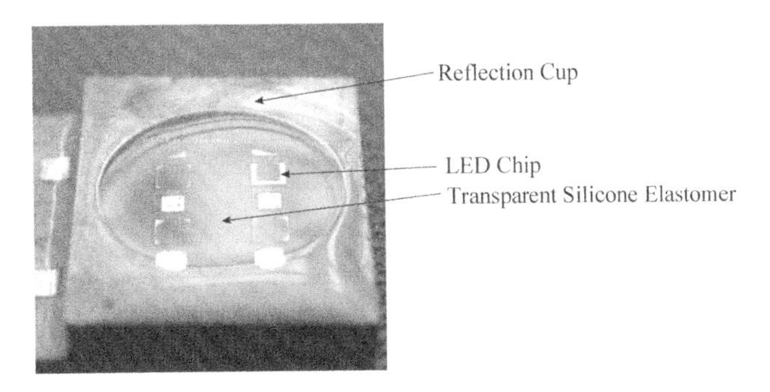

Figure 4.15 Packaged LED array.

Figure 4.16 Illuminated LED system with cooling solution.

(4) The reflection cup was fixed onto the circuit board, filling it with flexible and optically transparent silicone elastomer, then the chips were put into the oven to bake in order to cure the silicone elastomer. Figure 4.15 shows the LED modules after silicone elastomer curing. The lens and light reflector were installed according to the demands of optical design.

(5) The LED module like the one shown in Figure 4.15 cannot work stably by relying on the thermal dissipation structure on the copper board because a large amount of heat needs to be dissipated out. A cooling means is necessary to exchange the heat out to the ambient. Figure 4.16 shows the working LEDs array with cooling system added on the base.

4.4 Summary

In this chapter, issues in the thermal management of the LED in packaging level were discussed. Basic concepts of heat transfer were briefly introduced covering conduction, convention, radiation, and thermal resistance. Models for system level thermal resistance for typical LED packages were briefly discussed. Various ways to decrease thermal resistance in terms of packaging forms were discussed, including packaging materials. System in packaging (SiP) and chip on board (COB) were discussed and some examples were given.

References

[1] Arik, M., Petroski, J., and Weaver, S. (2001), "Thermal challenges in the future generation solid state lighting applications: light emitting diodes," *ASME/IEEE international packaging technical conference*, Hawaii, USA, 113–120.

[2] Arik, M., Petroski, J. and Weaver, S. (2002), "Thermal challenges in the future generation solid state lighting applications: light emitting diodes," *8th Intersociety conference on thermal and thermomechanical phenomena in electronic systems*, San Diego, CA, USA, IEEE, 113–120.

[3] Arik, M., and Weaver, S. (2004), "Chip scale thermal management of high brightness LED packages," *Proceeding of SPIE*, **5530**: 214–223.

[4] Arik, M. and Weaver, S. (2005), "Effect of chip and bonding defects on the junction temperatures of high-brightness light-emitting diodes," *Optical Engineering*, **44**(11): 111305-1-8.

[5] Holman, J.P. (2002), *Heat Transfer*, Boston, McGraw-Hill Press.

[6] Yang, S.M., and Tao, W.Q. (1998), *Heat Transfer*, Beijing, Higher Education Press. (In Chinese)

[7] You, S.M. and Lee, S. (2004), "High power microelectronics thermal management and packaging fundamentals," *Tutorial in Proceedings Intersociety Conference on Thermal and Thermomechanical Phenomena in Electronics Systems*, Las Vegas, Nevada, p. 4.

[8] Cheng, S.M., Huang, S.Y., Bai, C.Y., and Wei, B.W. (1990), Heat Transfer, Beijing, Higher Education Press. (In Chinese)

[9] Haque, S., Steigerwald, D., Rudaz, S., Steward, B., Bhat, J., Collins, D., Wall, F., Subramanya, S., Elpedes, C., and Elizondo, P. (2000), "Packaging challenges of high-power LEDs for solid state lighting," *www.lumileds.com/pdfs/techpaperspres/manuscript_IMAPS_2003.PDF*

[10] Krames M. R, Shchekin O. B, Mueller-Mach R, Mueller G.O., Zhou, L., Harbers G., and Craford M.G., (2007), "Status and future of high-power light-emitting diodes for solid-state lighting," *Journal of Display Technology*, **3**(2): 160–175.

[11] Daniel, A.S., Jerome, C.B., and Dave, C. (2002), "Illumination with solid state lighting technology," *IEEE Journal on Selected Topic in Quantum Electronics*, **18**(2): 310–320.

[12] Carl, Z. (2004), "New material options for light emitting diode packaging," *Proceedings of SPIE*, **5366**: 173–182.

[13] Fan, B.F., Zhao, Y., Xian, Y.L., and Wang, G. (2006), "Thermal simulation studies of high-power light-emitting diodes. In: Advanced LEDs for Solid State Lighting," *Proceedings of SPIE*, **6355**, 63550D-4.

[14] Liu, Z.Y., Liu, S., Wang, K., and Luo, X.B. (2009), "Status and prospects for phosphor-based white LED packaging," *Frontiers of Optoelectronics in China*. **2**(2): 119–140.

[15] Park, J. K., Shin, H. D., Park, Y. S., Park, S. Y., Hong, K. P., Kim, B. M. A suggestion for high power LED package based on LTCC. In: *Proceedings of the 56th Electronic Components and Technology Conference*. 2006, 1070–1075.

[16] Shin, M. W., Thermal design of high-power LED package and system. In: *Advanced LEDs for Solid State Lighting*. Gwangju: SPIE, 2006, 6355: 635509

[17] Yang, L., Jang, S., Hwang, W., Shin, M. Thermal analysis of highpower GaN-based LEDs with ceramic package. *Thermochimica Acta*, 2007, **455**(1–2): 95–99.

[18] Kim, K.M., Shin, S.H., Lee, Y.K., Choi, S.M., and Kwon, Y.S. (2008), "Aluminium-based packaging platform for LED using selectively anodising method," *Electronics Letters* **44**(1): 24–25.

5

Reliability Engineering of High Power LED Packaging

5.1 Concept of Design for Reliability (DfR) and Reliability Engineering

The concept of reliability engineering has been practiced in many industries in the past few decades. The aerospace industry is an example, but one where it is not practical to develop a mock-up due to the high cost of making even the prototypes. However, the development of the aerospace industry has matured to the stage that it has accumulated significant databases for materials, loading, components, sub-assemblies, and full vehicles so as to be able to design the vehicle by modeling and simulation. Design for X (DfX) has been widely used for this industry. X refers to almost everything such as manufacturing, assembly, reliability, durability, maintenance, and cost. Reliability engineering is one of the most important aspects of DfX, in which three major tasks are involved: design for reliability (DfR), reliability testing and data statistics analyses, and failure analysis [1]. Figure 5.1 schematically plots the relationship among these three tasks. In the DfR, both numerical methods and experimental validation methods are covered, including models for life prediction, models and simplified test vehicles (TVs) for material selection, processes development, and structural optimization. Numerical methods include the finite element method (FEM), boundary element method, coupled FEM, various validation tools such as strain gages, test chips, and optical measurement methods. Reliability testing and statistics data analyses include various accelerated testing methods associated with screening components/modules/systems by much more severe tests. Due to the fact that new generation chips appear every three to six months and even the harshest tests may last one or two months, new model based accelerated testing methods are desirable, which are still in the research stage. This second task will focus on the data analysis. The third task is concerned about failure analysis including fundamental failure mechanisms and their analysis methods by SEM, TEM, AFM and so on. This approach has also been successfully developed for the automobile industry. In the past 25 years, the integrated circuit (IC) industry has begun to systematically develop DfR based methodology for reliability problems. Here in this

LED Packaging for Lighting Applications: Design, Manufacturing and Testing, First Edition. Sheng Liu and Xiaobing Luo.
© 2011 Chemical Industry Press. All rights reserved. Published 2011 by John Wiley & Sons (Asia) Pte Ltd.

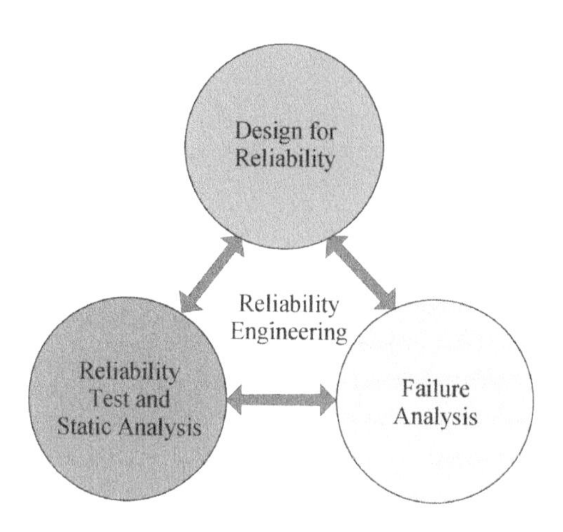

Figure 5.1 Schematic of reliability engineering.

chapter, we intend to present a platform for the LED industry with the focus on LED modules and application products.

5.1.1 Fundamentals of Reliability

Reliability is defined as the probability that a system will meet specified performance criteria for a predefined period of time while operating within specified environmental conditions [2].

(i) The Reliability Function

The reliability function $R(t)$ is defined as the probability that a device will function for a given period of time t under specific operating conditions. The formula is:

$$R(t) = P(T \geq t) \tag{5.1}$$

$R(t)$ ranges between 0 and 1:

$$0 \leq R(t) \leq 1 \tag{5.2}$$

(ii) The Cumulative Failure Probability

The cumulative failure distribution function $F(t)$ is the probability that failure occurs before time t. The formula is as follow:

$$F(t) = P(T \leq t) \tag{5.3}$$

Obviously,

$$F(t) + R(t) = 1 \tag{5.4}$$

(iii) Failure Distribution Density

Failure density function $f(t)$ is defined as the failure probability of the electronic components at the time t.

$$0 \leq f(t) \leq 1 \tag{5.5}$$

It is necessary that:

$$\int_0^\infty f(t)\mathrm{d}t = 1 \tag{5.6}$$

The cumulative failure distribution function in the time interval from zero to t can be expressed as:

$$F(t) = \int_0^t f(t)\mathrm{d}t \tag{5.7}$$

And the reliability function is given by:

$$R(t) = \int_t^\infty f(t)\mathrm{d}t \tag{5.8}$$

(iv) The Failure Probability

The failure probability $\lambda(t)$ is defined as the probability that failure occurs during time t. It is also known as the instantaneous failure rate or the hazard function.

$$\lambda(t) = \frac{f(t)}{R(t)} \tag{5.9}$$

5.1.2 Life Distribution

Life is a quantitative characterization of the reliability of electronic devices. Since reliability is a statistical concept, in a particular individual electronic device, it is difficult to evaluate the exact value of life before failure occurs. But if we know the failure rate $\lambda(t)$ of a number of electronic devices, we can obtain some life characteristic values. The mean time to failure is the average working or storing time of a device before its failure; it is usually presented as Mean Time to Failure (MTTF):

$$u = \int_0^\infty tf(t)\mathrm{d}t \tag{5.10}$$

$$u = \int_0^\infty R(t)\mathrm{d}t \tag{5.11}$$

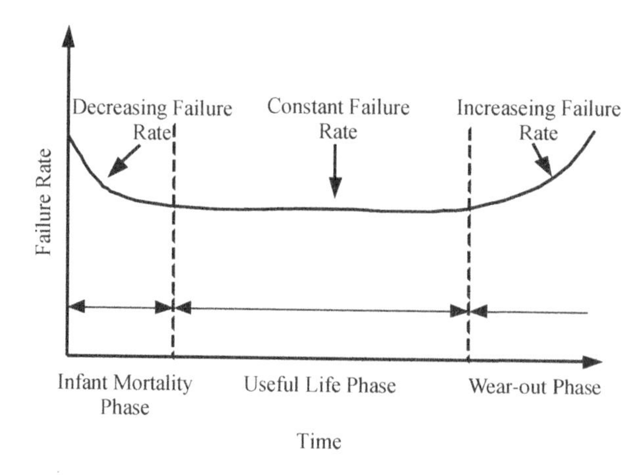

Figure 5.2 Typical failure rate curve.

The service life of LEDs can be defined by MTTF. As to the lighting purposes, it generally refers to the operating time during which the output luminous flux of LEDs attenuates to 70% of the initial value for high power packaging, which are commonly accepted [3]. The failure rate of LED devices can be generally divided into three periods: the infant mortality phase, the useful life phase and the wear-out phase. The bathtub reliability curve is a sequential combination of these three phases as shown in Figure 5.2.

The infant mortality phase is characterized by failure occurring in the early stage of the product utilization and the failure rate decreases rapidly with the extension of working time. The reason which causes the early failure is largely due to manufacturing defects or improper handling or installation. The useful life phase is characterized by a low and stable failure rate which is very close to a constant. Failures during this period are random. The wear-out phase is characterized by a marked increasing failure rate. It is largely due to wear-out or physical degradation of the devices.

Parametric failure and catastrophic failure are two failure modes of LED devices. Catastrophic failure is defined as when the LED device will not illuminate due to the change of key electrical or optical parameters. Parametric failure is defined as when the change of key parameters from their initial value exceeds a certain amount. The normal and slight parameter changes which do not affect LED devices operation should not be recognized as failure.

One always desires to use a single mathematical model to express the failure rate of electronic devices throughout the whole life. Here we present some common distribution functions, such as the exponential distribution function, normal distribution function, the Weibull distribution function, and lognormal distribution function.

(i) Exponential Life Distribution

The major characteristic of exponential life distribution is that its failure probability is a constant which is very convenient for the calculation. The hazard rate function is:

$$\lambda(t) = \lambda = \text{Constant} \tag{5.12}$$

The failure density function of a normal distribution is:

$$f(t) = \lambda \exp(-\lambda t) \quad (5.13)$$

The cumulative failure distribution and the reliability function are given by:

$$F(t) = 1 - \exp(-\lambda t) \quad (5.14)$$

$$R(t) = \exp(-\lambda t) \quad (5.15)$$

The MTTF is:

$$MTTF = \frac{1}{\lambda} \quad (5.16)$$

(ii) Normal Life Distribution

Normal distribution function is also known as Gaussian distribution or the error distribution function. The normal distribution is two parameters distribution: the mean, u and the standard deviation, σ. The probability density function is:

$$f(t) = \frac{1}{\sqrt{2\pi\sigma^2}} \exp\left[\frac{-(x-u)^2}{2\sigma^2}\right] \quad (5.17)$$

The cumulative failure distribution function and the reliability function are given by:

$$F(t) = \int_0^t \frac{1}{\sqrt{2\pi\sigma^2}} \exp\left[\frac{-(x-u)^2}{2\sigma^2}\right] dx \quad (5.18)$$

$$R(t) = 1 - \int_0^t \frac{1}{\sqrt{2\pi\sigma^2}} \exp\left[\frac{-(x-u)^2}{2\sigma^2}\right] dx \quad (5.19)$$

The MTTF is:

$$MTTF = u \quad (5.20)$$

(iii) Weibull Life Distribution

The Weibull distribution is widely used in the analysis of the failure distribution of semiconductor devices. The probability density function is as follows:

$$f(t) = \frac{m}{T}\left(\frac{t}{T}\right)^{m-1} \exp\left[-\left(\frac{t}{T}\right)^m\right] \quad (5.21)$$

The cumulative failure distribution function and the reliability function are given by:

$$F(t) = 1 - \exp\left[-\left(\frac{t}{T}\right)^m\right] \quad (5.22)$$

$$R(t) = \exp\left[-\left(\frac{t}{T}\right)^m\right] \quad (5.23)$$

The MTTF is:

$$MTTF = T \cdot \Gamma\left(1 + \frac{1}{m}\right) \quad (5.24)$$

$$\Gamma(n) = \int_0^\infty e^{-x} x^{n-1} dx \quad (5.25)$$

where m is the shape parameter, and T is scale parameter. When the shape parameter m is less than 1, the failure rate is decreasing, which can model the early failure period. When the shape parameter m equals 1, the Weibull distribution is changed to an exponential distribution with failure rate $1/T$, which can represent the useful life period. When the shape parameter m is greater than 1, it has an increasing failure rate, which is suitable for modeling wear-out period.

(iv) Lognormal Life Distribution

The failure density function of lognormal distribution is given by:

$$f(t) = \frac{1}{\sigma t \sqrt{2\pi}} \exp\left[\frac{-[\ln(t-u)]^2}{2\sigma^2}\right] \quad (5.26)$$

The lognormal cumulative distribution is obtained by transformation from standard normal tables. The transformation is:

$$Z = \frac{\ln t - u}{\sigma} \quad (5.27)$$

The MTTF is:

$$MTTF = \exp\left(u + \frac{\sigma^2}{2}\right) \quad (5.28)$$

5.1.3 Accelerated Models

MIL–HDBK–217, Telcordia SR–332, RAC's PRISM, JEDEC, SAE's reliability prediction method, and the CNET reliability prediction method, and so on have long been used for reliability prediction [4]. However, due to many reasons such as inappropriate field loading, and unreliable material databases and constitutive models, the effect of manufacturing induced defects on the life prediction, lack of the understanding of failure modes and failure criteria, the accuracy of the prediction has been low, and often with one order in difference for the life prediction as compared to the test data and field life. This is particular true for high power LED, as it is so new, and so many companies are rushing to join this booming business, that the reliability research is far behind the development of the industry.

Due to the facts mentioned above, traditional testing standards themselves have been challenged [4–12]. Those standards suffer from various disadvantages and that was why physics of failure based modeling approach has been suggested [12]. Therefore, those traditional testing standards may still be used as screening testing methods. The challenge is how to use accelerated testing to predict the reliability of LED devices quantitatively. As discussed by Xie and Pecht [13], neither the Arrhenius model, which considers the environmental condition of only constant temperature, nor the exponential distribution model developed on the basis of constant failure rate is suitable for LEDs reliability prediction. It is believed that the physics of failure based modeling approach represents the right direction to follow.

Unfortunately in the current industry, testing is still the most commonly used method of high-power LED reliability evaluation. By accelerated test, the life test of luminous semiconductor devices can be finished within several weeks or months. Usually, after testing life characteristics under high stress, life under normal stress can be derived using accelerated models to be described below. Generally, as pointed out previously, failure criterion is that the optical attenuation achieves 70% of the initial value.

When LEDs are used in a system such as in a road light or a tunnel light, we require them operate for a long time without much maintenance. It is partially true for tunnel lighting which is on 24 hours a day and seven days a week. Reliability of these systems evolves into more requirements such as reliability, safety, maintainability, and supportability, requiring advanced prognostics health management techniques [6,7]. Sensors and/or other leading indicators such as a material microstructure evolution in the device or system itself are needed. In the LED light fixture design, advanced sensors are being considered to provide a health monitoring for the LED road light systems [14].

In this chapter, we will still present those statistics based models, which can provide a reference to those practicing engineers and researchers. Thousands of samples are needed for testing, which are expensive, time consuming, and often cannot check out those failure modes in the field. At the same time, we will present what we feel is essential for the physics of a failure based modeling approach. With better and more complete material databases, sensor based loading monitoring, fundamental understanding of failure mechanisms, robust numerical modeling, and accelerated reliability evaluation models we will achieve the final designation for robust prediction of the actual life.

(i) Arrhenius Equation

The Arrhenius model can be used with great success if the key factor for failure is temperature. This empirically based model is expressed in the following form:

$$k = Ae^{\left(\frac{-E_a}{RT}\right)} \tag{5.29}$$

where k is the rate coefficient, A is a constant to be determined, E_a is the activation energy, R is the universal gas constant (8.314×10^{-3} kJ/mol·K), and T is the temperature (in Kelvin). The acceleration factor between accelerated stress and use environment is given by:

$$AF = e^{\frac{E_a}{R}\left(\frac{1}{T} - \frac{1}{T_{ref}}\right)} \tag{5.30}$$

where AF is acceleration factor, and T_{ref} represents use environment. Note that the only parameter unknown in Equation 5.30 is the activation energy which can be calculated by running multiple stress conditions.

(ii) Peck's Model

The Peck's model is widely adapted if the accelerated stresses are temperature and humidity. The Peck's model is expressed as following form:

$$AF = \left(\frac{RH}{RH_{ref}}\right)^{-c} e^{\frac{E_a}{R}\left(\frac{1}{T} - \frac{1}{T_{ref}}\right)} \tag{5.31}$$

where RH is the relative humidity in percentage, c is RH inverse power law coefficient, E_a is the activation energy, R is the universal gas constant (8.314×10^{-3}kJ/mol·K), and T is the temperature (in Kelvin). The interaction between the relative humidity and temperature is ignored in the model.

(iii) Eyring Model

A powerful model used for multiple accelerated stresses is the Eyring model. The Eyring model usually includes temperature and other relevant non-thermal stresses. The model for temperature and two additional stresses takes the general form:

$$AF = \left(\frac{T}{T_{ref}}\right) e^{\left[\frac{E_a}{R}\left(\frac{1}{T} - \frac{1}{T_{ref}}\right) + \left(B + \frac{C}{T}\right)(S_1 - S_{1ref}) + \left(D + \frac{E}{T}\right)(S_2 - S_{2ref})\right]} \tag{5.32}$$

where S_1 and S_2 are the functions of the stresses, E_a is the activation energy, R is the universal gas constant (8.314×10^{-3} kJ/mol·K), and T is the temperature (in Kelvin). B, C, D, E are constants related to stress. The interaction terms between the temperature and other stresses are presented by parameters C and E. The simplified Eyring model with temperature, humidity, and voltage stresses is given by:

$$AF = \left(\frac{V_{ref}}{V}\right)^{-n} \left(\frac{RH_{ref}}{RH}\right)^{-c} e^{\left[\frac{E_a}{R}\left(\frac{1}{T} - \frac{1}{T_{ref}}\right)\right]} \tag{5.33}$$

where c is the RH inverse power law coefficient, and n is the voltage inverse power law coefficient.

(iv) Coffin–Manson Model

For failure of solder joint or other metals under the thermo-mechanical fatigue in which the key factors are the temperature and frequency, the most widely used model is the Coffin–Manson model which is expressed in the following form:

$$AF = \left(\frac{\Delta T_{\text{ref}}}{\Delta T}\right)^{-n} \tag{5.34}$$

where ΔT is entire temperature cycle range within which a device operates, and n is material dependent inverse power law coefficient.

$$N_f = Af^{-\alpha}\Delta T^{-\beta}G(T_{\max}) \tag{5.35}$$

where N_{f} is the number of cycles to fail, f is the cycling frequency, α is cycling frequency exponent, β is the temperature range exponent, and $G(T_{\max})$ is the Arrhenius term evaluated at the maximum temperature.

5.1.4 Applied Mechanics

(i) Thermo-elasticity: Thermally Induced Strains and Stresses

When the temperature change is small, the expansion and contraction deformation of a substance can be considered as linearly proportional to the temperature change. This proportionality is expressed by the coefficient of thermal expansion (CTE). For the thermo-elastic problem the deformation due to thermal change will disappear when temperature change vanishes. The thermal strain for an isotropic material can be expressed as:

$$\varepsilon_x^{th} = \varepsilon_y^{th} = \varepsilon_y^{th} = \alpha(T - T_0) \tag{5.36}$$

$$\gamma_{xy}^{th} = \gamma_{yz}^{th} = \gamma_{zx}^{th} = 0 \tag{5.37}$$

where α is the coefficient of thermal expansion and the unit is $1/°\text{C}$, T is the current temperature, T_0 is the reference temperature. ε^{th} and γ^{th} represent normal and shear thermal strains respectively. The thermal shear components γ^{th} are zero because the thermal expansion is a volumetric expansion.

Due to mechanical constraints, the non-uniform temperature distribution and CTE mismatch between different materials of packaging, thermal stress is generated in the LED device inside the bulk materials and along interfaces. For an isotropic linearly elastic material the thermal stress and thermal strain can be related by generalized Hooke's law:

$$\varepsilon_x = \frac{\partial u}{\partial x} = \frac{1}{E}\left[\sigma_x - \mu(\sigma_y + \sigma_z)\right] + \alpha(T - T_0) \tag{5.38}$$

$$\varepsilon_y = \frac{\partial v}{\partial x} = \frac{1}{E}\left[\sigma_y - \mu(\sigma_z + \sigma_x)\right] + \alpha(T - T_0) \tag{5.39}$$

$$\varepsilon_z = \frac{\partial w}{\partial x} = \frac{1}{E}\left[\sigma_z - \mu\left(\sigma_x + \sigma_y\right)\right] + \alpha(T - T_0) \tag{5.40}$$

$$\gamma_{xy} = \frac{\tau_{xy}}{G} \tag{5.41}$$

$$\gamma_{yz} = \frac{\tau_{yz}}{G} \tag{5.42}$$

$$\gamma_{zx} = \frac{\tau_{zx}}{G} \tag{5.43}$$

where E is Young's modulus, μ is Poisson's ratio, and G is shear modulus. The components σ_x, σ_y, σ_z are normal stresses, and τ_{xy}, τ_{yz}, τ_{zx} are shear stresses. ε_x, ε_y, ε_z represents normal strain, and γ_{xy}, γ_{yz}, γ_{zx} represents shear strain. u, v, w are displacement component functions.

(ii) Hygro-elasticity-Moisture Induced Strains and Stresses

Moisture diffusion modeling can be based on the analogy between mass diffusion and heat conduction [15]. Transient moisture diffusion is assumed to be modeled by Fick's law of diffusion as described below:

$$\frac{\partial^2\theta}{\partial^2 x} + \frac{\partial^2\theta}{\partial^2 y} + \frac{\partial^2\theta}{\partial^2 z} = \frac{1}{D}\frac{\partial\theta}{\partial t} \tag{5.44}$$

where θ is moisture concentration, and D is the moisture diffusivity of respective materials. As the moisture concentration is discontinuous across the material interface, relative moisture concentration, w defined with Equation 5.45, could remain continuous across the multi-material interface.

$$w = \theta/\theta_{sat}, \quad 1 \geq w \geq 0 \tag{5.45}$$

where θ_{sat} is the saturated moisture concentration of materials. By using relative moisture concentration the Fick's law of diffusion can be reconstructed as:

$$\frac{\partial^2 w}{\partial^2 x} + \frac{\partial^2 w}{\partial^2 y} + \frac{\partial^2 w}{\partial^2 z} = \frac{1}{D}\frac{\partial w}{\partial t} \tag{5.46}$$

The hygroscopic swelling linearly increases with the moisture content. The mismatch of the coefficients of moisture expansion (CME) causes hygroscopic swelling stress in the packaging. This is analogous to the expansion induced by a mismatch of CTEs. Therefore, the hygroscopic stress could be obtained as:

$$\sigma = E\beta(\theta - \theta_{ref}) \tag{5.47}$$

where β is the coefficient of moisture expansion, θ is the current moisture concentration, and θ_{ref} is the reference moisture concentration.

(iii) Hygro-thermal-elasticity: Combined Temperature and Moisture Induced Strains and Stresses

In both manufacturing processes and reliability testing, temperature and moisture can occur individually in sequence or occur at the same time, resulting in very complicated stresses, which could cause significant yield and reliability problems, such as popcorning issue [15,16] in IC plastic packaging, BGA packaging, and reflow induced fall off of PC lenses in LED packaging [17,18]. The combined strain can be expressed as:

$$\varepsilon = \alpha \Delta T + \beta \Delta C \tag{5.48}$$

where α and β can be temperature dependent and care must be exercised to correctly predict quantitatively when water and vapor can co-exist along partially or fully delaminated interfaces during reflow and other complicated loading conditions.

(iv) Interfacial Facture Mechanics

Delamination and cracking will occur at the interfaces in high power LED packaging modules due to the mismatch of the coefficients of thermal expansion and the possible existence of moisture, as mentioned above, between different materials which could result in high interfacial stress concentration. Cracks may form in the interface from manufacturing defects, material flaws, damages caused by applied loads induced by temperature cycling, and moisture penetration. Those cracks may then propagate through the interface and influence the thermal and optical performance of LED packaging modules. Therefore, it is necessary to consider mechanisms of the interfacial fracture at the interfaces under thermal stress, hygrostress, or hygro-thermal conditions.

Cracks can be divided into Mode I, Mode II, and Mode III in the theory of fracture mechanics, which are shown in Figure 5.3. Mode I crack is open mode crack, and the load is perpendicular to the direction of crack propagation. Upper and lower crack surfaces open along the direction of the load and the crack is extended along the crack face. Mode II crack is sliding mode and Mode III is shearing mode. Mode II crack is controlled by shear stress which is parallel to the crack plane and perpendicular to the crack front line. The tearing shear stress is parallel to the crack plane and the crack line for Mode III.

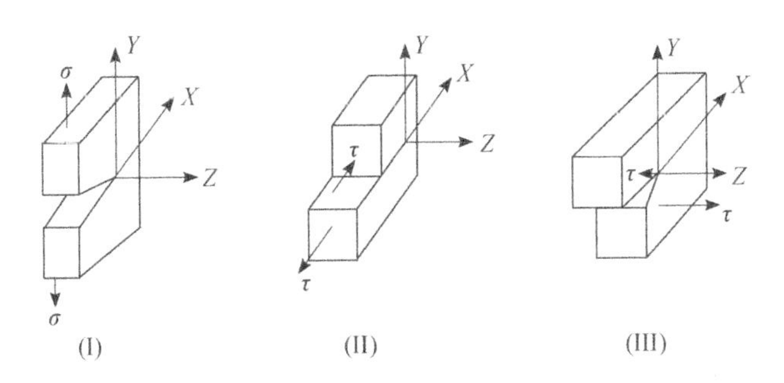

Figure 5.3 Typical crack modes.

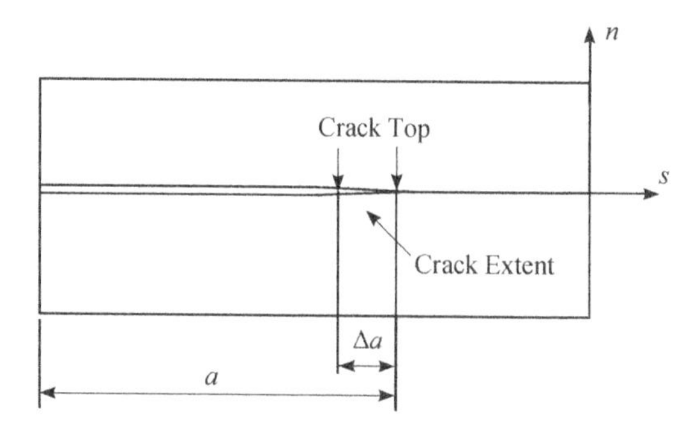

Figure 5.4 Interfacial fracture mechanics modeling.

Energy release rate G as a generalized crack driving force is an important physical quantity. The energy release rate may be calculated in two ways: three energy release rates corresponding to three modes and total energy release rate as a function of phase angles [19–23]:

$$G = G(\psi_{\text{local}}) \tag{5.49}$$

$$\psi_{\text{local}} = \tan^{-1}\left[\frac{\sigma_{\text{sn}}}{\sigma_{\text{nn}}}\right] \tag{5.50}$$

where ψ_{local} is the local phase angle, n and s denote normal and sliding direction respectively, σ_{nn} and σ_{sn} are normal shear stresses at the crack tip associated with a delamination length of a, as shown in Figure 5.4.

The strain energy release rates for Mode I and Mode II fracture modes, respectively, can be expressed based on the linear elastic fracture mechanics as:

$$G_{\text{I}} = \lim_{\Delta a \to 0}\left\{\frac{1}{2\Delta a}\int_0^{\Delta a}[u_{\text{n}}^{+}(a+\Delta a) - u_{\text{n}}^{-}(a+\Delta a)]\sigma_{\text{nn}}\text{d}s\right\} \tag{5.51}$$

$$G_{\text{II}} = \lim_{\Delta a \to 0}\left\{\frac{1}{2\Delta a}\int_0^{\Delta a}[u_{\text{s}}^{+}(a+\Delta a) - u_{\text{s}}^{-}(a+\Delta a)]\sigma_{\text{sn}}ds\right\} \tag{5.52}$$

where Δa is the crack extension, $u_{\text{n}}^{+}, u_{\text{s}}^{+}, u_{\text{n}}^{-}, u_{\text{s}}^{-}$ are the displacements of the upper and lower surfaces associated with a delamination length of $a + \Delta a$, respectively, as shown in Figure 5.4.

Ideally, the above integral should be calculated by two solutions, that is, one before the crack growth with a crack of a and the second solution after the crack growth for an amount of Δa. In terms of the finite element framework, it has been demonstrated that only one run can generate very accurate results for the strain energy release rates as long as the finite element mesh is reasonable, even with a fairly coarse mesh [24–26]. The demonstrated examples covered isotropic, orthotropic, anisotropic, long fiber reinforced composites, and woven composites by the first author [19] and many others. In practice, it is reasonable to use finite increment Δa to obtain accurate results for the strain energy release rate and its components. Therefore, numerically simple formulae are shown as follows:

$$G_{\text{I}} = \frac{1}{2\Delta a}\left[u_{\text{n}}^{+}(a) - u_{\text{n}}^{-}(a)\right]f_{\text{n}} \tag{5.53}$$

$$G_{\text{II}} = \frac{1}{2\Delta a}\left[u_{\text{s}}^{+}(a) - u_{\text{s}}^{-}(a)\right]f_{\text{s}} \tag{5.54}$$

where the f_{s} and f_{n} are nodal forces at the crack tip in s (sliding) and n (normal) directions respectively, u_{s} and u_{n} are relative displacements for a pair of nodes right ahead of the crack front in both tangential and normal directions respectively.

It is noted that no actual crack growth is needed if the growth is not implemental. Therefore, one step is needed for the strain energy release rate calculation. It should be noted that finite elements before and after the crack tip should have the same size in the crack direction in order to simplify the energy release rate calculation [25]. It was once noted that the energy release rate components are not converged for an opening crack tip when the finite element mesh is sufficiently small, which has been demonstrated numerically by Sun [27]. Therefore, when the crack closure technique is used, a reasonable mesh is desirable as long as the components of the energy release rate are needed. Fortunately, the total energy release rate can achieve a converged value even with a sufficiently fine mesh near the crack tip, making it a useful fracture quantity [27]. The total energy release rate is the sum of its components:

$$G_{\text{total}} = G_{\text{I}} + G_{\text{II}} \tag{5.55}$$

$$G_{\text{total}} = G(\psi_{\text{local}}) \tag{5.56}$$

The phase angel is defined as:

$$\tan\psi = \sqrt{\frac{G_{\text{II}}}{G_{\text{I}}}} \tag{5.57}$$

or approximately calculated by crack tip opening displacement in terms of tangential and normal displacement components:

$$\tan\psi = \frac{\Delta u_{\text{s}}}{\Delta u_{\text{n}}} \tag{5.58}$$

The first author's previous work [19] should be referred to for three-dimensional delamination crack front.

5.2 High Power LED Packaging Reliability Test

5.2.1 Traditional Testing Standards, Methods, and Evaluation

The defects may be induced at the early stage of product design and manufacturing of the light emitting diodes. Therefore, the detection of potential defects within the LED devices through suitable reliability tests is very important for new LED product development. The MTTF and failure rate of LED products can also be obtained through these tests. However there is not yet a unified standard for reliability testing in the high power LED industry. The LED companies use the referential reliability testing standards described for electronic devices by those screening testing standards mentioned in Section 5.3. The environmental reliability tests are used to simulate the operating environment of LEDs. Typical environmental reliability tests which can be used for LED packaging are listed as follows:

(1) Room Temperature Operating Life Test (RTOL)
(2) Low Temperature Operating Life Test (LTOL)
(3) High Temperature Operating Life Test (HTOL)
(4) Wet High Temperature Operating Life Test (WHTOL)
(5) High/Low Temperature Storage Test
(6) Temperature Cycle Test
(7) Pulse Life Test
(8) Resistance to Soldering Heat Test
(9) Thermal Shock
(10) Mechanical Shock
(11) Vibration Test
(12) Salt Atmosphere (Corrosion Test)
(13) Dust Test

Some important environmental reliability tests done by the Cree and Philips Lumileds Lighting Corporations for high power LEDs are shown in Table 5.1. It can be seen that most of the environmental reliability conditions are similar except for the low temperature operation life test and thermal shock test. In addition, autoclave for 96 hours and precondition at 60 °C and 60% RH for 120 hours prior to soldering reflow tests is done instead of the wet high temperature operation life test for the Luxeon K2 [28,29].

The performance of LED devices is always evaluated by measurement of thermal, electrical, and optical parameters, which will be discussed in some detail in Chapter 7, but briefly discussed here. Thermal parameters of LEDs include junction temperature, case/shell temperature, and the thermal resistance. The junction temperature is the key parameter to evaluate the quality of LEDs which greatly influences the optical and electrical performance of LEDs. Thermocouples can be used to measure the thermal parameters of LED devices and an infrared thermometer can be used for non-contact measurement. The optical parameters of LEDs include luminous flux, correlated color temperature, radian flux, chromaticity coordinates, color rendering index, and emission wavelength. The output optical parameters of LEDs can be measured using photometric and colorimetric parameters (PMT), spectroradiometer, and CCD spectrometer. Electrical parameters of LEDs include leakage current, forward current, forward voltage, power dissipation, and the current flow caused by generation/recombination of carriers. The temperature, optical, and electrical performances are dependent on each other. For example, the temperature of LEDs is related to the forward voltage, luminous flux, and dominant wavelength.

The performance degradation phenomena for LED devices include (1) light output decrease; (2) leakage and generation-recombination current increase; (3) increase in series resistance and temperature of the devices; and (4) modifications of spectral properties.

Table 5.1 Some environmental reliability test conditions of Cree–XLamp and Luxeon–K2

Test Method	Stress Conditions	
	Cree–XLamp	Luxeon–K2
RTOL	Ambient Temperature: 45 °C Forward Current: Maximum in datasheet Test Period: 1008 hours	Ambient Temperature: 55 °C Forward Current: Maximum in datasheet Test Period: 1000 hours
HTOL	Ambient Temperature: 85 °C Forward Current: Maximum in datasheet Test Period: 1008 hours	Ambient Temperature: 85 °C Forward Current: Maximum in datasheet Test Period: 1000 hours
LTOL	Ambient Temperature: − 40 °C Forward Current: Maximum in datasheet Test Period: 1008 hours	Ambient Temperature: − 55 °C Forward Current: Maximum in datasheet Test Period: 1000 hours
WHTOL	Ambient Temperature: 85 °C Forward Current: Maximum in datasheet Humidity: 85% relative humidity (RH) Time: 1008 hours (cycles)	Precondition at 60 °C and 60% RH prior to reflow soldering, 120 hours Autoclave: 121 °C, 100% RH, 15 psig, 96 hours
Thermal Shock	Temperature Range: − 40 °C to 125 °C Dwell Time: 15 minutes Transfer Time: <20 seconds Cycles: 200 cycles	Temperature Range: − 40 °C to 110 °C Dwell Time: 15 minutes Transfer Time: <20 seconds Cycles: 1000 cycles
Mechanical Shock	Shock: 1500 G Pulse Width: 0.5 ms Direction: 5 each, 6 axis (30 total)	Shock: 1500 G Pulse Width: 0.5 ms Direction: 5 each, 6 axis
Salt Atmosphere	Ambient Temperature: 35 °C Salt Deposit: 30 g/m^2/day Test Period: 48 hours	Ambient Temperature: 35 °C Test Period: 48 hours

The failure criteria for the LED packaging modules must be established before the reliability testing. Currently, the failure criteria are mainly based on the optical degradation or being unable to be illuminated, which are widely used by industry. The failure criteria for Cree–XLamp and Luxeon–K2 are listed in Table 5.2. The failure criteria for Cree-XLamp are that luminous flux degradation is more than 15% for InGaN LEDs and more than 25% for AlInGaP LEDs. It is noted that some companies use different criteria. For instance, the failure criterion for the Luxeon-K2 is that the degradation of optical output is 50% instead of 70% [28,29].

It should be pointed out that little information on the fundamental understanding of materials, microstructures, thermal degradation, and mechanics based cracks, delamination growth, and accelerated life prediction for high power LEDs is available to the community. For instance, the cap fall off is a simple thermo-mechanical issue, but the final performance does get reflected optically, resulting in significant light output loss.

Table 5.2 Environmental reliability test failure criteria of Cree–XLamp and Luxeon–K2

Test Method	Cree–XLamp	Luxeon–K2
RTOL HTOL LTOL WTOL	Forward Voltage Shift: >200 mV Luminous Flux Degradation InGaN LEDs: >15% AlInGaP LEDs: >25% Forward or Reverse Leakage: >10 μA Catastrophic Failure	Luminous Flux Degradation: >50% Catastrophic Failure
Thermal Shock Mechanical Shock Salt Atmosphere	LEDs no longer Illuminated after Test	LEDs no longer Illuminated after Test

5.2.2 Methods for Failure Mechanism Analysis

It is necessary to perform a detailed failure analysis to determine the root causes of the failure of LED devices. The methods for failure mechanism analysis which could be used for LED devices are listed as follows.

(i) Optical Microscopic Analysis

Optical microscopy is one of the main tools for LED packaging failure analysis. It can be used to observe and analyze the breakdown phenomena of the chip under electrical stress, wire failure, cracks, stains, and scratches of the LED chip, the corrosion of metal and the cracks in the silicone glue, and so on.

(ii) Infrared Imaging System

An infrared imaging system provides a non-contact temperature measurement method for a given region of the LED devices. Sometimes a local temperature in a small region will be much higher than average temperatures, called hot spots, due to defects generated in the packaging. These phenomena will directly affect the LED device's reliability and lifetime. An infrared imaging system is able to meet the requirements of non-contact temperature measurement under the testing or operating condition.

(iii) SEM Analysis

A scanning electron microscope (SEM) has a series of advantages such as high resolution, high magnification, large depth of field, and a strong three-dimensional sense, which can be used to observe fine structures that cannot be seen under an optical microscope. In the failure analysis of LED packaging, a scanning electron microscope can be used to observe short circuits, open circuits, electro-migration, oxide layer and corrosion of metal contact and wire bonding. It can also be used to observe the dislocations and other defects of epitaxial layer of the LED chip, voids, cracks, debonding, and so on.

(iv) Acoustic Microscope Analysis

A C mode scanning acoustic microscope (C–SAM) is an excellent tool for a non-destructive failure analysis of packaging. It can provide a fast and comprehensive image for the critical packaging defects and can identify three-dimensional aspects of those defects in packaging. In the process during which the packaging undergoes accelerated environment (such as high temperature, high humidity), C–SAM can also be used for continuous monitoring of the packaging.

(v) Nano X-Ray CT [30]

One of the newly developed tools used is computed tomography (CT) to observe tiny hidden defects and three-dimensional imaging in a very fine resolution. Figure 5.5 presents the CT "cross-section" with details and non-connected balls. It is believed this system is going to be

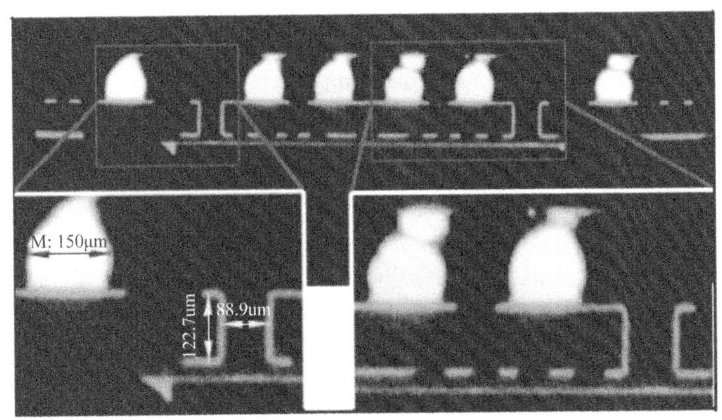

(a) Cross section image

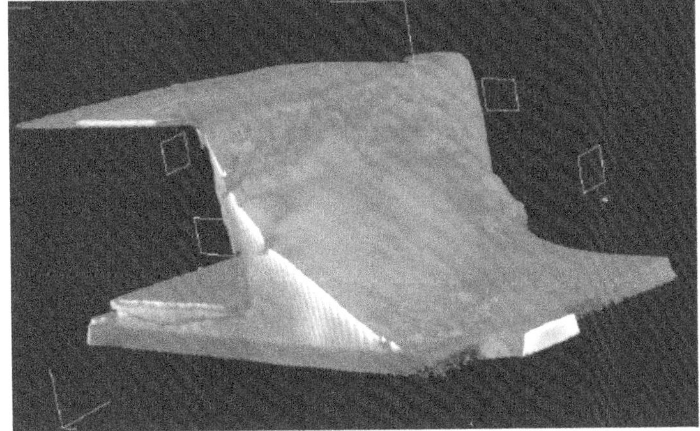

(b) Three-dimensional image of a solder joint

Figure 5.5 CT and its cross section and three-dimensional images [30].

able to detect a number of issues, such as the damage shapes of flip-chip LED lenses, and interface issues in LEDs.

5.2.3 Failure Mechanisms Analysis

(i) Deformation, Strain, and Stress

During the process of manufacturing many kinds of defects will be induced due to improper manufacturing and installment processes. During the LED chip die attach process the dalamination of the die attachment may be induced by using improper process parameters. Defects such as electrode striping, damage of chip, breakage of gold wire may be induced during the wire bonding process of LED packaging.

(ii) High Temperature and Its Induced Stresses

The diffusion velocity, concentration, and life of the current carrier as well as the forbidden gap of semiconductor LEDs are all related to temperature. Poor heat dissipation will lead to an increase in the temperature of LED packaging modules which will affect the performance of LED devices. The degradation of LED devices due to the thermal stress is listed as follows:

(1) The increase in temperature will affect the band-gap of LEDs and change the color of LED's light-emitting. This phenomenon is known as colors drift/bleaching [31]. It is found that the emission photon's wavelength of the LED module will change with electromagnetic radiation, temperature, and stress. The degree of color drift/bleaching caused by the temperature is more significant [32].
(2) The phosphor conversion technology is generally used in white light LEDs currently. With a rise in temperature, the conversion efficiency of phosphor will decrease [33]. The high temperature will change the stimulated radiation wavelength of the phosphor resulting in color drift/bleaching and thus affect the optical performance of LED modules. In addition, the high temperature will accelerate the aging of phosphor to shorten its lifetime, which will consequently lead to a decline of life of LED modules.
(3) The rise of temperature will lead to a decline in the optical transparency of the polymer lens, resulting in a decrease of luminous flux output [34]. The extremely high temperature can cause yellowing of the optical lens which is made of PC, PMMA, and silicone gel. As shown in Figure 5.6, the relative ratio between the intensity of the yellow and blue peaks decreases from 0.45 to 0.25 as a consequence of thermal stress (200 °C), thus determining a significant shift of the light emitted toward the blue. This may contribute to the decay of the conversion efficiency of the phosphor, degradation of the transmittance of the lens and uniform browning of the white epoxy material as shown in Figure 5.7 [34].
(4) The temperature has significant impact on the lifetime of LED modules. The experiments have shown that an LED chip's lifetime is reduced sharply with a rise in the junction temperature [35]. Figure 5.8 shows that the light output of LEDs decreases with time under different ambient temperatures. Figure 5.9 shows that LED lifetime is a function of junction temperature.
(5) Due to the mismatch of thermal expansion coefficients in different LED packaging materials, a lager thermal stress is likely to be induced, which will also have an

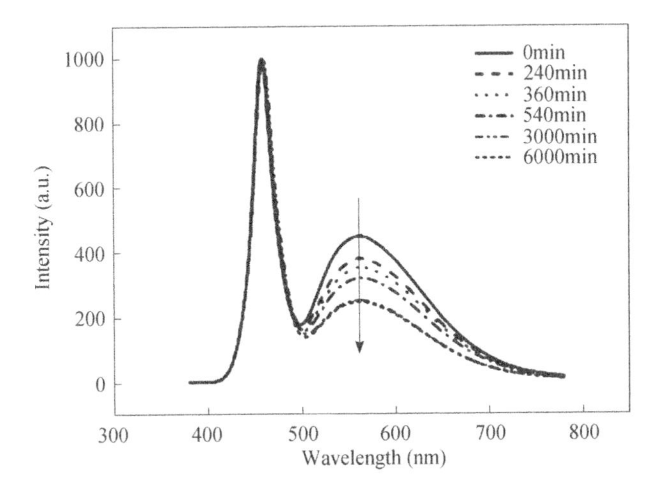

Figure 5.6 EL spectra measured during stress at 200 °C on one of the analyzed white LEDs [34]. (Reproduced with permission from M. Meneghini, L.R. Trevisanello, G. Meneghesso and E. Zanoni, "A review on the reliability of GaN-based LEDs," *IEEE Transactions on Device and Material Reliability*, **8**, 2, 323–331, 2008. © 2008 IEEE.)

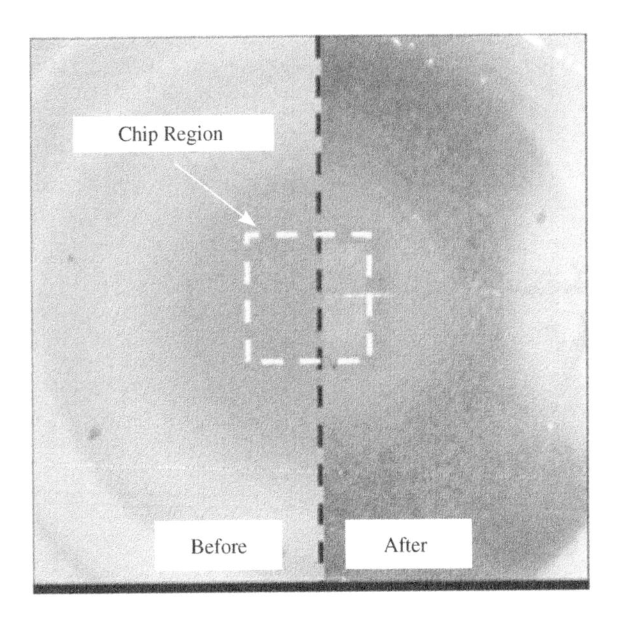

Figure 5.7 Micrograph of the emissive region of one of the analyzed samples, taken before and after stress [34]. (Reproduced with permission from M. Meneghini, L.R. Trevisanello, G. Meneghesso and E. Zanoni, "A review on the reliability of GaN-based LEDs," *IEEE Transactions on Device and Material Reliability*, **8**, 2, 323–331, 2008. © 2008 IEEE.)

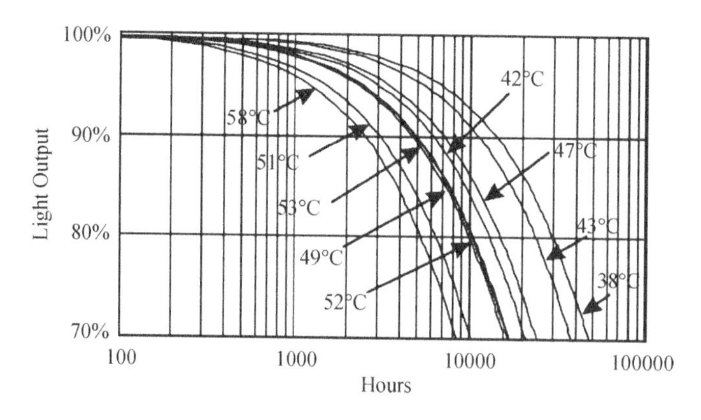

Figure 5.8 Light output as a function of time for high-power white LEDs operated at various ambient temperatures [35]. (Reproduced with permission from N. Narendran, and Y.M. Gu, "Life of LED-based white light sources," *Journal of Display Technology*, **1**, 1, 167–171, 2005. © 2005 IEEE.)

unfavorable impact on LED packaging modules. When the strain/stress induced is high, various damage modes will be caused to the devices [33], and cracking and delaminations will likely occur. Even if the thermal stress does not reach the damage criteria, it will result in deformation of the packaging lens, seriously affecting LED lighting output [36].

To sum up, the temperature and induced strain/stress have significant effects on the performance and reliability of LED modules. In order to meet the lighting demands, LED modules are often provided with high power and high packaging density, which leads to a difficult problem of LED heat removal. Therefore, effective thermal management is very important for the usage of LED modules.

(iii) Moisture and Moisture Induced Stresses (Hydrostresses)

LED packaging contains polymer materials such as silicone gel, epoxy resin, molded, or potted plastics. These polymer materials are capable of absorbing moisture due to hydrophilic groups in a lot of polymers, especially when the cure is not complete. In addition, the moisture can also penetrate into the LED packaging through some pores which are inevitably brought into the polymer materials during the preparation process. It is inevitable for the LED module to be

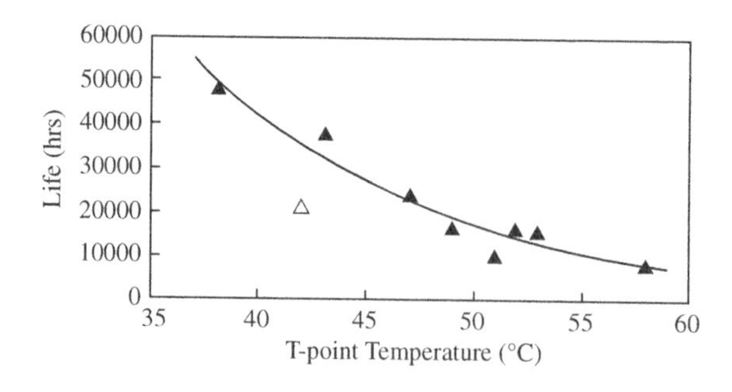

Figure 5.9 Life as a function of T-point temperature [35].

exposed to certain high humidity environments when it is in normal operation, or testing, or manufacturing. The LED road light is a typical example in such environments. The absorbed moisture will do harm to the LED module's performance, as explained below.

(1) The LED packaging module will expand after absorbing an amount of moisture. There will be great stresses due to the hydro expansion coefficient mismatch between the packaging materials. During the solder reflowing process there will be a large pressure due to the moisture evaporation under the reflow peak temperature over 200 °C which is much higher than the boiling point of water (100 °C) at standard atmosphere pressure;
(2) The moisture will reduce the bonding strength between different materials [37,38], and cracking and delamination will likely be induced in the LED packaging. Severe cracking and delamination will bring serious damage and even failure to LED devices. Cracking and delamination which appear at the interfaces between the chips and the heat sink will lead to an increase of the contact thermal resistance and thus increase the junction temperature of LED devices [17,39]. Cracking and delamination which appear at the interfaces between the lens and silicone gel or between the chip and silicone gel is harmful to the light extraction of LED packaging [17,18];
(3) Moisture penetrating into the LED packaging module will corrode the metallic materials in the packaging structure, reducing the lifetime and performance of the entire packaging structure [40]; and
(4) Moisture penetrating into the LED packaging module will affect the efficiency of phosphor, absorbing and scattering light resulting in acceleration of degradation of luminous efficiency.

(iv) Electrical Stress

Figure 5.10 shows the normalized light output degradation of LEDs under forward current and reverse-bias stress. It was found that the slow formation of point defects, which enhance non-radiative recombination and low-bias carrier tunneling [41].

The gold wire of the LED packaging will be fused due to high energy electrical transient currents. Figure 5.11 shows a typical fused-wire catastrophic failure. Figure 5.12 displays a wire bond lift-off due to electro-migration and/or electrolytic corrosion (if humidity is also involved), which has been observed after decapsulation of the LED device [40].

The failure mechanisms, failure location and acceleration factors of LEDs are shown in Table 5.3.

5.3 Rapid Reliability Evaluation

The life of a high power LED is more than 100,000 hours in theory and its service life can reach more than 50,000 hours generally. Most of the key reliability tests usually need more than one thousand hours [42–44]. For the reliability test, the test time of a few weeks to a few months is appropriate. In the long run, if a virtual reliability assessment could be done with realistic considerations of major factors, a significant saving will be achieved, which could reduce the time of screening to a very short period. In particular, the time for each new generation of LED chips can be as short as a few months and the time needed for reliability assessment should be significantly reduced. For those extremely harsh environments, many tests need to be done,

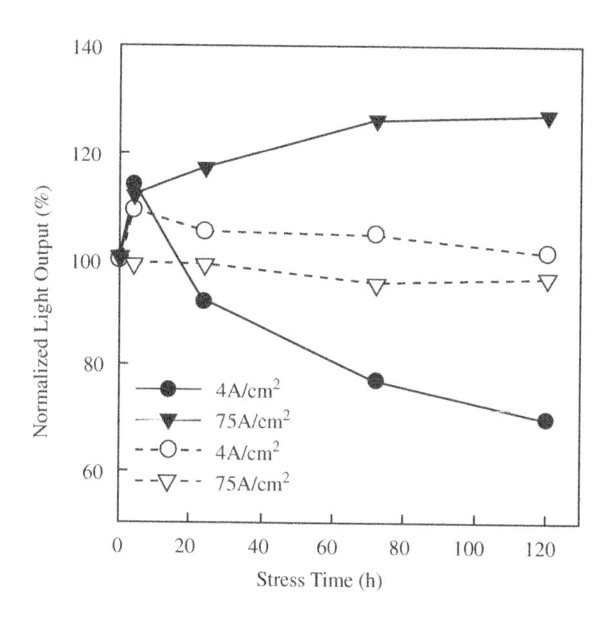

Figure 5.10 Normalized light output of the LEDs as a function of time under forward-current stress (solid lines) and reverse-bias stress (dotted lines) [41].

but most small vendors cannot afford this due to the time and funding pressure. It is therefore highly desirable to enhance our modeling capability.

As mentioned before, manufacturing induced defects can affect both the yield and the long term reliability, which has been proven to be true for both IC and MEMS packaging. It is also true for LED packaging and a coupled modeling of stress build-up during the manufacturing is also desirable.

The rapid reliability evaluation technology has been developed to simulate those traditional reliability tests, with the hope of replacing them in the future. More and more attention has been

Figure 5.11 Fused gold wire failure.

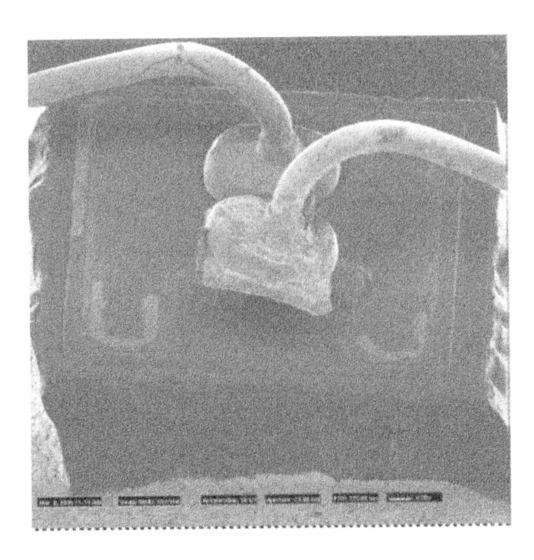

Figure 5.12 LED bond wire liftoff, including the pad metal [40]. (Reprinted from *Microelectronics and Reliability*, **46**, P. Jacob, A. Kunz and G. Nicoletti, "Reliability and wearout characterisation of LEDs," 1711–1714, 2006, with permission from Elsevier.)

paid to the accelerated life test and reliability rapid evaluation. This trend can be summarized as follows:

(1) Robust material constitutive models that can deal with dependence of time, loading rate, and history of those soft or low temperature materials;
(2) Coupled analysis and co-design modeling and simulation capabilities;
(3) Fundamental understanding of failure mechanisms and failure modes and advanced failure analysis tools;
(4) Use sensors to real-time monitor the parameters which are sensitive to the health monitoring system for modules and systems; and
(5) Integrated consideration of manufacturing processes simulation, packaging co-design advisor, and rapid reliability qualification tools, coupled with new development of validation tools. See Figure 5.13.

Table 5.3 Failure mechanisms and accelerating factors of LEDs

Failure Mechanism	Failure Location	Accelerating Factors
Electro-migration	Metallization	Current Temperature
Corrosion	Metallization	Current Temperature Humidity Salt Environment
Delamination	Bonding Interfaces	Temperature Humidity
Degradation	Phosphor and Encapsulate	Temperature Humidity
Degradation	LED Chip	Current Temperature
Disconnection	Wire Bond Electrode	Vibration

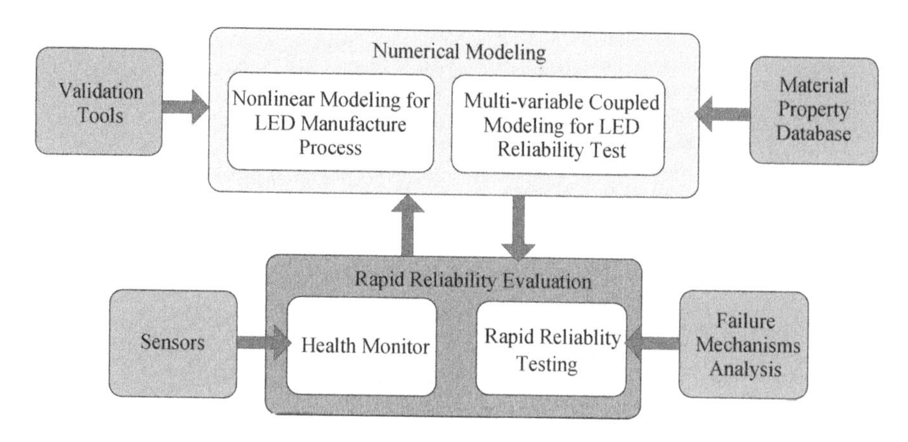

Figure 5.13 An integrated platform for coupled manufacturing, rapid reliability qualification, and packaging co-design for LED modules and systems.

5.3.1 Material Property Database

The performance and reliability of LED packaging are basically determined by materials, loading history, processes, and structure. In order to determine the aging and failure mechanisms of high power LED packaging under multiple stress conditions (such as stress, heat, humidity, current, and so on.), a material property database for LED packaging materials including thermal property parameters, optical property parameters, mechanical property parameters, constitutive equations, and interfacial strength should be established.

A database of exact material properties of LED packaging is also needed for precision numerical modeling. Fortunately a lot of data has been accumulated in the past many years [45–48]. However, we are still far from being satisfied: more testing of material properties are needed to build a robust database.

(i) Thermal Property Parameters

Generally speaking, thermal properties are a function of temperature. The basic thermal properties and measuring instruments are listed in Table 5.4. The thermal conductivity, specific heat, and CTE parameters of some LED packaging materials are listed in Table 5.5.

Table 5.4 Thermal property parameters and measuring instruments

Thermal Property Parameter	Measuring Instrument
Melting Point	DSC, DTA
Softening Point	DSC, DTA
Glass Transition Temperature	DSC, DMA, DTA, TMA
Pyrolysis Temperature	TG
Specific Heat	DSC
Thermal Conductivity	Thermal Conductivity Meters
Coefficient of Thermal Expansion	TMA
Thermophysical and Heat Transfer Properties of Phase Change	DSC, DMA, TG, DTA

Table 5.5 Thermal property parameters of typical LED packaging materials

	Thermal Conductivity (W/m °C)	Specific Heat (J/kg °C)	CTE (ppm/°C)
GaN	130-140	490	5.6
Sapphire	35-46	730	7.9
Cu	393	390	16.5
Silicone Glue	1.8	92	300
Epoxy Lens	0.2	350	45
Epoxy + Ag	2.45	250	38
Al	216	940	23
Plastic	0.24	110	24
Die Attachment	57.1 @ − 25 °C 56.2 @ 25 °C 53.3 @ 60 °C 52.9 @ 100 °C 52.8 @ 150 °C	248	27

(ii) Mechanical Property

It is critical to investigate the effects of mechanical properties on LED failures for accurate reliability assessments. The mechanical properties of materials include static mechanical properties, dynamic mechanical properties, and bi-material interfacial strength.

(1) Static and quasi-static mechanical properties
Mechanical behaviors and important mechanical parameters such as Young's modulus, Poisson's ratio, yield strength, strain-stress curves, creep or stress relaxation curves need be obtained.

Mechanical properties may be a function of one or more independent variables such as temperature, humidity, or the direction in the material. Therefore, mechanical measurements should be conducted with different strain rates under different temperature and humidity conditions. Constitutive equations can be constructed based on the test curves and are used to describe material behaviors. A simple constitutive equation for solid materials is the Hookean linear elasticity model. Plastic flow will occur at the end of elastic deformation. Deformations of solid materials may be time-dependent, temperature dependent, and history dependent and highly nonlinear. The constitutive equations of many real materials resemble some combination of elastic and plastic responses, even viscous responses. The basic mechanical properties of some LED packaging materials are listed in Table 5.6.

(2) Dynamic mechanical property
Epoxy materials including PC, transparent silicone resin, and solders in LED packaging are viscoelastic materials. The viscoelastic property cannot be measured under quasi-static conditions, because the response stresses or strains are constant or change too slowly to be detected. When the materials are subjected to dynamic mechanical loading, the sample deforms under the impact or cycle loading. By comparing the strain response lag to the applied force, the viscoelastic properties of the material can be determined. Stress analysis needs the frequency-dependent dynamic module. The viscoelastic motions with various frequencies within the whole temperature range could be written as:

Table 5.6 Young's modulus, Poisson's ratio, and the density of LED packaging materials

	Young's Modulus (GPa)	Poisson's Ratio	Density (Kg/m^3)	Temperature (°C)
GaN	210–295	0.31	6150	
Sapphire	400	0.22	3965	
Cu	128	0.26	8950	
Die Attachment	55.8	0.255	7400	−25 °C
	48.1			25 °C
	41.3			60 °C
	35.3			100 °C
	30.3			150 °C
Silicone Glue	6.1×10^{-4}	0.34	1200	20 °C
	6.4×10^{-4}	0.31		65 °C
	7.0×10^{-4}	0.28		100 °C
	7.3×10^{-4}	0.26		125 °C
Epoxy Lens	0.31	0.45	980	
Epoxy + Ag	7.6×10^{-3}	0.269	3800	
Al	69	0.33	2700	
Plastic	3.1	0.42	1300	

$$E = E' + iE'' \tag{5.59}$$

where E' is storage modulus which represents the rigidity of materials, and E'' is loss modulus.

$$\tan\delta = \frac{E'}{E''} \tag{5.60}$$

where $\tan\delta$ is the loss factor.

Typical dynamic mechanical properties as a function of both frequency and temperature for a silicone resin are shown in Figure 5.14.

(3) Interfacial strength

The mechanical integrity of many electronic devices and their components are determined by the strength of the interfaces between dissimilar materials. Therefore, a test of interfacial strength is critically important to the design for reliability of these devices. The important interfaces of the LED packaging include:

(1) GaN/Silicone
(2) Phosphor/Silicone
(3) PC/Silicone
(4) Copper/Silicone
(5) GaN/Solder
(6) Copper/Solder

Failure of these interfaces will result in thermal and optical performance degradation of LED devices. For example, the CTE mismatch between the silicone resin and other packaging materials for LED packaging is inevitable during thermal cycling. It would generate stress concentration at the interface and may lead to delamination [17,50].

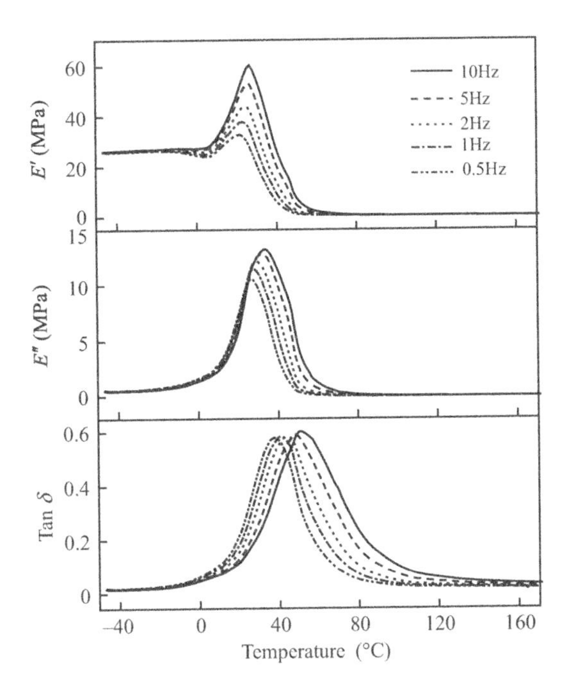

Figure 5.14 Storage modulus (E'), loss modulus (E''), and loss factor (tanδ) as functions of temperature for a silicone resin at 0.5, 1, 2, 5, and 10 Hz frequencies [49]. (Reproduced with permission from Q. Zhang, X. Mu, K. Wang *et al.*, "Dynamic mechanical properties of the transparent silicone resin for high power LED packaging," *International Conference on Electronic Packaging Technology & High Density Packaging (ICEPT)* 2008. © 2008 IEEE.)

The method based on fracture mechanics is known as damage tolerant design [51]. It assumes that some detectable cracks/flaws exist in the structure and then one can predict the probability of crack propagation during processing and operation cycles. The interfacial strength of bimaterials can be studied by experimental and finite element methods [16].

(4) Measurement equipment and methods
There are many standardized test methods to measure mechanical properties of materials, such as documents published by ASTM International. Static mechanical properties and interfacial strengths can be measured by universal testing machines. A six-axis submicron test machine for small samples (as shown in Figure 5.15) was first invented by the first author group at Wayne State University, see Lu *et al.* [52]. Dynamic mechanical properties of materials can be measured by plate impact experiments or dynamic mechanical analysis using a dynamic mechanical analyzer.

(iii) Moisture Absorption Characteristics

As mentioned previously, LED packaging materials such as die attachment, silicone, PC, and epoxy are very sensitive to moisture and the performance of LED packaging modules is severely influenced by moisture penetration. It was found that the luminous flux output of the LED packaging modules will decrease by nearly 11.05% and the optical efficiency will decrease by

Figure 5.15 Six-axis submicron tester.

approximately 10.47% under the ambient condition of 85 °C/RH 85% for 100 hours in the authors' lab. If delamination occurs, significant reduction can be generated, as described before.

The diffusivity coefficient, coefficient of moisture expansion (CME), and saturated humidity ratio of LED packaging materials should be measured. The moisture property parameters of some LED packaging materials are listed in Table 5.7 [39].

(iv) Optical Property

The optical performance of LEDs is influenced by the properties of packaging materials of transparent silicone resins and PC lens. Compared to epoxy resin, silicone has excellent light resistance characteristics. It has been found that silicone can eliminate the yellowing which occurs in LED packaging with conventional epoxy, as it encapsulates and enhances the optical extraction efficiency. The optical properties of transparent silicone and PC include transmittance, absorption rate, and refractive index, and so on. Similar to the mechanical properties, the optical properties will also be affected by environmental conditions such as temperature, moisture, and corrosive gases. Table 5.8 lists some optical properties. If aging of the material occurs, its transmittance, absorption rate, and refractive index will be changed severely.

5.3.2 Numerical Modeling and Simulation

The reliability of LED devices can be modeled with a nonlinear multi-variable coupling technology. The loading may include thermal, moisture, static or dynamic forces, and

Table 5.7 Moisture property parameters of LED packaging materials

	Moisture Diffusion Coefficient (m^2/s)	Saturated Humidity Content (kg/m^3)	Coefficient of Moisture Expansion (m^3/kg)
Plastic	1.886×10^{-12}	4.88	0.210×10^{-3}
Die Attachment	1.25×10^{-11}	6.20	0.445×10^{-3}

Table 5.8 Optical properties of LED packaging materials

	Refractive Index	Optical Absorption (/mm)	Scatter Coefficient (/mm)
Lens	1.586	–	–
Silicone	1.45–1.54	–	–
Phosphor	1.65	1.5	5
p-GaN	2.43	8	–
MQW	2.61	8	–
n-GaN	2.39	8	–
GaN	2.42	8	–
Sapphire	1.71	–	–

electrical loading. Modeling in general is a very challenging subject and the authors of this book tend to refer those colleagues in the community to another book specifically on modeling and simulation for general packaging and interconnecting including IC packaging, namely MEMS Packaging and Nano Packaging and Interconnects, co-authored by Sheng Liu and Yong Liu [53].

(i) Temperature and Thermostress Modeling

The processes of thermal stress build-up are (1) the transient thermal modeling, (2) the sequential thermal and stress coupling modeling as shown in Figure 5.16.

With a transient thermal model, it is found that it takes half an hour for the LED chip 's temperature to increase from an initial temperature of 25 °C to about 90 °C when the power is on, and it takes 20 minutes to cool down when the power is off, as shown in Figure 5.17.

It is also notable that the highest thermal mechanical stress (more than 60 MPa) is at the interface of the die attachment and Cu base due to the mismatch of CTE at the highest working temperature of 90 °C, as shown in Figure 5.18.

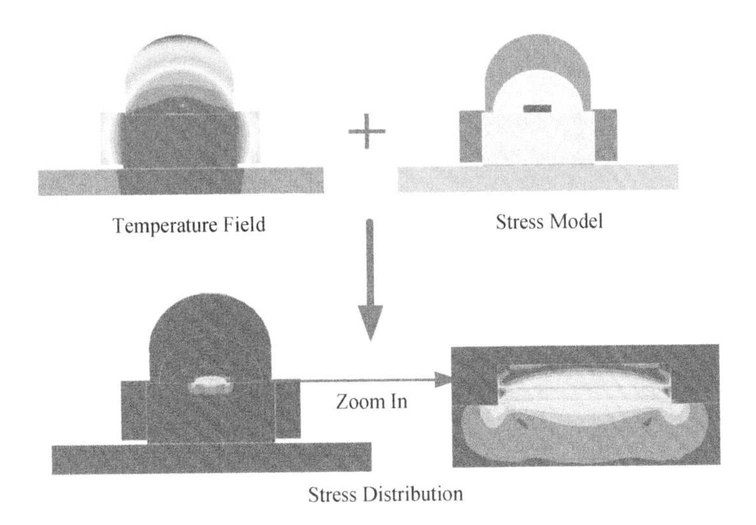

Figure 5.16 Sequential thermal stress coupling modeling.

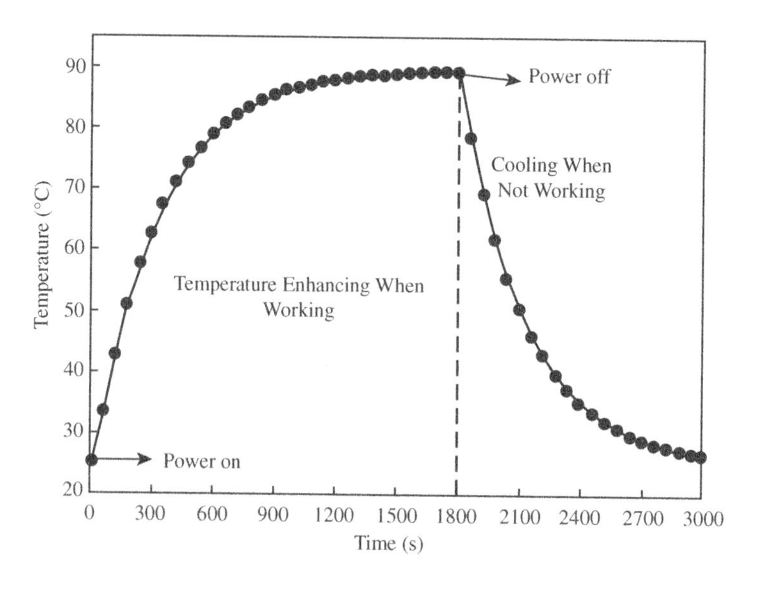

Figure 5.17 LED chip temperature profile.

Therefore, it could be inevitable to generate cracking at the interface of the die attachment and Cu base if hard die attach material is used. It is notable that the cracks could be further developed to become delaminations when there are some voids and short pre-cracks at the interface, because the energy release rate increases with the temperature when the working temperature rises and the length of delamination grows, as shown in Figure 5.19. It is noted that the delamination could be unstable as the energy release rate is bigger with a longer delamination size, as is common in fracture mechanics [54].

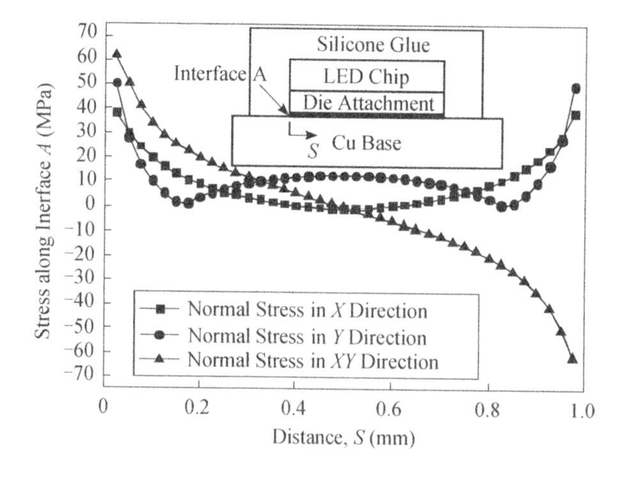

Figure 5.18 Thermo-mechanical stresses along one interface.

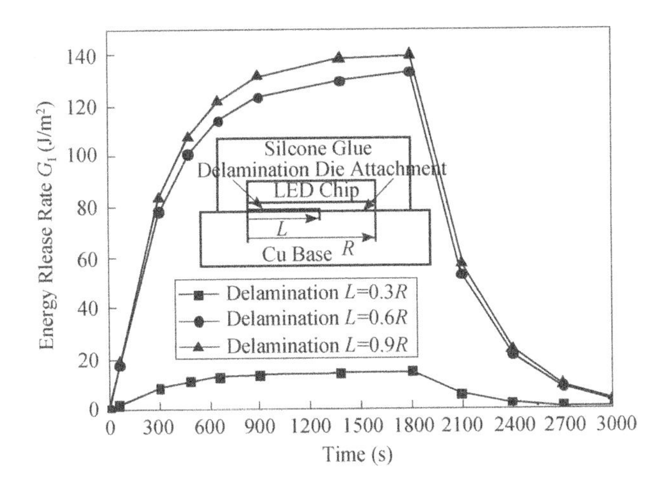

Figure 5.19 Energy release rates as a function of time for three different delamination sizes.

(ii) Modeling for Moisture and Induced Hygrostress

When LED packaging is exposed to a moisture-heavy environment as in the case of combined temperature and moisture loading, and harsh field applications, the LED packaging absorbs moisture quickly. The moisture distribution in the LED packaging with processing time is shown in Figure 5.20. It can be seen that the relative moisture concentration reaches 90% in most of the LED's components after only 200 hours in our research. Moisture penetration at the position of A, B, and C in the LED packaging is shown in Figure 5.21.

In die attachment, the hygroscopic stress evolution with exposed time is shown in Figure 5.22. After 700 hours, the maximum von Mises stress reaches 19.6 MPa at die

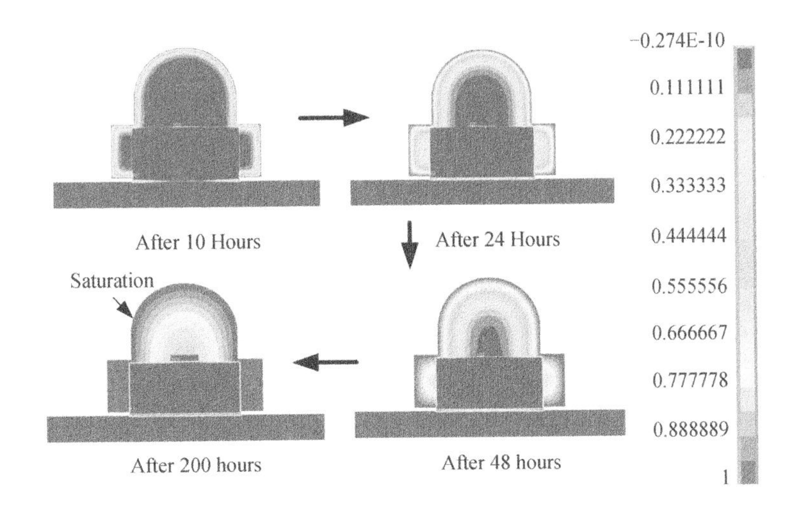

Figure 5.20 Process of moisture penetration in the LED packaging.

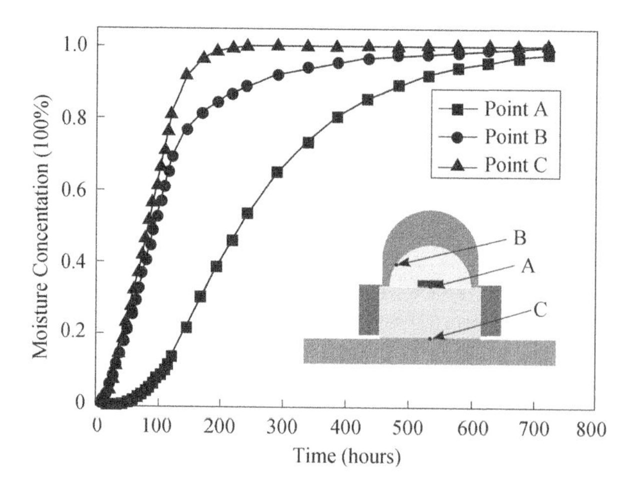

Figure 5.21 Moisture penetration in LED packaging.

attachment interface. If there is any defect such as a crack existing at the interface in general, it is possible for a short crack to develop into a long crack or delamination which is harmful to thermal management and long term reliability, as will be seen in a following section.

(iii) Effect of Defects on Thermal Resistance

In our research, it is found that the thermal resistance of the die attachment will be enhanced greatly when there is any delamination at the interface of the die attachment and Cu base, which is shown in Figure 5.23a. It can be seen from the figure that thermal resistance increases dramatically when the size of the delamination area reaches 90% of the die attachment's area. Although the moisture is harmful to LED packaging, the thermal resistance effect of

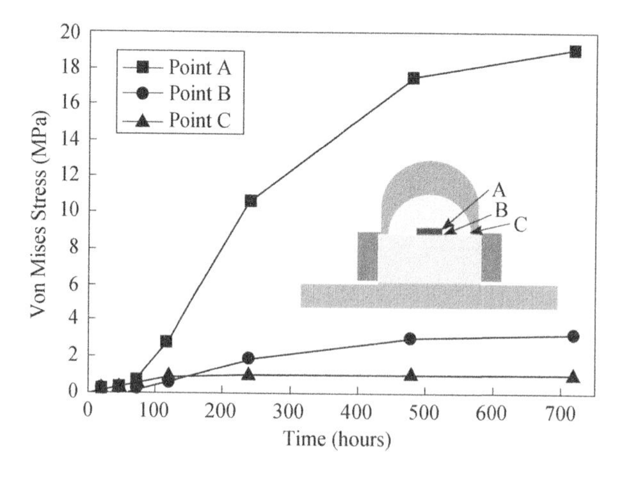

Figure 5.22 Hygroscopic stress evolution in LED packaging.

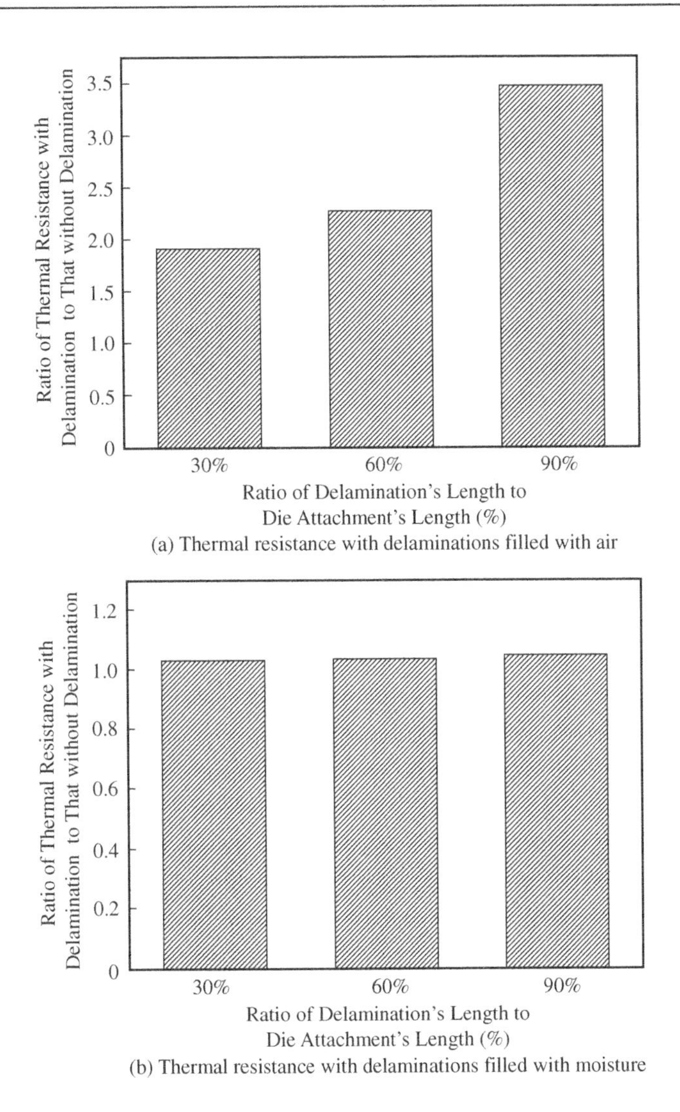

Figure 5.23 Effects of delamination on thermal resistance.

delamination filled with moisture is less than that of delamination filled with air as shown in Figure 5.23b.

Another important defect studied in our research is voids in LED packaging. During LEDs' manufacturing process, it is inevitable to generate voids in the solders or air bubbles in epoxy glue or silicones due to bad mixing or even insufficient vacuum removal of air bubbles, which are common even in a lot of commercial products. There are many kinds of voids in die attachment, but only two kinds of voids were studied in this study as shown in Figure 5.24. The effect of voids on thermal resistance is shown in Figure 5.25. From the figure, it is evident that voids affect the thermal resistance greatly, and the voids filled with air would enhance the thermal resistance by six times when the area of voids reaches 30% of the area of

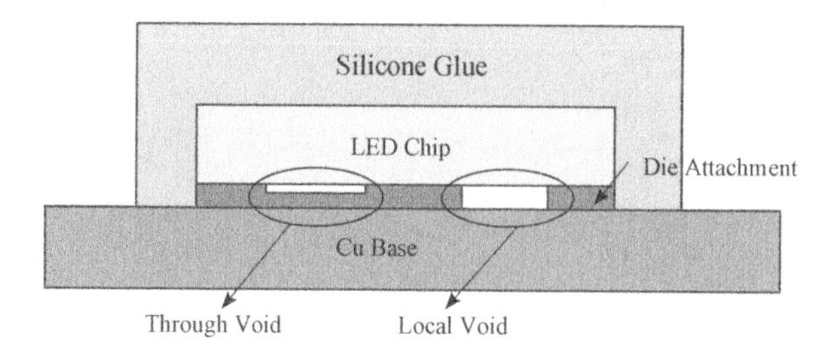

Figure 5.24 Possible voids in die attachment.

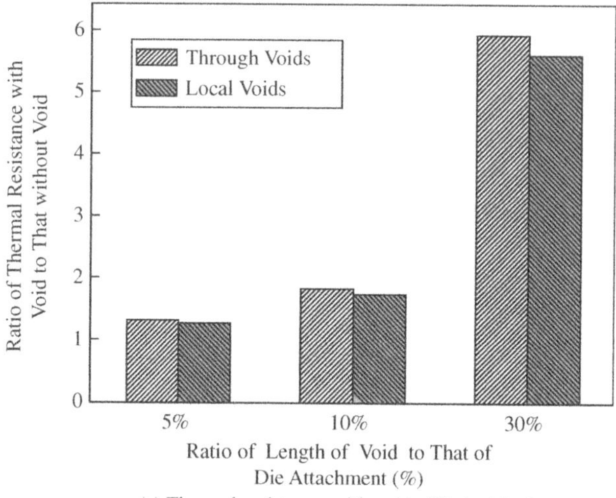

(a) Thermal resistance with voids filled with air

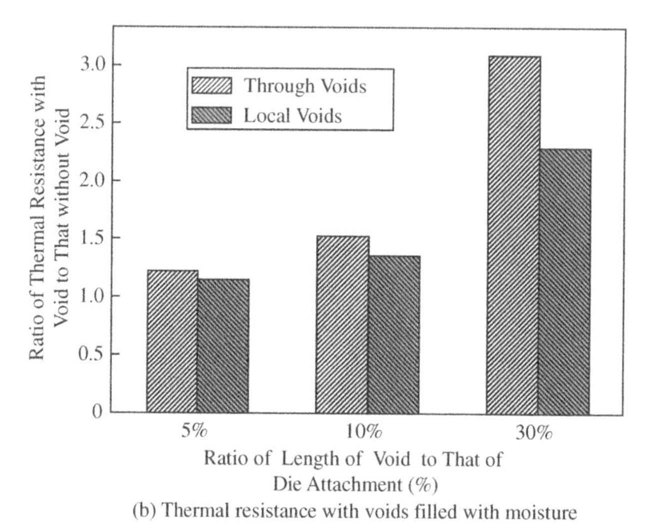

(b) Thermal resistance with voids filled with moisture

Figure 5.25 Effects of different depth voids on the thermal resistance of die attachment.

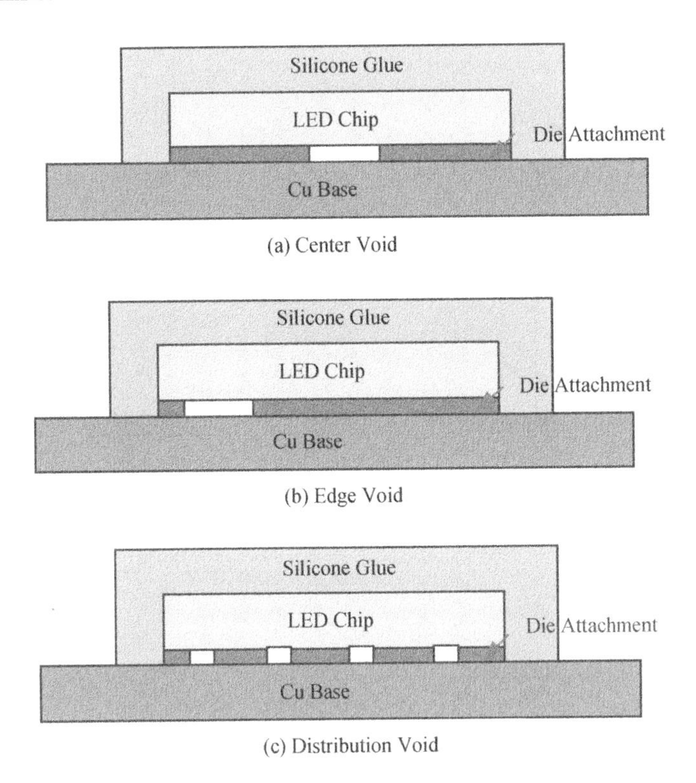

Figure 5.26 Different location voids: (a) center; (b) edge; and (c) distribution.

die attachment. It is noticeable that the effect on the thermal resistance of voids filled with air is more than that of voids filled with moisture.

In fact, the voids in the LED are also distributed in different places. But only three locations are considered in this study, as shown in Figure 5.26. The effect of voids on thermal resistance is shown in Figure 5.27. It is obvious that the thermal resistances change greatly with different locations of voids. The voids at the edge affect the thermal resistance most, and the voids at the edge which are filled with air could enhance the die attachment's thermal resistance by almost six times.

(iv) Effect of Defects on LED Light Extraction Efficiency

The effects of defects on the light extraction efficiency are studied with the Monte Carlo ray tracing method. As shown in Figure 5.28, the LED's relative light extraction efficiency decreases when the ratio of delamination length to chip size increases. However, the extent of the decrease is small, and only less than 4% of lights are lost when the ratio of delamination length to chip size reaches 65%. The slow decrease might be caused by random scattering of phosphor. The lights refracted into the phosphor are scattered by the phosphor and about 60% of the lights were scattered backwards [55], so that the light loss caused by the scattering of the phosphor originally existing at the air gap is probably similar to the loss caused by total

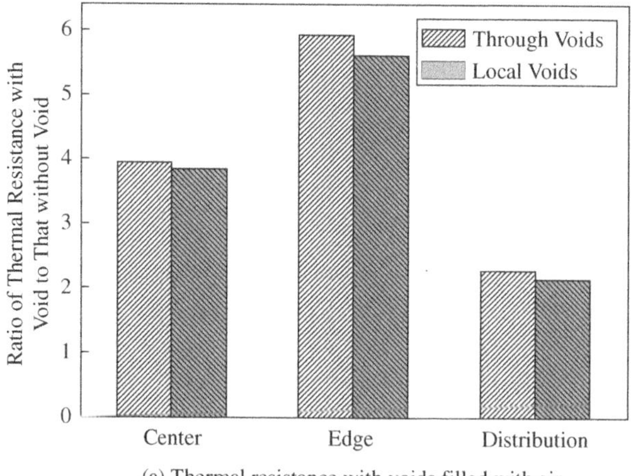

(a) Thermal resistance with voids filled with air.

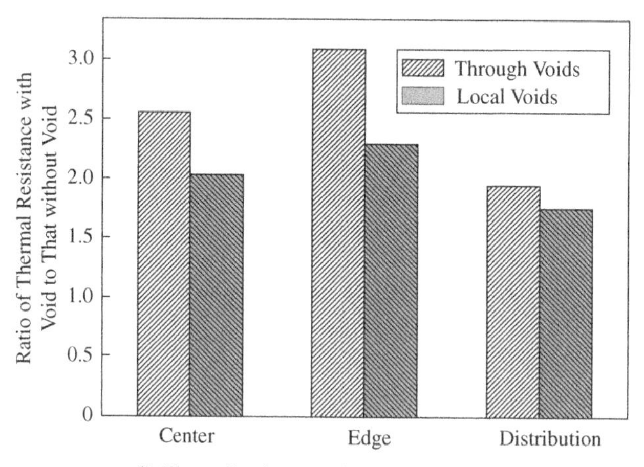

(b) Thermal resistance with voids filled with moisture.

Figure 5.27 Effects of different location voids on thermal resistance of die attachment.

reflection at the air delamination interface. It is also found that when the phosphor's refractive index increases from 1.65 to 1.8, the LED's relative light extraction efficiency decreases faster and by more than 7%, and lights are lost when the ratio of delamination length to chip size reaches 65%.

But, in another condition where delamination exists at the interface of the phosphor layer and silicone glue, the effect of delamination on light extraction efficiency is significant. As shown in Figure 5.29, the LED's relative light extraction efficiency decreases fast when the ratio of delamination length to chip length increases, and more than 28% of lights are lost when the ratio of delamination to phosphor rises to 65%. Comparing delamination filled with different media, it is found that the effect on the light extraction efficiency of delamination filled with air is more than that of delamination filled with moisture.

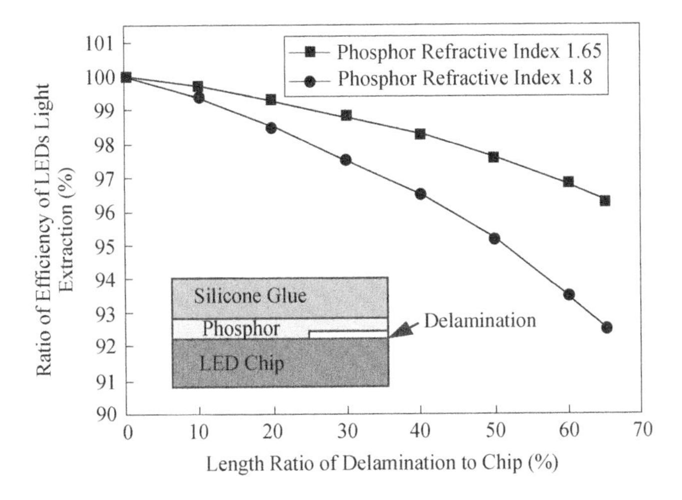

Figure 5.28 Effects of delamination on light extraction efficiency.

(v) Modeling of Manufacturing Processes: Wire Bonding Process

Wire bonding is one of the main processes of LED packaging. Impropriate bonding parameters may lead to reliability problems of the interconnect of LEDs, such as bond pad cratering, peeling, and cracking below the bond pad, especially when there are defects induced in epitaxial growth and chip manufacturing. In this section, a numerical simulation is presented to investigate the stress and strain distribution on the electrode structure of LED devices during the impact and ultrasonic vibration stages of the wire bonding process.

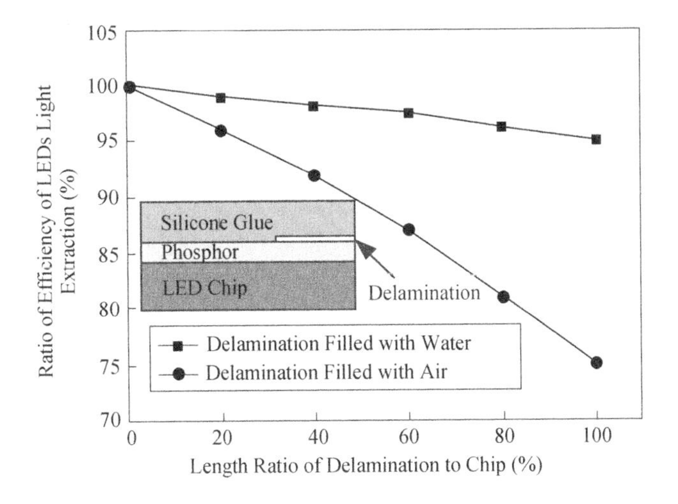

Figure 5.29 Effects of delamination on phosphor top interface on light extraction efficiency.

Parametric studies are also carried out to examine the effects of the amplitude of the ultrasonic vibration, the friction between the free air ball (FAB) and bond pad, and the tilt of the bond pad on the stress level, and on the potential of structural defects in the ohmic contact layers of wire bonding interconnection.

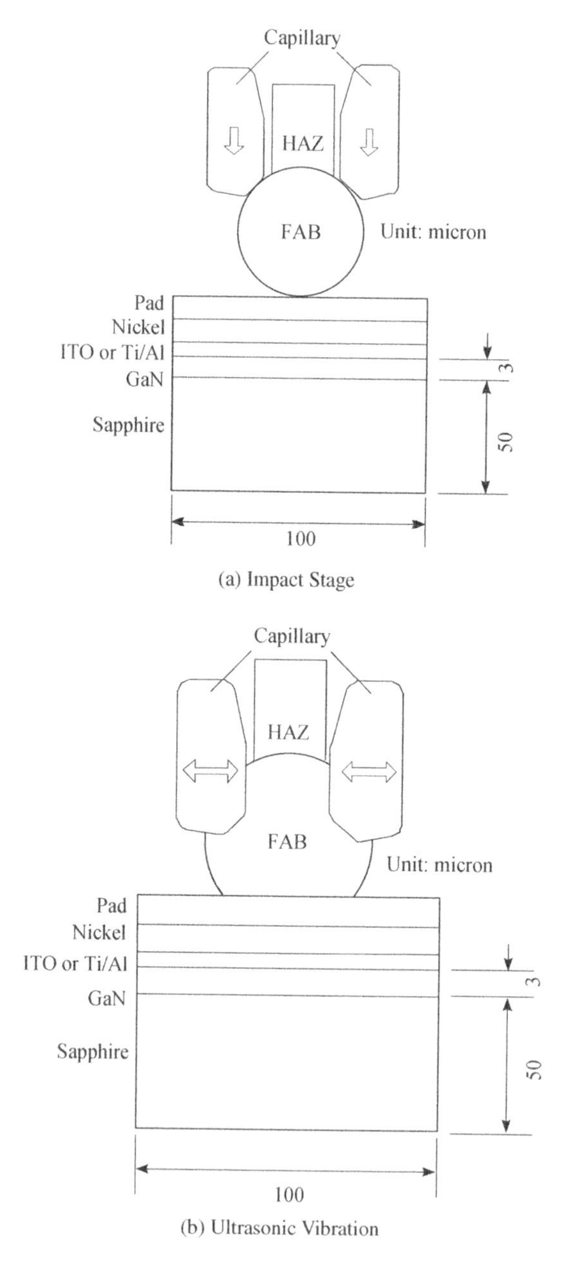

Figure 5.30 Schematic of wire bonding on LED chip: (a) impact stage, and (b) horizontal ultrasonic vibration.

The wire bonding process of high power LEDs is studied with a nonlinear finite element framework. Typically the whole wire bonding process consists of four stages: the wire tip heated to generate a FAB, the *z*-motion of capillary and FAB, the impact of the FAB with bond pad, and the input of ultrasonic wave energy. In our model, the whole wire bonding process on the bond pad of the LED was simplified to consist of the impact stage and ultrasonic vibration stage, as shown in Figure 5.30. The model involves the capillary, the gold ball, the heat affect zone (HAZ), GaN layer, and the sapphire substrate. The active layer of the LED is not considered. The thicknesses of the GaN layer and sapphire substrate are 3 μm and 50 μm respectively. The length of the structure is 100 μm.

The *p*-type electrode structure of the LED chip consists of the ITO layer as the ohmic contact layer, which is covered by the nickel layer and the gold bond pad. The thickness of ITO layer is assumed to be 0.1 μm in *p*-type electrode structure. The thicknesses of the nickel layer and gold bond pad are chosen as 0.2 μm and 1 μm respectively.

The geometry of capillary is shown in Figure 5.31. A specific set of parameters, $d_1 = 30\ \mu m$, $d_2 = 40\ \mu m$, $d_3 = 90$, $\alpha = 3°$, $\beta = 90°$, $R_1 = R_2 = 2\ \mu m$, and $R_3 = 20\ \mu m$, are applied. The diameter of gold wire used in the package of high power LEDs is 25 μm. Generally the diameter of the FAB must be 1.5–4 times bigger than that of gold wire. The diameter of the FAB is chosen to be 50 μm in present work.

The actual wire bonding process is rather complex and modeling cannot solve every detail of the bonding process. In the present study, in order to conduct an effective numerical simulation, the following assumptions are made:

(1) The thermal stress induced by the difference of thermal expansion coefficients between the different metal layers of the electrode is not considered;
(2) The heat and temperature induced by plastic deformation and friction between the FAB and bond pad is not included in this study; and

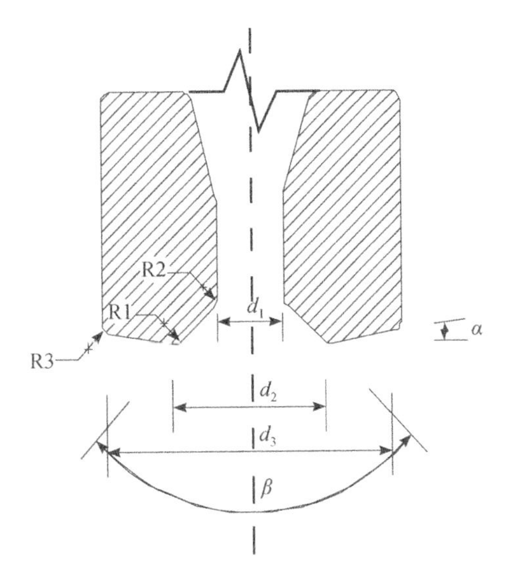

Fig. 5.31 Geometry of capillary.

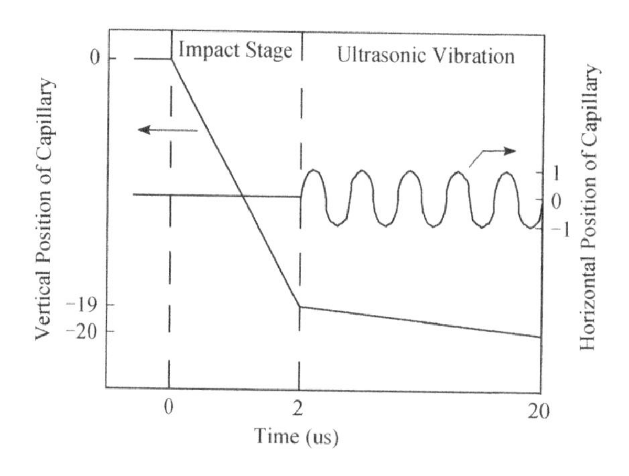

Fig. 5.32 Position of capillary during wire bonding process.

(3) The capillary is assumed to be a rigid body due to much higher Young's modulus and hardness and the inertia force is not considered [56,57].

Two-dimensional plane strain analyses were carried out for simplicity. Since the bonder capillary is considered to be a rigid body due to high hardness, this leads to the rigid and elastic-plastic contact pair between capillary and FAB. While the contact surfaces between FAB and gold bond pad are a nonlinear contact pair with consideration of the dynamic friction. The bottom of the sapphire is fixed and two sides are constrained in horizontal direction.

Figure 5.32 shows the vertical and horizon position of the capillary over the wire bonding process. Two phases are defined in Figure 5.32, the first phase includes the contact impact with strain hardening and the second phase deals with horizontal ultrasonic vibration. In the impact stage, the capillary is supposed to move along a distance of 19 μm within duration of 2 μs. In the ultrasonic vibration stage, the amplitude of horizontal movement cycle of the capillary is assumed as 1 μm while the frequency is set to be 100 kHz. During the ultrasonic vibration stage, the displacement of capillary is supposed to be 1 μm to exert a bond force to the gold ball within duration of 20 μs [56,57].

The material parameters are listed in Table 5.9. FAB, bond pad, nickel, titanium, and aluminum layers are nonlinear (bi-linear) materials; the other materials are considered to be linear elastic [56–60].

(1) Stress Distribution during Wire Bonding Process

Distributions of von Mises stress along the middle of the ohmic contact layer from the center to the right edge at different times are shown in Figure 5.33. The vertical dots lines in Figure 5.33a represent the contact edge between the FAB and bond pad.

At the impact stage, the von Mises stress is increased with time and the maximum stress occurs close to the contact edge. The maximum stresses of 211 MPa in the ITO layer occur at the end of the impact stage. While at the ultrasonic vibration stage, the von Mises stress appears periodically. The level of von Mises stresses is much higher than those in the impact stage. It can be concluded that the defects such as peeling and cracks may be more possibly induced at the ultrasonic vibration stage during the wire bonding process.

Table 5.9 Material parameters

Material	Modulus (GPa)	Poisson Ratio	CTE (ppm/°C)	Yield Strength (MPa)
ITO	160	0.335	8.6	–
Titanium	40	0.36	8.6	–
Aluminum	70	0.35	23	400
Nickel	219	0.31	13.4	620
Gold Pad	30	0.44	14.2	110
HAZ	35	0.44	14.2	135
FAB	30	0.44	14.2	110
Sapphire	400	0.22	7.9	–
GaN	295	0.31	5.6	–

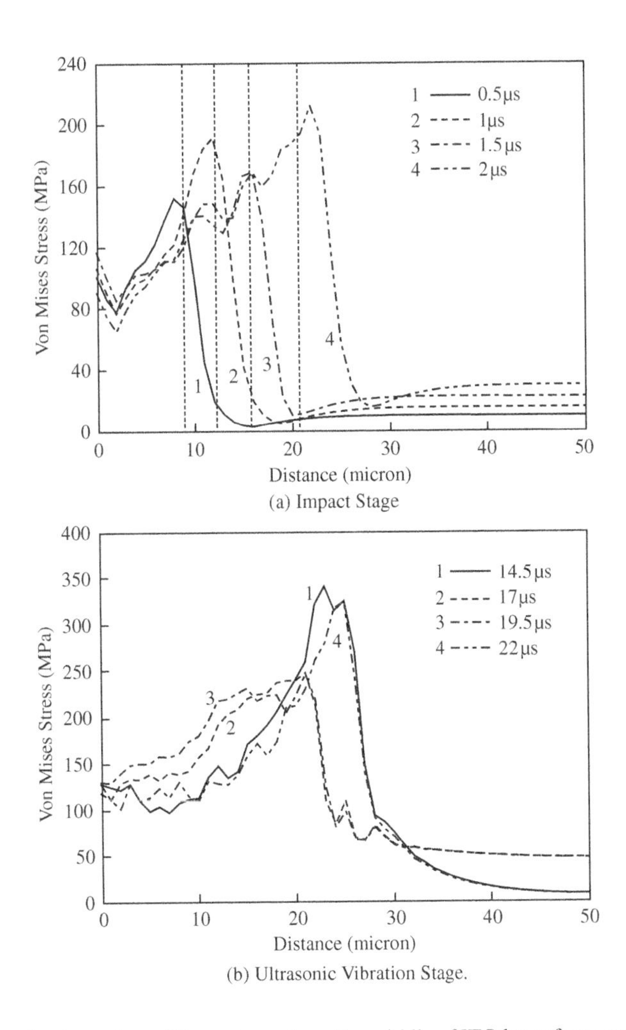

Figure 5.33 Distributions of von Mises stress along the middle of ITO layer from center to right edge at different times in *p*-type electrode structure.

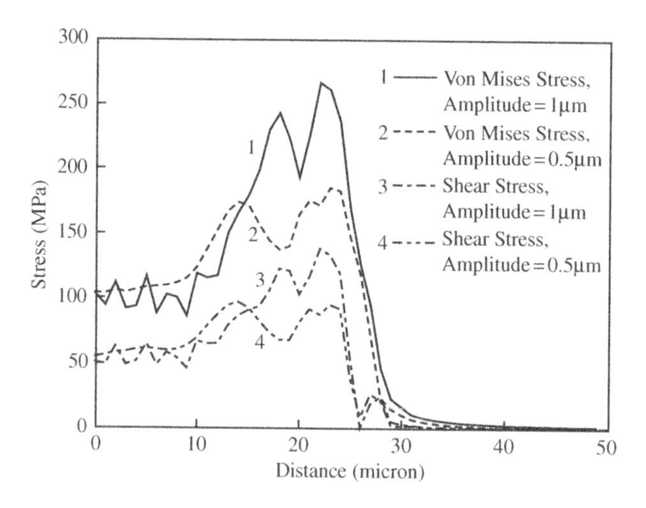

Figure 5.34 Distributions of von Mises and shear stresses in the bond pad from center to right edge with different amplitudes at the end of ultrasonic vibration stage in the *p*-type electrode structure.

(2) Effects of Amplitude of Ultrasonic Vibration

Effects of the ultrasonic vibration amplitude on the stress distributions in the bond pad and ohmic contact layers during the wire bonding process are discussed in this section. The results are listed in Figures 5.34 and 5.35.

Figure 5.34 shows that the stresses in the bond pad increase greatly when the amplitude vibration changes from 0.5 μm to 1 μm. In the *p*-type electrode structure, the maximum von Mises stress increases from 185 MPa to 266 MPa and shear stress increases from 94 MPa to 138 MPa.

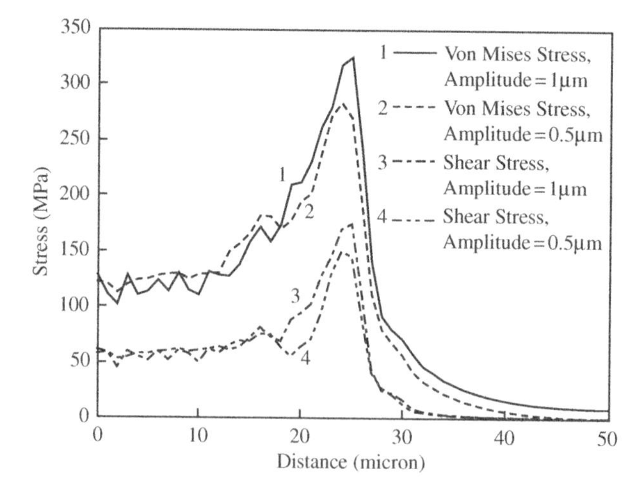

Figure 5.35 Distributions of von Mises and shear stresses in the ITO layer from center to right edge with different amplitudes at the end of ultrasonic vibration stage in the *p*-type electrode structure.

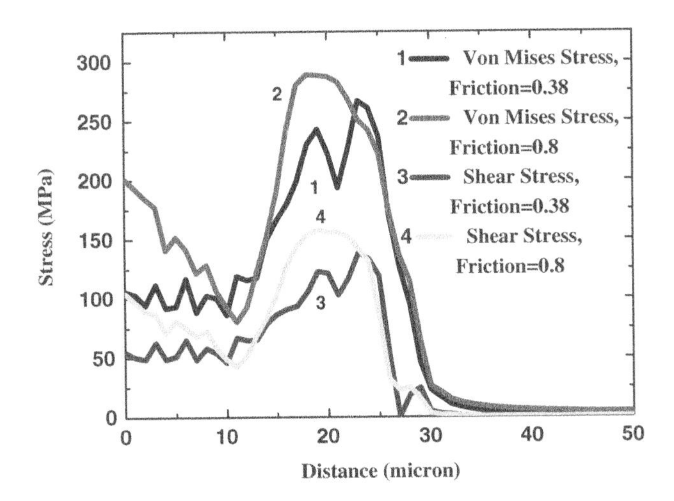

Figure 5.36 Distributions of von Mises and shear stresses in the bond pad in the *p*-type electrode structure with different friction coefficient at the end of ultrasonic vibration stage.

Figure 5.35 shows that the stresses on the ohmic contact layer also increase with the greater ultrasonic wave amplitude. In the *p*-type electrode, the maximum von Mises stress increase from 283 MPa to 324 MPa and shear stress increase from 149 MPa to 174 MPa.

The above results show that the amplitude of the ultrasonic wave has significant influence on the stresses in the bond pad and ohmic contact layers under the bond pad. It is essential to control the ultrasonic wave energy input to avoid the defects which may be induced by too high energy input during wire bonding process.

(3) Effects of Friction between FAB and Bond Pad

Friction at the interface between the FAB and bond pad is a complicated multi-physics process; bonding occurs when enough energy is available to overcome the active energy of barrier and surface oxidation [56,59]. In this study, the effects of the friction coefficient on the stress distribution were studied by using two different friction coefficients 0.38 and 0.8 in the numerical analysis. The modeling results are listed in Figure 5.36 to Figure 5.37.

From Figure 5.36, we can see that in the *p*-type electrode structure, the maximum von Mises stress in the bond pad of high friction case is 288 MPa, greater than the value of lower friction case which is 266 MPa. The maximum shear stresses in the bond pad of high and lower frication cases are 157 MPa and 138 MPa respectively.

A higher friction coefficient results in greater stress and a larger area of higher stress level on the bond pad, which may be good for the wire bonding process. However, this will result in higher stresses in the ohmic contact layer under the bond pad and cracks and delaminations may be induced there. From Figure 5.37, we can see that the maximum von Mises stress and shear stress in the ITO layer increase from 324 MPa to 336 MPa and from 174 MPa to 183 MPa respectively.

(4) Effects of Bond Pad Tilt

The tilt of the bond pad is induced during the die attach process of LED chip. A schematic of the tilt of the bond pad and LED chip is shown in Figure 5.38. The tilt angle of the LED

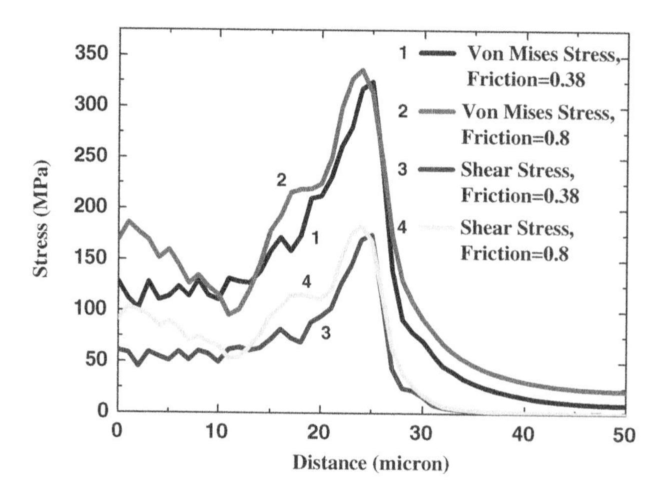

Figure 5.37 Distributions of von Mises and shear stresses in the ITO layer in the *p*-type electrode structure with different friction coefficients at the end of ultrasonic vibration stage.

chip and the bond pad is set to be 1° in this analysis. The effects of the tilt of bond pad on the stress distribution in different metal layers were studied. The maximum von Mises stresses in different layers of both types electrode structures at both impact and ultrasonic stages of wire bonding are listed in the Table 5.10. The effects of bond pad tilt are dramatic at the ultrasonic vibration stage, while at the impact stage the influences of bond pad tilt is less significant. At the end of the ultrasonic vibration stage, the maximum von Mises stresses at the right part of the different metal layers of the tilt case increase, compared to the flat case, while the values at the left part of the metal layer decrease. The tilt of the bond pad induces a more rigorous condition for the metal layer under the bond pad during the wire bonding process. Therefore, the tilt of the pad must be avoided through using more suitable process parameters in the LED chip die attach process.

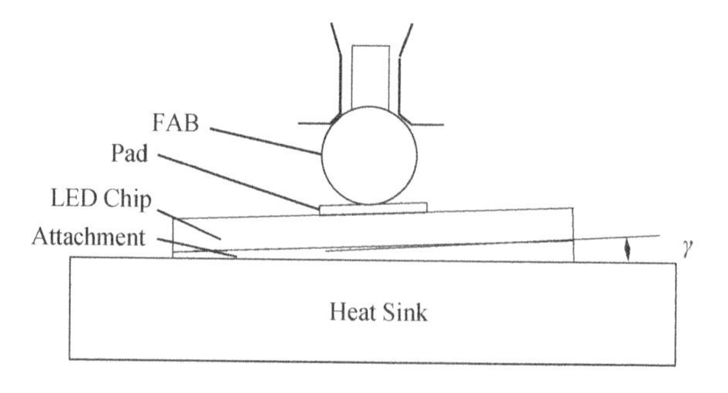

Figure 5.38 Schematic of LED chip tilt.

Table 5.10 Maximum von Mises stress on *p*-type electrode structure (MPa)

Time	Location	Flat		Tilt (1°)		Change	
		L	R	L	R	L	R
2 μs	Pad	110	110	110	110	0	0
	Ni	232	227	247	238	+ 15	+ 11
	ITO	222	220	226	236	+ 4	+ 16
22 μs	Pad	288	286	237	411	− 51	+ 125
	Ni	387	426	284	532	− 103	+ 106
	ITO	266	344	279	396	+ 13	+ 52

5.4 Summary

In this chapter, the concept of design for reliability (DfR) and reliability engineering were introduced for LED packaging. Fundamentals of reliability, life distribution, and accelerated models were first discussed. Due to the fact that physics of failure becomes essential, applied mechanics for both bulk materials and interfaces was introduced. High power LED packaging reliability tests were introduced along with various traditional testing standards, methods, and evaluation. Existing methods for failure mechanism analysis were also presented and examples of failure mechanisms/modes were provided. Finally, a model-based rapid reliability evaluation approach was introduced. A material property database was identified as being essential for reliable modeling. Detailed moisture/temperature dependent properties and rate dependent properties were the required properties for reliable modeling. Some examples were provided. In particular, a coupled manufacturing and reliability modeling may be essential to evaluate the effect of the process window on the yield and long term reliability.

References

[1] Lau, J.H. and Pao, Y.H. (1999), *Solder Joint Reliability of BGA, CSP, Flip Chip, and Fine Pitch SMT Assemblies*, New York, McGraw-Hill.

[2] Merrill, L.M. and Cyril, A.D. (1989), *Electronic Materials Hand Book Volume 1 Packaging*. ASM International.

[3] Alliance for Solid-State Illumination Systems and Technologies (ASSIST) (2005), "*ASSIST Recommends: LED Life for General Lighting*," *http://www.lrc.rpi.edu*

[4] Pecht, M. and Nash, F.R. (1994), "Predicting the reliability of electronic equipment," *Proceedings of the IEEE*. **82** (7): 992–1004.

[5] Pecht, M. and Kang, W.C. (1988), "A critique of Mil-Hdbk-217E reliability prediction methods," *IEEE Transactions on Reliability* **37**(5): 453–457.

[6] Pecht, M. (2009), "A new perspective on electronic product reliability," *Invited Talk*, Wuhan National Laboratory for Optoelectronics, China, May, 2009.

[7] Lall, P., Bhat, C., Hande, M.H., More, V.K., Vaidya, R.H. and Pandher R.J. (2008), "Interrogation of system state for damage assessment in lead-free electronics subjected to thermo-mechanical loads," *IEEE 58th Electronic Components & Technology Conference*, Lake Buena Vista, FL, USA, ECTC, 918–929.

[8] Bowles, J.B. (1992), "A survey of reliability prediction procedures for microelectronic devices," *IEEE Transactions on Reliability* **41**(1): 2–12.

[9] Dylis, D.D. and Priore, M.G. (2001), "A comprehensive reliability assessment tool for electronic systems," *Proceedings of the Annual Reliability and Maintainability Symposium*, IEEE, Philadelphia, PA, USA, RAMS, 308–313.

[10] Edson, B. and Tian, X. (2004), "A prediction based design-for-reliability tool," *Proceedings of the Annual Reliability and Maintainability Symposium*, IEEE, Los Angeles, CA, RAMS, 412–417.

[11] Leonard, C.T. and Pecht, M. (1989), "Failure prediction methodology calculations can mislead: use them wisely, not blindly," *Proceedings of the Aerospace and Electronics Conference*, IEEE, Dayton, OH, USA, NAECON, 1887–1892.

[12] Hansen, C.K. (1995), "Reliability prediction and simulation for a communications satellite fleet," *Proceedings of the Annual Reliability and Maintainability Symposium*, IEEE, Washington, DC, USA, RAMS, 152–158.

[13] Xie, J. S. and Michael, P. (2003), "Reliability prediction modeling of semiconductor light emitting device," *IEEE Transactions on Device and Materials Reliability* **3**(4): 218–222.

[14] Liu, S., Luo, X.B., Chen, H., Cao, G., Wang, X.P., Cao, W., Wang, K. and Jin, C.X. (2009), A wireless intelligent control system for high power LEDs, *China Patent Application:* 200910038193.9.

[15] Liu, S. and Mei, Y.H. (1995), "Behavior of delaminated plastic IC package subjected to encapsulation cooling, moisture absorption, and wave soldering," *IEEE Transactions on Component, Packaging, and Manufacturing Technology* **18**(3): 634–645.

[16] Liu, S., Zhu, J.S., Hu, J.M. and Pao, Y.H. (1995), "Investigation of crack propagation in ceramic/conductive epoxy/glass systems," *IEEE Transactions on Components, Packaging, and Manufacturing Technology Part A* **18** (3): 627–633.

[17] Tan, L.X., Li, J., Liu, S., Gan, Z.Y. and Wang, K. (2007), "Effect of defects on thermal and optical performance of high power LEDs," *4th China International Forum on Solid State Lighting*, Shanghai, P.R. China, China SSL, 304–309.

[18] Chen, Z.H., Li, J., Wang, K. and Liu, S. (2008), "The simulation of interfacial stress in high brightness light emitting diodes," *5th China International Forum on Solid State Lighting*, Shenzhen, P.R. China, China SSL, 349–353.

[19] Liu, S. (1992), "Damage Mechanics of Cross-ply Laminates Resulting from Transverse Concentrated Loads," *Ph.D. dissertation, Mechanical Engineering Department, Stanford University*, Aug. 1992.

[20] Liu, S., Mei, Y.H. and Wu, T.Y. (1995), "Bimaterial interfacial crack growth as a function of mode-mixity," *IEEE Transactions on Components, Packaging, and Manufacturing Technology. Part A* **18**(3): 618–626.

[21] Wang, J.J., Lu, M.F., Ren, W., Zou, D.Q. and Liu, S. (1999), "A study of the mixed-mode interfacial fracture toughness of adhesive joints using a multi-axial fatigue tester," *IEEE Transactions Electronics Packaging Manufacture Technology* **22**(2): 166–184.

[22] Suo, Z. and Hutchinson, J.W. (1989), "Sandwich test specimens for measuring interface crack toughness," *Materials Science and Engineering A* **107**: 135–143.

[23] Suo, Z. (1990), "Delamination specimens for orthotropic materials," *Journal of Applied Mechanics* **57**: 627–634.

[24] Jih, C.J. and Sun, C.T. (1992), "Evaluation of a finite element based crack closure method for calculating static and dynamic strain energy release rates," *Engineering Fracture Mechanics* **37**: 313–322.

[25] Rybicki, E.F. and Kenninen, M.F. (1977), "A finite element calculation of stress intensity factors by a modified crack closure integral," *Engineering Fracture Mechanics* **9**: 931–938.

[26] Liu, S., Kutlu, Z. and F.K. Chang (1991), "Matrix cracking and delamination in laminated polymeric composites resulting from transversely concentrated loadings," *Proceeding of the 1st International Conference on Deformation, Fracture. Composites*, 30/1–30/7

[27] Sun, C.T. and Manoharan, M.G. (1987), "Strain energy release rates of an interfacial crack between two orthotropic solids," *Proceeding of American Society for Composites*, 2nd Technical Conference, 49–57.

[28] *Cree X Lamp LED Reliability Datasheet, Cree, Inc., www.cree.com/xlamp.*

[29] *Luxeon-K2 Reliability Datasheet RD06, Philips Lumileds Lighting Company, www.luxeon.com.*

[30] Wolter, K.J., Oppermann, M. and Zerna, T. (2008), "Nano packaging a challenge for non-destructive testing," *10th Electronics Packaging Technology Conference*, Singapore, EPTC, 873–878.

[31] Monemar, B. (1974), "Fundamental energy gap of GaN from photoluminescence excitation spectra," *Physical Review* **10**(2): 676–681.

[32] Christensen, A., Nicol, D., Ferguson, I. and Graham, S. (2005), "Thermal design considerations in the packaging of GaN based light emitting diodes," *Proceedings of the SPIE*, San Diego, California, USA, SPIE, 594118.1–594118.14.

[33] Mehmet, A. and Stanton, W. (2004), "Chip scale thermal management of high brightness LED packages," *Proceedings of the SPIE*, Denver CO, SPIE, 214–223.

[34] Meneghini, M., Trevisanello, L.R., Meneghesso, G. and Zanoni, E. (2008), "A review on the reliability of GaN-based LEDs," *IEEE Transactions on Device and Material Reliability* **8**(2): 323–331.

[35] Narendran, N. and Gu, Y. M. (2005), "Life of LED-based white light sources," *IEEE/OSA Journal of Display Technology* **1**(1): 167–171.

[36] Hsu, Y.C., Lin, Y.K., Chen, M.H., Tsai, C.C., Kuang, J.H., Huang, S.B., Hu, H.L., Su, Y.I. and Cheng, W.H. (2008), "Failure mechanisms associated with lens shape of high-power LED modules in aging test," *IEEE Transactions on Electron Devices* **55**(2): 689–694.

[37] Teh, L.K., Wong, C.C. and Mhaisalkar, S. (2004), "Characterization of nonconductive adhesives for Flip-Chip interconnection," *Journal of Electronic Materials* **33**(4): 271–276.

[38] Lu, M.G., Shim, M.J. and Kim, S.W. (2001), "Effects of moisture on properties of epoxy molding compounds," *Journal of Applied Polymer Science* **81**(6): 2253–2259.

[39] Hu, J.Z., Yang, L.Q. and Shi, M.W. (2007), "Mechanism and thermal effect of delamination in light-emitting diode packages," *Microelectronics Journal* **38**: 157–163.

[40] Jacob, P., Kunz, A. and Nicoletti, G. (2006), "Reliability and wearout characterization of LEDs," *Microelectronics Reliability* **46**: 1711–1714.

[41] Cao, X.A., Sandvik, P.M., LeBoeuf, S.F. and Arthur, S.D. (2003), "Defect generation in InGaN/GaN light-emitting diodes under forward and reverse electrical stresses," *Microelectronics Reliability* **43**: 1987–1991.

[42] Chen, C.H., Tsaiand, W.L. and Tsai, M.Y. (2008), "Thermal resistance and reliability of low-cost high-power LED packages under WHTOL test," *International Conference on Electronic Materials and Packaging*, Taipei, China, EMAP, 271–276.

[43] Grillot, P.N., Krames, M.R., Zhao, H. and Teoh, S.H. (2006), "Sixty thousand hour light output reliability of AlGaInP light emitting diodes," *IEEE Transactions on Device and Materials Reliability* **6**(4): 564–574.

[44] Shen, H., Pan, J. and Feng, H. (2007), "Accelerated life test for high-power white LED based on spectroradiometric measurement," *Proceedings of SPIE*, Beijing, China, SPIE, 684104.1–684104.9.

[45] Qian, Z.F., Lu, M.F. and Liu, S. (1997), "Constitutive modeling of polymer films from viscoelasticity to viscoplasticity," *ASME Manufacturing Science and Engineering* **6**(1): 377–382.

[46] Ren, W., Qian, Z.F. and Liu, S. (1998), "Thermal-mechanical creep of two solder alloys," *IEEE 48th Electronic Components & Technology Conference*, Seattle, USA, ECTC, 1431–1437.

[47] Wang, J. J, Zou, D.Q. and Liu, S. (1997), "Evaluation of interfacial fractures toughness of a bi-material system under thermal loading conditions," *ASME Manufacturing Science and Engineering* **6**(1): 411–421.

[48] Zhu, F.L., Zhang, H.H., Guan, R.F. and Liu, S. (2007), "The effect of temperature and strain rate on the tensile properties of a $Sn_{99.3}Cu_{0.7}$ (Ni) lead-free solder alloy," *Microelectronic Engineering* **84**: 144–150.

[49] Zhang, Q., Mu, X., Wang, K., Gan, Z. Y., Luo, X. B. and Liu, S. (2008), "Dynamic mechanical properties of the transparent silicone resin for high power LED packaging," *International Conference on Electronic Packaging Technology & High Density Packaging*, Shanghai, China, ICEPT-HDP, 1–4.

[50] Karlicek, R.F., Jr., (2005), "High power LED packaging," *Conference on Lasers & Electro-Optics*, Baltimore, Maryland, CLEO, 337–339.

[51] Kanninen, M.F. and Popelar, C. H. (1985), *Advanced Fracture Mechanics*, New York, Clarendon Press.

[52] Lu, M.F., Ren, W., Liu, S. and Shangguan, D. K. (1997), "A unified multi-axial sub-micron fatigue tester with applications to electronic packaging materials," *Electronic Components and Technology Conference*. San Jose, California, USA, ECTC, 144–148.

[53] Liu, S. and Liu, Y. (2009), *Modeling and Simulation for Packaging Assembly: Manufacture, Reliability and Testing*, New York, John Wiley and Sons.

[54] Wang, J.J., Lu, M.F., Zou, D.Q. and Liu, S. (1998), "Investigation of interfacial fracture behavior of a flip-chip package under a constant concentrated load," *IEEE Transactions Electronics Packaging Manufacture Technology* **21**(1): 79–87.

[55] Narendran, N., Gu, Y., Freyssinier-Nova, J.P. and Zhu, Y. (2005), "Extracting phosphor-scattered photons to improve white LED efficiency," *Physical Status Solid (A)* **202**(6): 60–61.

[56] Yeh, C.L. and Lai, Y.S. (2006), "Comprehensive dynamic analysis of wire bonding on Cu/low-k wafers," *IEEE Transactions on Advanced Packaging* **29**(2): 264–270.

[57] Yeh, C.L. and Lai, Y.S. (2006), "Transient simulation of wire pull test on Cu/low-k wafers," *IEEE Transactions on Advanced Packaging* **29**(3): 631–638.

[58] Liu, Y., Irving, S. and Luk, T. (2004), "Thermo-sonic wire bonding process simulation and bond pad over active stress analysis," *Proceedings of 54th Electronic Components and Technology Conference*, USA, ECTC, 383–391.

[59] Liu, Y., Allen, H., Luk, T. and Irving, S. (2006), "Simulation and experimental analysis for a ball stitch on bump wire bonding process above a laminate substrate," *Proceedings of 56th Electronic Components and Technology Conference*, USA, ECTC, 1918–1923.

[60] Medvedovski, E., Alvarez, N., Yankov, O. and Olsson M. *K*. (2008), "Advanced indium-tin oxide ceramics for sputtering targets," *Ceramics International* **34**(5): 1173–1182.

6

Design of LED Packaging Applications

6.1 Optical Design

6.1.1 Introduction of Light Control

With the continuous increase in the LED's lumen efficiency and luminous flux of one single module, the LED has quickly entered diversified markets from the traditional application of indicators, decorative lighting to general lighting, such as backlighting, display, projectors, automotive lighting, road lighting, indoor lighting, landscape lighting and so on, with a rapid expansion of market share year after year (Figure 6.1). The LED has many advantages such as energy-saving, environmental friendliness, long lifetime, small size, and so on. With its cost continuously reducing, it is reasonable for us to believe that solid state lighting based on the high-power LED will replace the traditional luminaires in a growing number of applications and set the world lighting trend [1–3].

(i) Light Pattern Design of LED Lighting

Light pattern control and color control of LED luminaires are two key factors in the optical design of the products in which the LED is applied. At present, circular symmetrical lens are adopted in most high-power LED packaging and the light pattern is circular, as shown in Figure 6.2. However, rectangular or sub-rectangular or elliptical light patterns are demanded on many LED lighting occasions, such as backlighting, projectors, automotive lighting, road lighting, and so on. If the circular light pattern is directly adopted in lighting, the ideal lighting performance cannot be guaranteed. A comparison of road lighting performance among different LEDs is shown in Figure 6.3, in which Figure 6.3a is the LED road lighting performance with a circular light pattern, while Figure 6.3b is the LED road lighting performance with a rectangular light pattern. It is very simple even at the first sight to recognize which is better.

At the same time, the light distribution curve of most high-power LEDs is a Lambertain curve; there are also other types of light distribution curves, such as batwing curve, side-emitting curve,

LED Packaging for Lighting Applications: Design, Manufacturing and Testing, First Edition. Sheng Liu and Xiaobing Luo.
© 2011 Chemical Industry Press. All rights reserved. Published 2011 by John Wiley & Sons (Asia) Pte Ltd.

Figure 6.1 Applications of high power LEDs. *(Color version of this figure is available online.)*

and so on, as presented in Figure 6.2. An LED with Lambertain light distribution is provided with a large beam divergence angle and nonuniform illumination distribution.

Therefore, it is difficult for some lamps (such as the MR16 lamp, down lamp, auto headlights, and so on) to meet the demands of light beam convergence and light pattern uniformity. Therefore, LED luminaires are usually demanded to re-distribute the spatial distribution of LED's light energy through the secondary optical system with a particular design and to accurately control the light pattern shape and uniformity of luminaires, thus achieving high-quality lighting.

There are three main methods for the secondary optical system to control LED's light-emitting: reflection, refraction, and scattering, as shown in Figure 6.4. The following is an introduction to these basic optical concepts.

(1) Law of reflection

The incident ray, reflection ray, and projection point normal are in the same plane; the absolute values of the incident angle and reflection angle are equal but with an opposite sign; that means that the incident ray and the reflection ray are at each side of the normal respectively. The law of reflection can be expressed in the following equation as:

$$I = -I'' \tag{6.1}$$

For the rough interface, when a beam of parallel incident light projects on it, the reflected light will no longer be parallel, resulting in an irregular diffuse reflection. However, as to any tiny reflecting surface on the rough surface, it still complies with the law of reflection.

(2) Law of refraction (Snell's Law)

The incident ray, refraction ray, and projection point normal are in the same plane; the ratio of the incident angle's sinusoid to the reflected angle's sinusoid is not related to the size of the incident angle, but related to the nature of the two mediums. As to the light, with a certain wavelength, this ratio is a constant under certain temperature and pressure

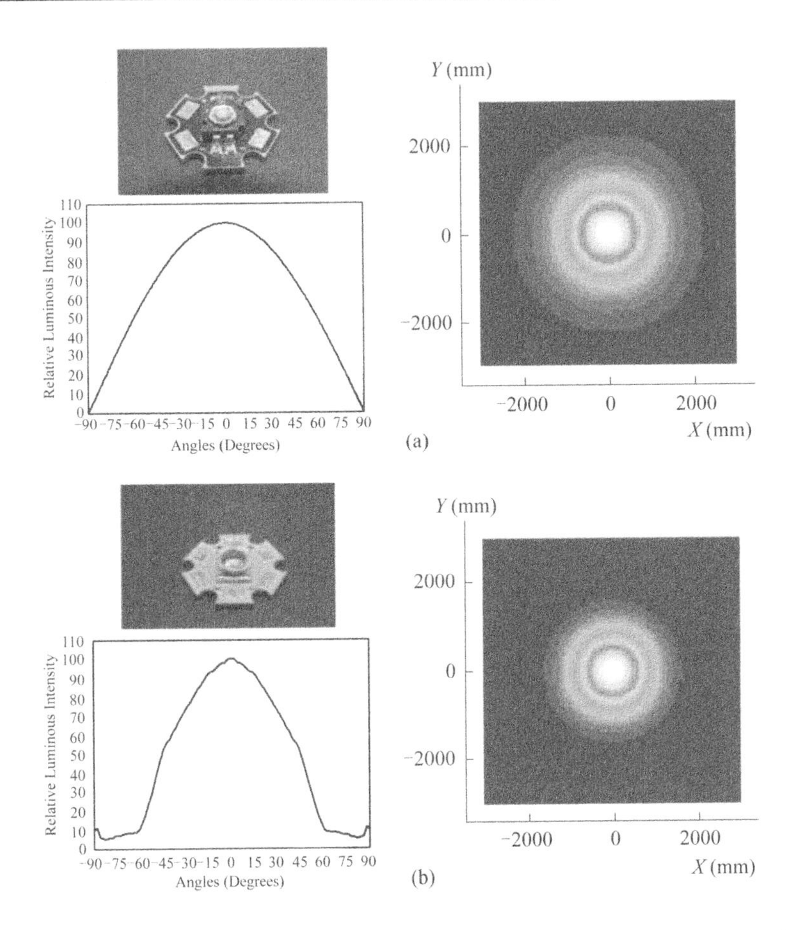

Figure 6.2 Luminous intensity distributions curves (radiation patterns) and light patterns (at 1 meter away) of several typical high power LEDs: (a) Lumileds Lambert; (b) Cree–XLamp.

Figure 6.3 A comparison of road lighting performances of two LED road lights with circular light pattern and rectangular pattern respectively. *(Color version of this figure is available online.)*

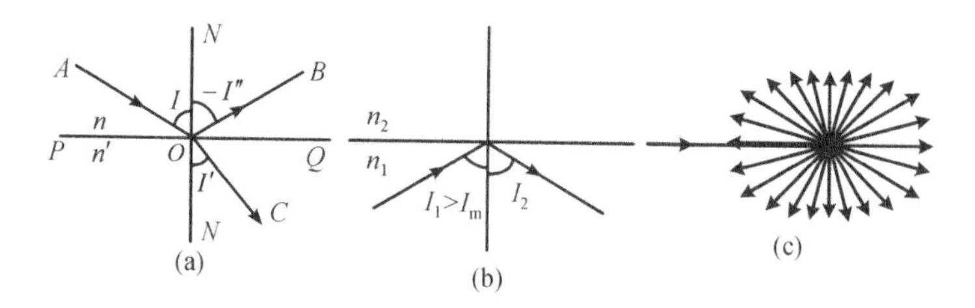

Figure 6.4 Schematic of (a) law of reflection and refraction; (b) total internal reflection; and (c) scattering.

conditions; it is equivalent to the ratio of the refractive index of the medium in which the refractive ray exists (n') to the refractive index of the medium in which the incident ray exists (n); it can be expressed in the following equation as:

$$n \sin I = n' \sin I' \tag{6.2}$$

(3) Total internal reflection
When light is emitting from the optically denser medium toward the optically thinner medium ($n_2 < n_1$), according to the law of refraction, there exists the following equation:

$$\frac{\sin I_1}{\sin I_2} = \frac{n_2}{n_1} < 1. \text{ If } \sin I_1 > \frac{n_2}{n_1}, \ \sin I_2 > 1,$$

which of course is meaningless. In fact, there is no refraction ray at this time; the incident rays are reflected back to the optically denser medium. This phenomenon is known as the total internal reflection (TIR). The incident angle which complies with the formula, $\sin I_m = \frac{n_2}{n_1}$ is called the critical angle, and the corresponding angle of refraction is $I_2 = 90°$.

(4) Scattering
When the incident ray is entering a medium, the inhomogeneity of the medium (that is, when the suspended particles of different refractive indices exist in the medium) will lead to the light radiating in all directions. Even if we are not facing the direction of the incident light, we can also see the light clearly; this phenomenon is known as the scattering of light.

The reflection control of the LED beam in LED lighting devices is generally achieved through the reflectors. The reflector mainly plays the role of convergence of LED beams and strengthens their directivity, such as in flashlights, down lamp, and so on. With the three-dimensional spatial distribution of various reflectors, large-scale lighting can be achieved through angle combination, such as road lights. In addition, the specially designed reflectors with a shaped surface (freeform surface for instance) can be provided with a convergent non-circular symmetrical light pattern, such as in the case of low-beam headlights.

The refraction control of the LED beam in LED lighting devices is achieved by the lens. The lens can make LED beams converge and also make them diverge and the control is flexible. Especially in recent years, with the emergence of LED freeform surface lens technology, the light extraction efficiency of the lens has been further improved and the control of light energy is more accurate. Therefore, it is applied on more and more occasions such as the MR16 lamp, road light, and backlighting, and so on.

Among LED luminaires, the most widely used scattering component is the diffuser plate. The diffuser plate contains a lot of diffuse particles or diffuse structures, which can well scatter the incident rays, thus achieving uniform illumination on the diffuser plate surface. It is widely used in LED lighting devices such as backlighting, daylight lamp, and so on.

(ii) Color Design and Control of LED Lighting

Color control is another key factor in the optical design of the products in which high-power LED is applied. In the past ten years, the concept of "Green Lighting" has been gradually recognized. In this version, lighting products with high efficiency, long lifetime, safety, and stable performance can be utilized by a science-based lighting design so as to improve people's working, study, and living condition and quality, thus creating a highly efficient, comfortable, safe, economic, and favorable environment and achieving the lighting which embodies modern civilization. People are no longer satisfied with brightness alone, they pay more attention to the influence of color, color temperature, and color rendering, and so on, on people's physiology and psychology, by using different luminaires according to different occasions, to meeting people's demands and to improving the customer's lighting experience.

For example, in road lighting, when the light source has met certain criteria, the color temperature is neutral but inclining to warm white which makes the pedestrian feel comfortable; and it does not demand a high color rendering of the light source, with color rendering index (CRI) of only about 50. In indoor lighting such as the office and classroom lighting, the color temperature of the light source should be neutral but inclining to cool white in order to improve people's working and study efficiency, with the CRI of about 75. As to lighting with a high demand of color such as backlight source, the CRI should be over 90.

The solid state lighting represented by LED technology is the best carrier of Green Lighting. It is not only because that compared with traditional luminaires, LED luminaires are provided with some advantages such as high efficiency, long lifetime, no environmental hazard and so on, but also because it is provided with some incomparable advantages such as rich colors and high controllability (Figure 6.5).

Figure 6.5 Colorful LEDs. *(Color version of this figure is available online.)*

The following is a detailed description of various methods of LED luminaires' light pattern control such as reflectors, TIR lens, freeform surface lens, diffuser plate, and so on. Corresponding practical design examples will also be given in order to help the readers get a better understanding of the methods of LED light energy distribution control. At the same time, the color design and control of LED lighting products will also be introduced, including the multi-color mixing, the color temperature choosing, and so on.

6.1.2 Reflectors

The bottom diameter of most reflectors is smaller than that of the top diameter which enables the incident rays convergence (Figure 6.6). LED's light distribution curves are mostly a Lambertain curve, with its half intensity angle of 60°. Therefore, reflectors are suitable for designing the convergent beam in which its half intensity angle is less than 60°. The design of divergent LED lighting systems based on reflectors can be achieved by changing the LED light source with a large diverging angle (such as batwing light distribution) or the three-dimensional organization of more than one reflector. At present, the materials of reflectors used in the LED mainly include metals and plastics. The reflector's interior surface is coated with a metallic film (such as silver, aluminum, nickel, stannum, and so on) to form a reflecting surface; the reflectivity of the smooth reflecting surface is generally over 85%; some specially coated ones can even be over 95%; the reflectivity of specially roughed reflecting surface is generally between 75–85% (Figure 6.7).

(i) LED Collimation Reflectors

Collimation reflectors are mostly used on the occasions where convergent beam lighting is demanded, such as flashlight, search light, and so on. The interior surface of collimation reflectors are usually a paraboloid which is formed by a parabola turning 180° around the symmetric axis. According to the paraboloid's mathematical properties, the light emitting from the focus will radiate in the direction parallel to the symmetric axis after it is reflected by the paraboloid. Therefore, by placing the LED light source at the paraboloid's focus, the collimation beam can be obtained theoretically, as shown in Figure 6.8.

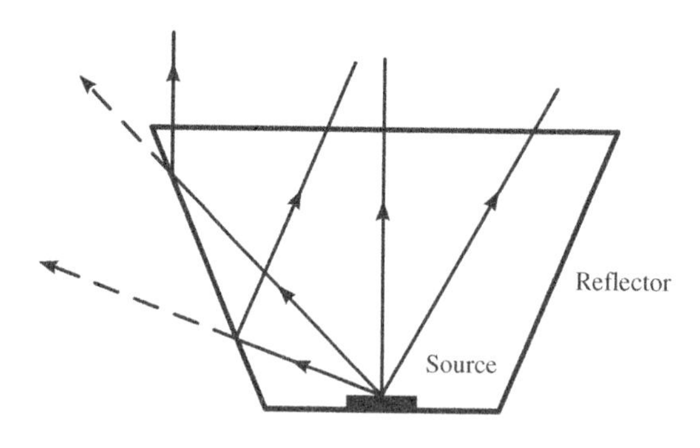

Figure 6.6 Schematic of a reflector.

Figure 6.7 LED reflectors in the market.

However, in practical applications, the beam with a small diverging angle which radiates from the LED with a Lambertain light distribution cannot incident at the reflector's interior surface limited by the height of the reflectors, but can exit out directly, making this part of light keep the original exit direction without collimation. The lower the reflector's height is, the more light exits directly. Therefore, as the design of obtaining the collimation beam through reflectors is limited in practice by the height of the reflector, it is not suitable for some occasions where the optical system is highly limited. According to the far-field condition, when the proportion of the reflector's height to the LED's edge length is less than 10, the LED light source is no longer regarded as a point light source, but an extended source, for example, 1 mm × 1 mm. The LED chip's edge is no longer at the paraboloid's focus, and the light emitting from this edge will be off the collimation direction after it is reflected by reflector. The collimation effect of reflectors is limited by its own height and the size of the LED light source, but for the lighting application which does not demand a high collimation (such as flashlight, spot light, down lamp, and so on), it is still an effective method.

(ii) Reflectors for Traditional LED Road Lighting

Currently, road lighting is a major application area of high power LEDs. The optical system of LED road lighting can be achieved through reflectors or lens. This section will introduce the

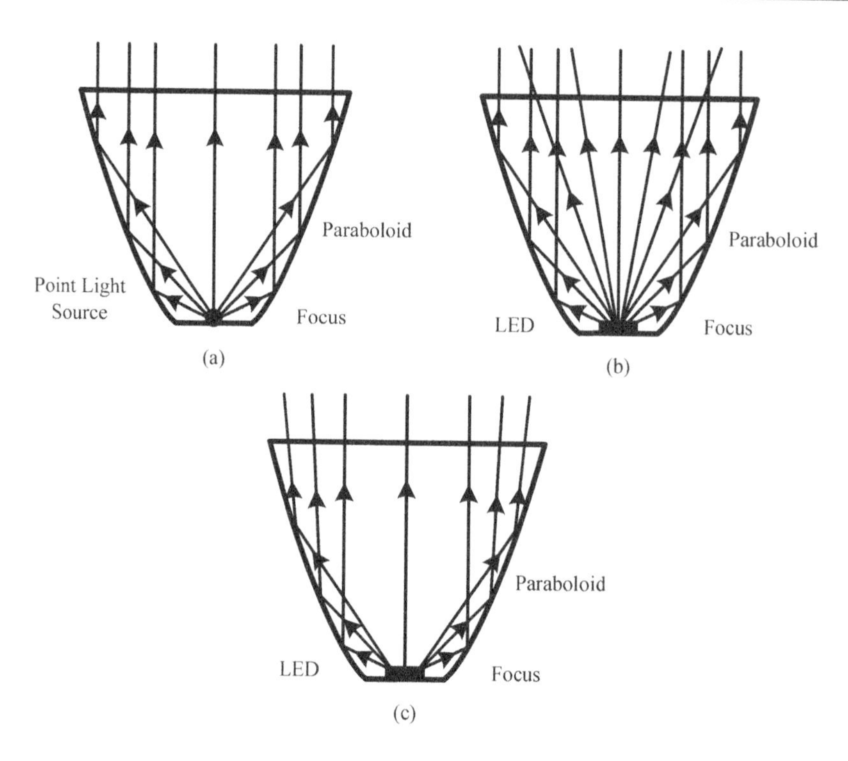

Figure 6.8 Schematic of collimation reflectors. (a) Collimation of paraboloid; (b) exit lights from LED with small diverging angles are out of control; and (c) effects of extended light source on the collimation of lights.

design based on reflectors and the design of lens used in LED road lighting will be introduced in the next section.

At present, the road light posts in urban collector roads and major roads are mainly 8 m, 10 m, and 12 m high and the ratio of road light space to the height is about three. Figure 6.9 presents a typical two-way four-lane collector road, with road lights on each side; the light post is 10 m high and the space between two lights is 30 m; the road is 20 m wide in total, including four motorways and two cycle ways. The most common 1W high-power LED packaging module based on hemispherical lens is set as the light source to design. Its light distribution curve is a Lambertain curve as shown in Figure 6.10a. The light energy distribution of this LED on the road when it is 10 m high is shown in Figure 6.10a; it is a circular light pattern with a diameter of approximately 22 m and the illumination distribution is nonuniform. Making a reference to the oblong road with single side 30 m × 10 m, this light pattern is not long enough in the direction along the road, easily resulting in nonuniform illumination. But at the same time it is too wide in the vertical direction of the road, causing light waste. Therefore, a Lambertain curve cannot be directly used in road lighting.

Turning the LED's installation angle can change its irradiation direction. The combination of LEDs with different projection angles can increase the light pattern's length, so that it is suitable for road lighting. However, as shown in Figure 6.10b, the direct three-dimensional

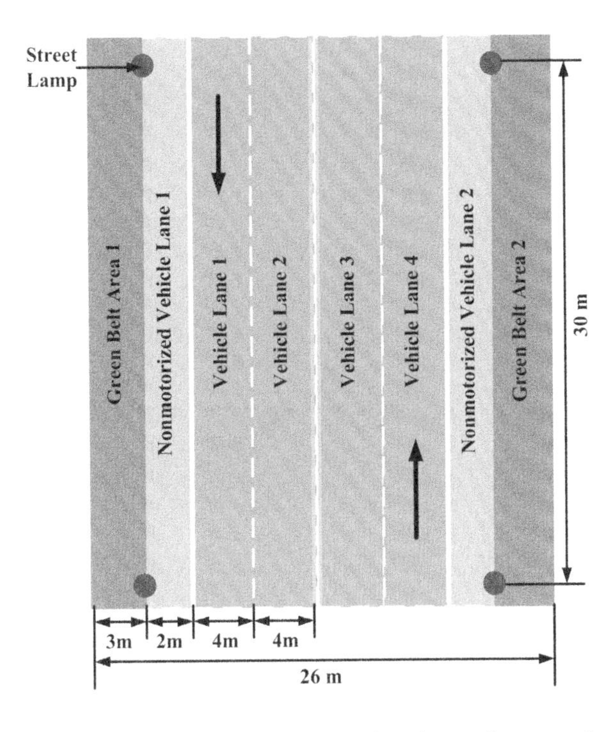

Figure 6.9 Schematic of two-way four-lane collector road.

arrangement of LED packaging modules will easily result in glare, light pollution, and great light waste in the vertical direction of the road. Therefore, it is necessary to control the light emerging angle of the LED through the reflectors.

The combined design method based on different high-power LED modules and the three-dimensional arrangement of reflectors is as follows: firstly, combine the LED modules which are with different special light distributions with different reflectors, adopt Monte Carlo ray tracing method to observe LED module groups' light energy distribution on the road after it is combined with reflectors, and record the modules whose light pattern's size and light energy distribution are reasonable. Secondly, choose different LED module groups according to different road designs; optimize the three-dimensional arrangement to allow every LED module group to control a lighting area on the road. Thirdly, simulate the combination of more than one LED module group, observe the overall road lighting performance, compare it with the indexes in road lighting standards, optimize the design, and eventually obtain the ideal LED road lighting design. The advantage of this design is its simplicity and by optimization, it can fulfill road lighting on different roads and with different lighting qualities.

Figure 6.11a is a reflector designed to be used in road lighting. The bottom diameter is 7 mm and the height is 10 mm; the sidewall is a straight wall and its inclination angle is 58°; the reflectivity of the interior surface is 85%. This reflector plays a significant role in controlling the LED beam and when it is 10 m high, the light pattern's diameter decreases to 7 m.

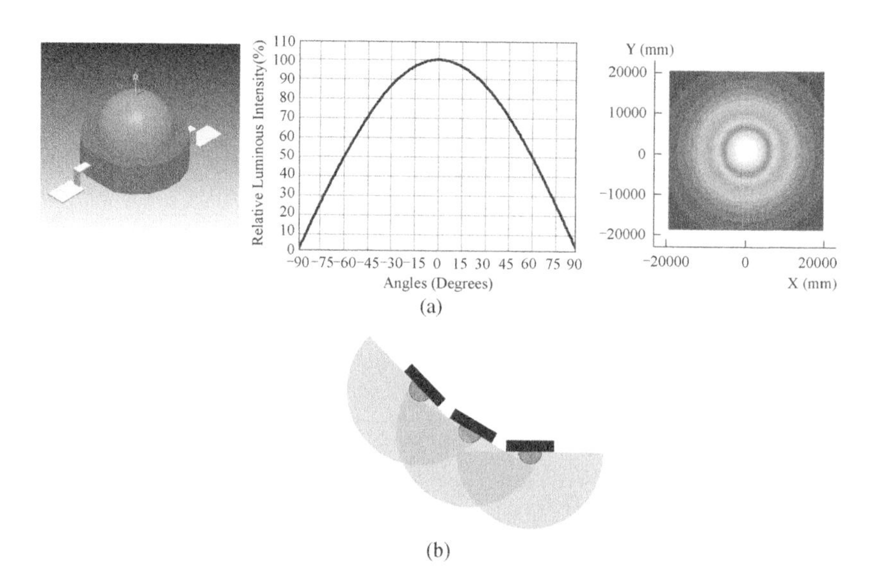

Figure 6.10 (a) Lambertian LED packaging module and its illumination performance at 10 meters high; and (b) the schematic of multi projection angles' three-dimensional combination of LED without reflectors.

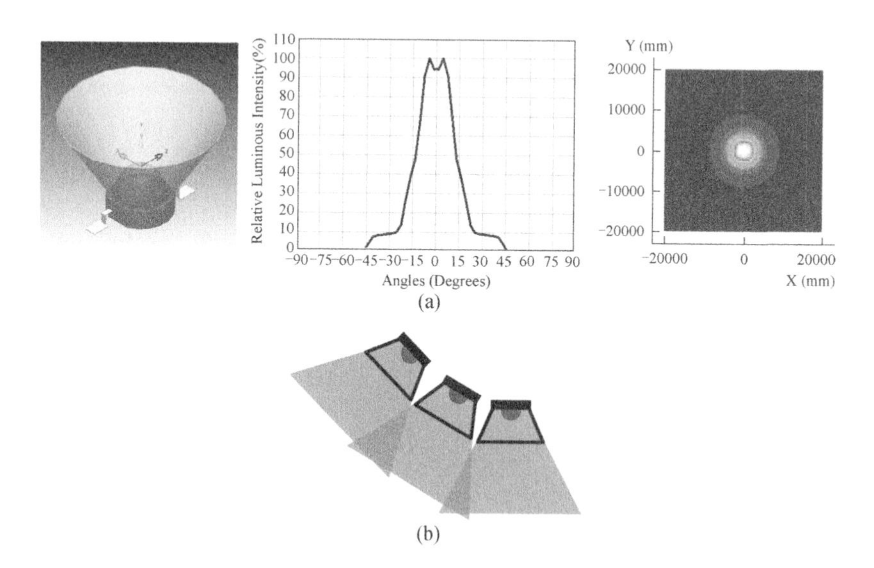

Figure 6.11 (a) Lambert LED packaging module with a reflector and its illumination performance at 10 meters high; and (b) the schematic of multi projection angles' three-dimensional combination of LED with reflectors.

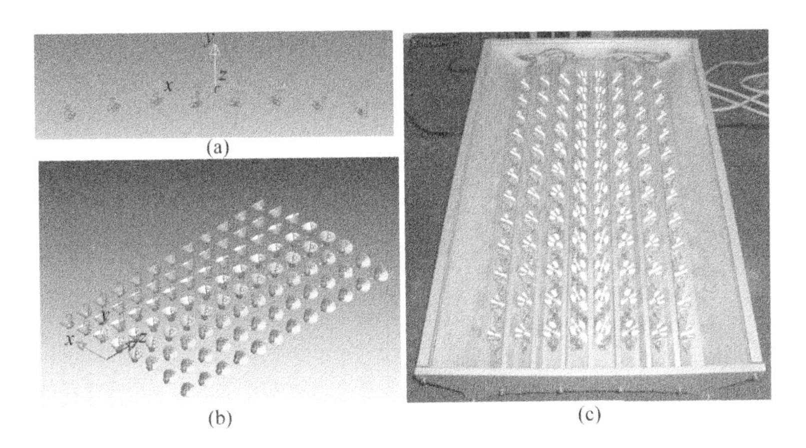

Figure 6.12 (a) Schematic of inclination angles of 8 LED module groups; (b) LED array for an LED road light consisting of 96 LEDs; and (c) the 96W LED road light.

A three-dimensional arrangement of the LED with reflectors is made to meet the demands that road lighting approximates a rectangular light pattern; it is divided into eight groups and their inclination angles are respectively ±10°, ±30°, ±40°, and ±45°, as shown in Figure 6.12. In order to meet the demand of illumination of collector road lighting, each group consists of 12 LED modules and this LED road light consists of 96 LEDs in total. The simulation illumination performance of the LED light is shown in Figure 6.13. When the lamp is 10 m high, an oblong light pattern which is 33 m long and 12 m wide is formed and the illumination distribution in the central area is uniform, which is suitable for collector road lighting. A prototype 96 W LED light fixture based on this design of reflector array is shown in Figure 6.12c. As shown in Figure 6.14, an LED road light developed by Hong Kong Polytechnic University also was designed by this method.

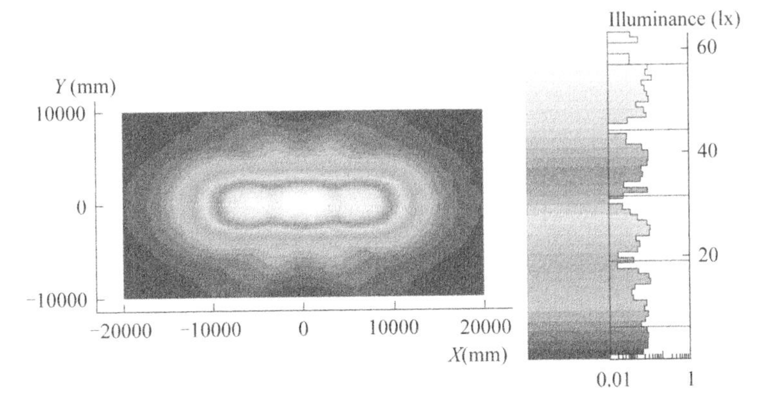

Figure 6.13 Numerical illumination performance of the 96W LED road light. *(Color version of this figure is available online.)*

Figure 6.14 An LED road light based on reflectors and 3D arrangement, developed by Hong Kong Polytechnic University [4].

(iii) Reflectors for Ultra-Long Span LED Road Lighting

At present, the bilaterally symmetrical light pattern is commonly adopted in urban expressways, major roads, and collector roads. The distance between two lights along the road is generally 30–60 m. The short distance between light posts is beneficial to improve the uniformity of road lighting. However, it will lead to an increase in the numbers of lights per unit-distance on the road. On many urban roads, light posts stand in great numbers, which limits the drivers' field of view to a large degree and affect their estimation and judgment of the roadside, thus increasing the probability of traffic accidents. In addition, with the increase in the lighting points on the road, night pedestrians and drivers cannot concentrate if there is an increase of glare, which is a potential safety hazard. Reducing the number of light posts can save construction, reduce repair, and beautify the city. Therefore, on the premise of reaching the road lighting standards, it is the road lighting development trend to increase the distance between two lights along the road and reduce the light posts per unit-distance on the road.

As to the general road light with ultra-long span lighting (the distance between two lamps >60 m), this assumption is not likely to be realized because of the limitation of the light source's luminescence properties [5,6]. However, as to the LED, this assumption is very likely to be realized because of the good directivity of the LED light sources. The design method of the reflector LED array integrated with a three-dimensional arrangement is also adopted in LED ultra-long span road lights. LED modules adopting different light distribution curves and reflectors with different beam control effects are combined to form different LED module groups, as shown in Figure 6.15. One may optimize the special inclination angle of every LED module group to make each LED module group control a lighting area on the road; more than one LED module groups' lighting areas are combined to form an even oblong lighting area along the road, as shown in Figure 6.16, thus meeting the demands of road lighting.

Simulated illuminance distribution on the road of the ultra-long span LED road light is shown in Figure 6.17. Table 6.1 presents a comparison of the simulated road lighting indices

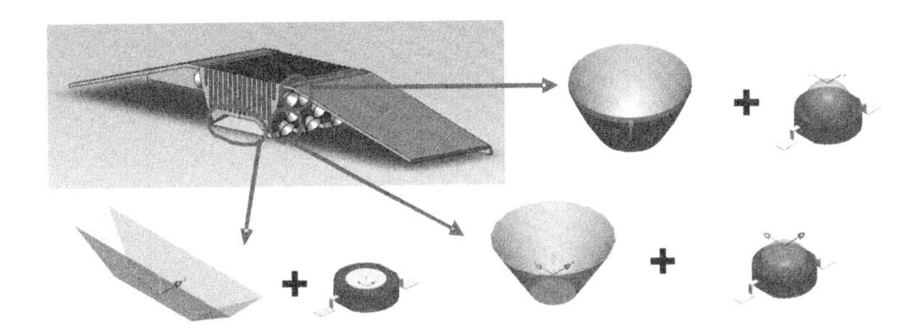

Figure 6.15 LED module groups' structural schematic of the ultra-long span LED road light.

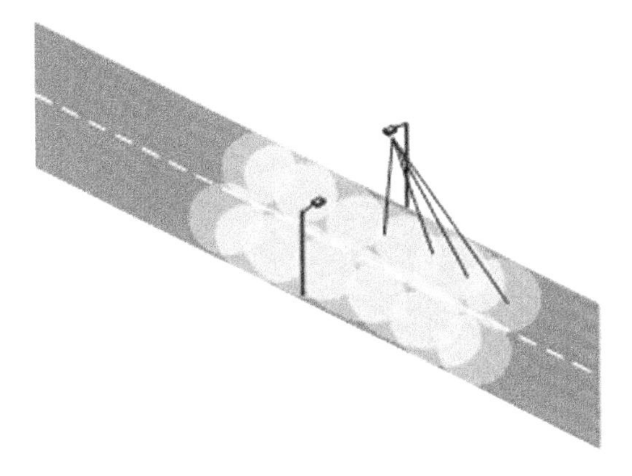

Figure 6.16 Schematic of combination effect of different LED module groups' light pattern on the road.

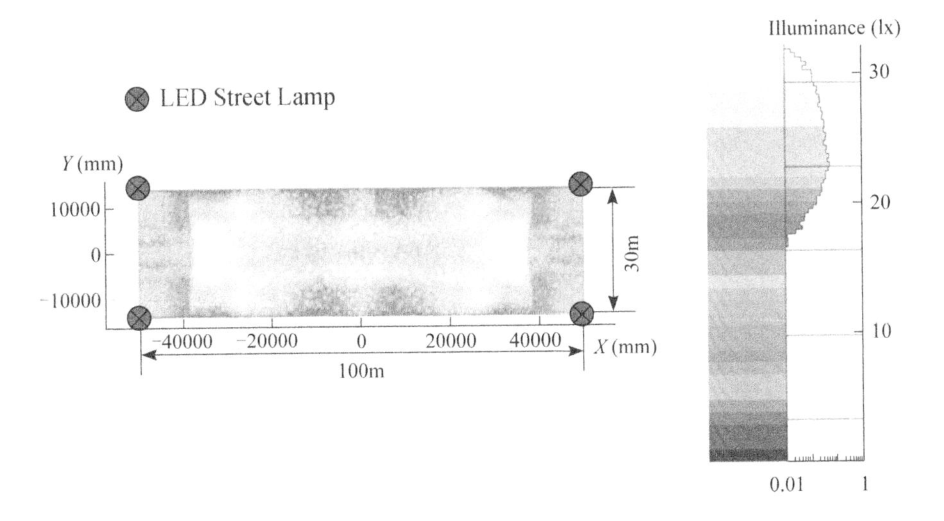

Figure 6.17 Schematic of simulated illuminance distribution on road of ultra-long span LED road lighting system (100 m × 30 m). *(Color version of this figure is available online.)*

Table 6.1 Comparison of the road lighting performance of ultra-long span LED road lighting system with the national standards

Grade	Road Type		Road Luminance			Road Illuminance
			Average Luminance L_{av}(cd/m^2)	Overall Uniformity U_o	Longitudinal Uniformity U_L	Average Illuminance E_{av}(lx)
I	Expressway and Major Road	National Standard	1.5/2.0	0.4	0.7	20/30
		Ultra-long Span LED Road light	1.6	0.64	0.74	24.0
Grade	Road Type		Road Illuminance Uniformity U_E		Glare Limitation Threshold Increment TI (%)	Surround Ratio S_R
I	Expressway and Major Road	National Standard	0.4		10	0.5
		Ultra-long Span LED Road light	0.64		9.3	0.56

with the national standards. Through the comparison, we find that the indices of the ultra-long span LED road lighting system all reach the vehicle road lighting standards in the *Design Standard of Urban Road Lighting CJJ45–2006*; some indices are even more favorable than those of the road lights with traditional light sources. All these prove that the ultra-long span LED road lighting system is not only possible, but can provide better road lighting through optimization design. Therefore, the ultra-long span LED road lighting system is one of the important development directions of road lighting in the future; integrated with intelligent control, it will provide the people with a totally new idea of road lighting design and road lighting with higher quality.

(iv) Reflectors for LED Headlamp

The LED has great freedom of design and the lamps are small in size. This is beneficial to achieve the streamline design of a headlamp which will reduce the wind resistance, reduce the gasoline consumption, and at the same time improve the body's aesthetic effect, making the headlamp look attractive. However, the optical design of the car's headlamps, especially for the lower-beam headlamp, is the most difficult in various car lights. The lower-beam headlamp should not only ensure that the driver can observe hazards 40 m before the car, but also ensure that drivers or pedestrians in the opposite direction do not suffer from glare; at the same time, it should provide enough brightness for pedestrians on the roadside and to illuminate signs. Therefore, there is a strict light distribution standard for the lower-beam headlamp. At present, the most common light distribution standard is ECE standard which mainly prescribes the light pattern's distribution 25 m away from the car lights, Figure 6.18.

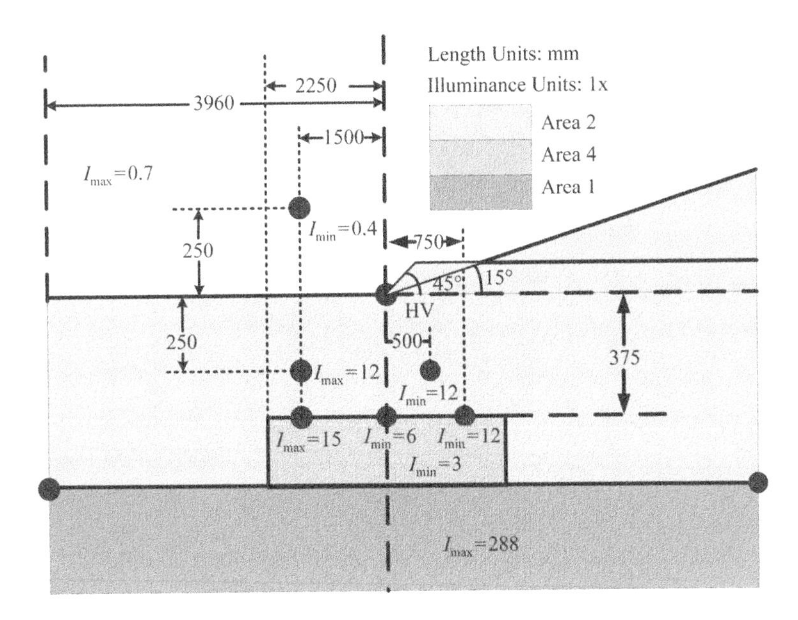

Figure 6.18 Schematic of ECE light distribution standard. *(Color version of this figure is available online.)*

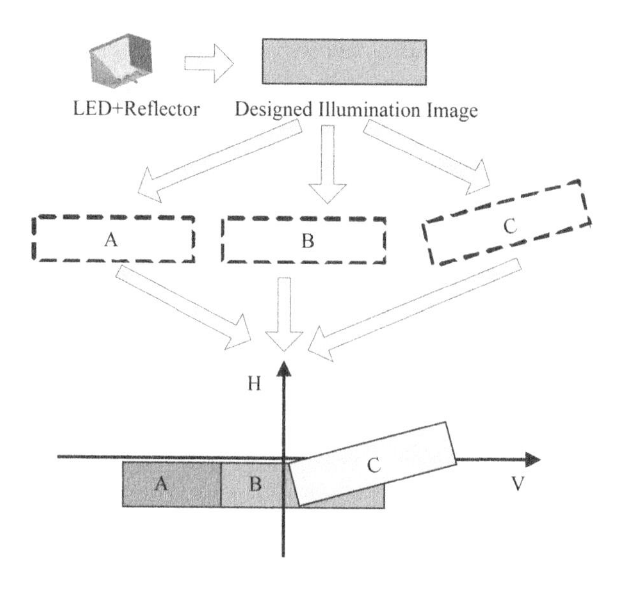

Figure 6.19 Schematic of realization of lower-beam headlamp's light distribution.

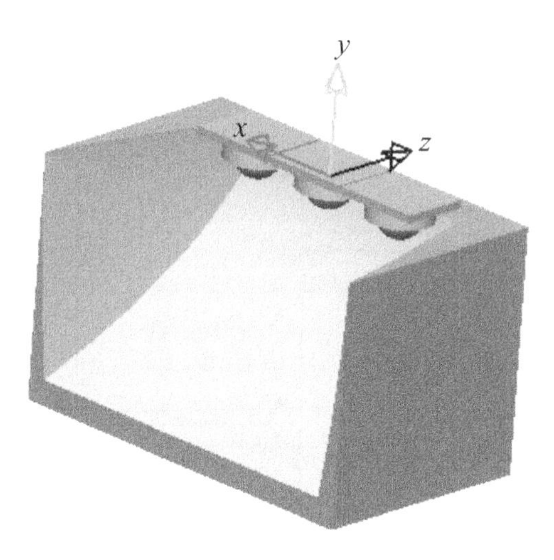

Figure 6.20 Schematic of a single reflector plus LED module.

Light distribution which meets ECE standard can be obtained by using reflectors. Usually, in order to make the design simple, reflector is designed which can generate an oblong light pattern with obvious cut-off line at first, then according to the combination of certain proportions and angles, superpose the light pattern to generate the prescriptive light pattern (Figure 6.19).

A single reflector plus LED module is shown in Figure 6.20. The light pattern generated by the single reflector plus LED module is shown in Figure 6.21.

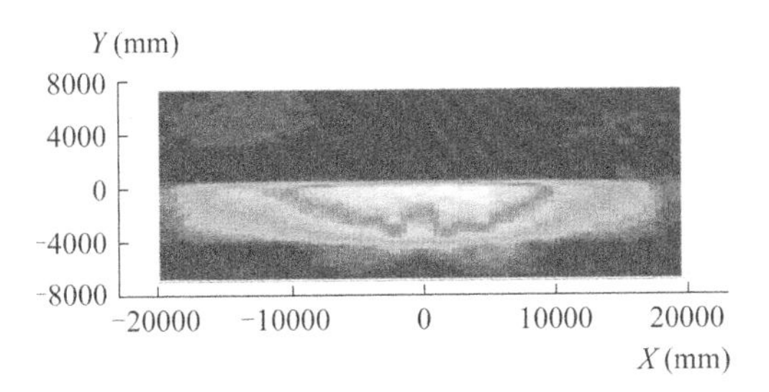

Figure 6.21 Light pattern generated by a single reflector.

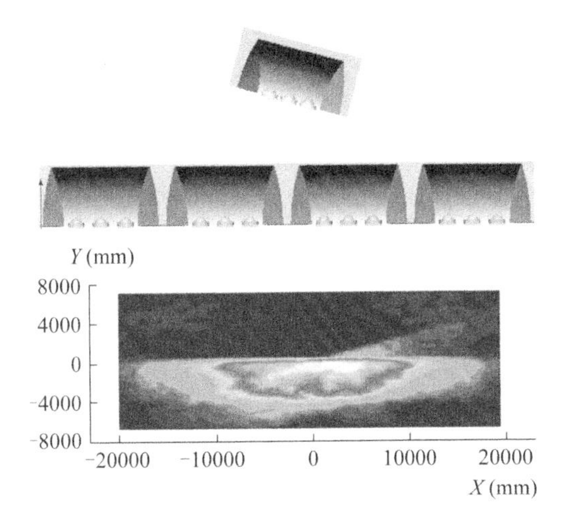

Figure 6.22 Lower-beam headlights and its simulated lighting result.

The lower-beam headlamp after the combination of more than one modules and its simulated lighting result are shown in Figure 6.22. It can be seen that on the right side of the light pattern there are obvious cut-off lines.

Currently, part of the lower-beam headlamp design not only adopts reflectors, but also employs the convergence lens in front of the reflectors. A single module can realize the light

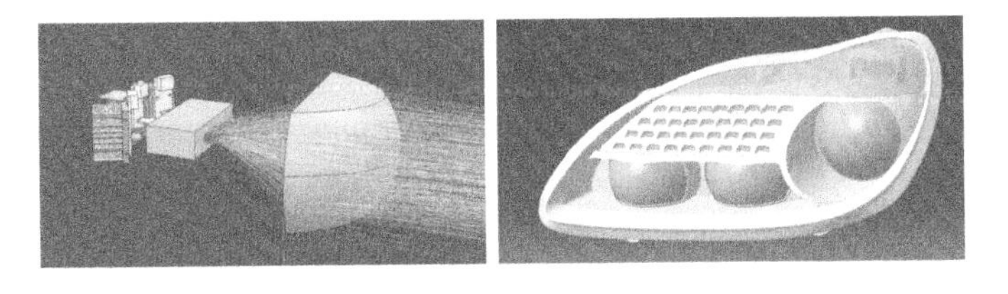

Figure 6.23 Lower-beam headlamp module and overall lamp adopting reflector plus lens method [7]. (Reproduced from U. Schlöder, "New optical concepts for headlamps with LED arrays," *Automotive Lighting Technology and Human Factors in Driver Vision and Lighting*, 2007-01-0869, 2007, SAE International.)

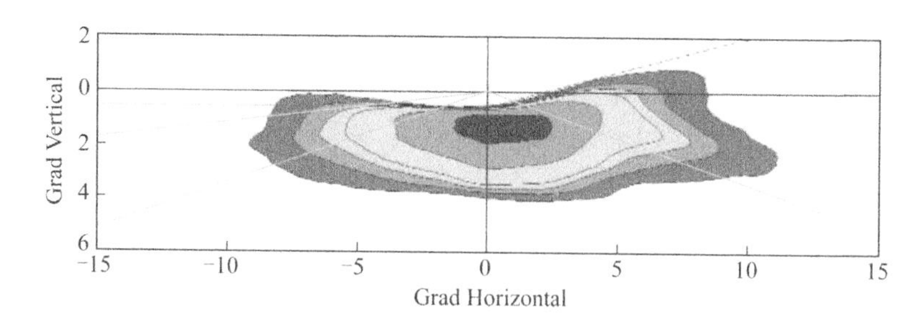

Figure 6.24 Simulated light pattern of the lower-beam headlights adopting reflector plus lens method [7]. (Reproduced from U. Schlöder, "New optical concepts for headlamps with LED arrays," *Automotive Lighting Technology and Human Factors in Driver Vision and Lighting*, 2007-01-0869, 2007, SAE International.)

pattern's shape which meets the normal requirement, only with a need to increase the modules so as to meet the demands of illumination. This method can achieve a better lighting performance; in addition, the lamp structure is more compact. The single module adopting this method and the overall lamp is shown in Figure 6.23. The simulated result of this lamp is shown in Figure 6.24.

6.1.3 Lenses

From the section above, we can see that reflectors play a role in converging the lights emitting from the LED. If we want to obtain a divergent LED lighting performance, such as road light, the three-dimensional arrangement of the reflectors becomes a necessity, and this makes the entire LED lighting system large in volume and the installation and maintenance is inconvenient. At the same time, because most LEDs are Lamberts or similar to Lambertian light distribution, the reflector can only control the rays with big emerging angles emitting from LED, but cannot control the rays with small emerging angle close to the symmetrical axis, leaving them radiating freely. Therefore, the reflector's ability of controlling the LED's overall beam is relatively weak.

As another major controlling means in LED lighting, the lens can not only generate a convergent beam but also a divergent beam. In addition, it can control effectively the light of various angles emitting from the LED. Furthermore, it has great freedom in design and the realization ways are flexible, especially with the rise of freeform lens, the designs of various noncircular light patterns are becoming more convenient. Therefore, the lens is more and more widely used in LED lighting, becoming a key means in optical design (Figure 6.25).

The materials commonly used in LED lens are polycarbonate (PC), polymethyl methacrylate (PMMA), and glass. PC is provided with very good toughness and good impact resistance; its light transmittance is about 90%; the refractive index is close to 1.586; the melting point is about 149 °C; the temperature for allowable thermal deformation is 130–140°, however its fatigue resistance is not very good. In LED lenses, PC is mainly used in primary lenses and as it is provided with weak flowing properties in injection molding, it is not suitable

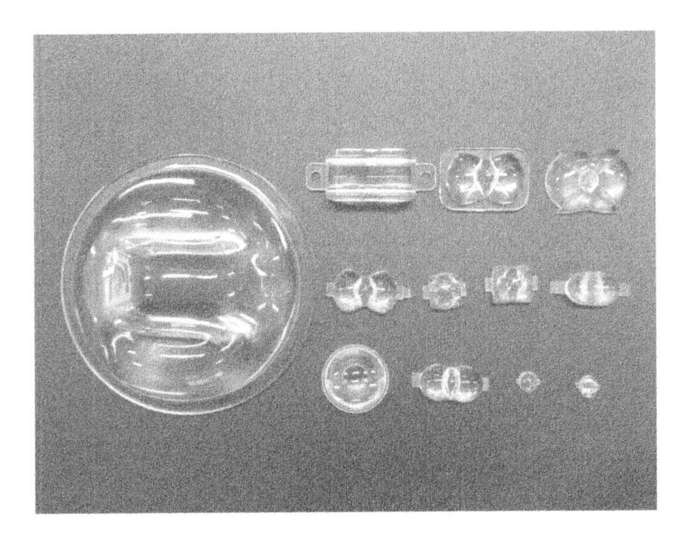

Figure 6.25 Various LED secondary lenses in the market.

for producing thick products. PMMA is a kind of plastic material to replace glass. Its light transmittance can reach 94%; the refractive index is close to 1.49; but its enduring temperature is relatively low, about 80–110 °C. In LED lenses, PMMA is mainly used in secondary lenses. Glass lenses are also widely used in high power LED lenses. It is characterized by high light transmittance, widely distributed refractive index (the common refractive index is about 1.50), and good temperature tolerance but has a complicated manufacturing process; therefore, it is hard to use to make more complicated lenses.

(i) LED Collimation Lens

As the optical system for LED collimation is axisymmetric, the real lens body can be obtained only by constructing a lens' cross section and rotating it 180°. A collimation lens's cross section is shown in Figure 6.26.

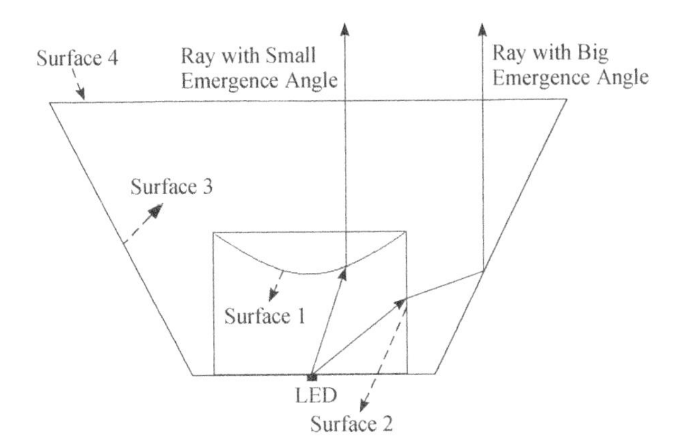

Figure 6.26 Schematic of an LED collimation lens' cross section.

In this collimation lens, there are four curved surfaces which control the beam. Curved surface 1 is to collimate the rays with small emerging angles. Curved surface 2 is a cylindrical surface which is to refract the rays with large emerging angles and reduce the lens' volume. The rays with large emerging angles which are refracted by the second curved surface into the lens will achieve total internal reflection in Curved surface 3 and radiate upward with collimation. Surface 4 is a plane and it will not change the exiting directions of the collimation rays. In other LED collimation lenses, the fourth surface could also be designed as a curved surface.

The design of the first curved surface and the third curved surface is the key point in this lens design. Since these two curved surfaces are freeform surfaces, it is necessary to calculate the point's coordinates on the cross section's contour line of the lens and then fit these points to form the lens' cross section. There are two major steps in calculating the contour lines of the first curved surface and the third curved surface.

(1) Construct the space coordinates of the points on the contour lines of the cross section
Firstly, set a seed point according to the size of the required lens. The incident ray of this point is on an upward longitudinal propagation after refraction or total internal reflection (the first curved surface controls the beams through refraction and the third curved-surface controls the beams through total internal reflection). The normal vector of this point can be obtained according to Snell's law.

Secondly, the tangent line of this point on the curve can be obtained based on the normal vector, and can then calculate the coordinates of the crossing point of the next ray radiating from the light source and the tangent line.

Thirdly, repeat the steps above until all the space coordinates on the curve have been obtained.

(2) Fit the points on the curves to obtain the contour lines of the first curved surface and the third curved surface

The construction process of the collimation lens is shown in Figure 6.27.

Figure 6.28a is a PMMA collimation lens and its ray tracing effect. We can see that one-side 0–90° rays almost radiate vertically upward after the refraction and reflection by the lens. Figure 6.28b is a UFO LED collimation lens in which the fourth surface is not a plane [8].

(ii) Axisymmetric Freeform Lens for LED MR16

One typical general lighting product of the LED is the LED MR16 lamp (Figure 6.29). MR16 lamps are typically used as outdoor spotlights or accent lighting for part of a room, in restaurants, museums, or retail displays. The light distribution of the LED is always similar to Lambertain which is not suitable for MR16 lamps illumination, therefore secondary optics design is essential. Freeform lens is an effective way to control lights accurately and to achieve high quality illumination with high light efficiency into main beam and uniformity [9].

The MR16 lamp is designed to form a circular and uniform light beam on the target for accent lighting. This freeform lens design belongs to the category of circularly prescribed illumination. As shown in Figure 6.30, the lens refracts the incident ray I into output ray O, and irradiates O at corresponding point Q on the circularly target plane. According to the energy mapping relationship, edge ray principle and Snell's law, the coordinates and normal vector of point P on the surface of the freeform lens is able to be calculated.

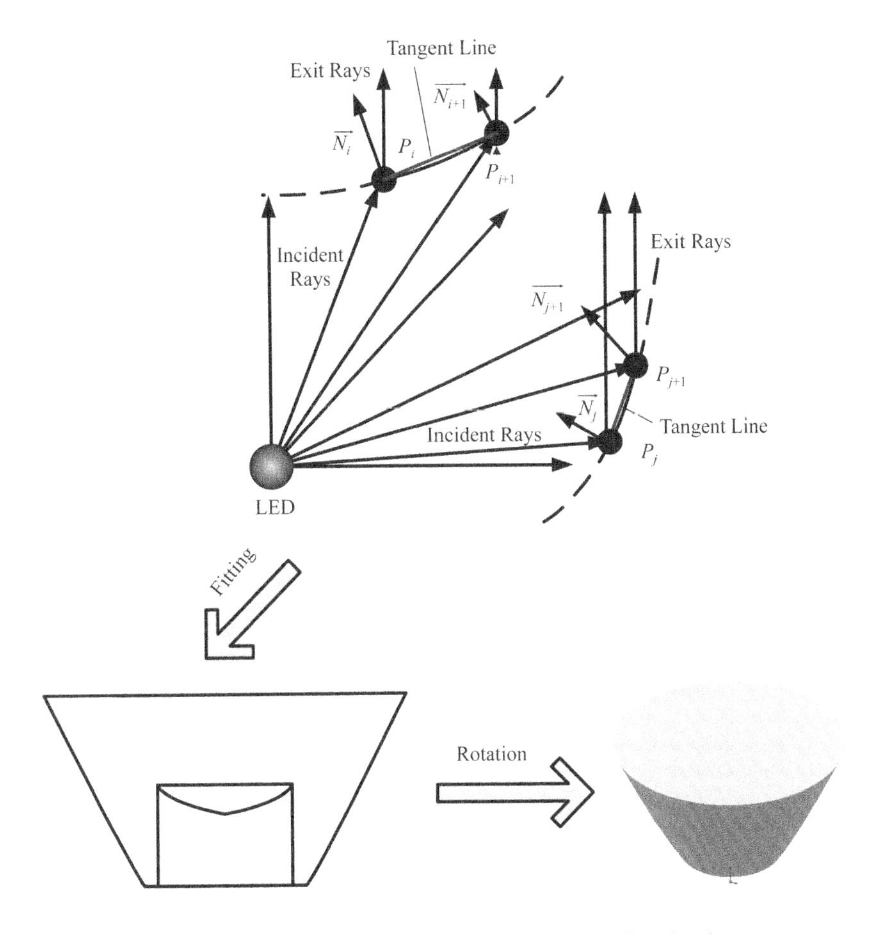

Figure 6.27 Construction process of an LED collimation lens.

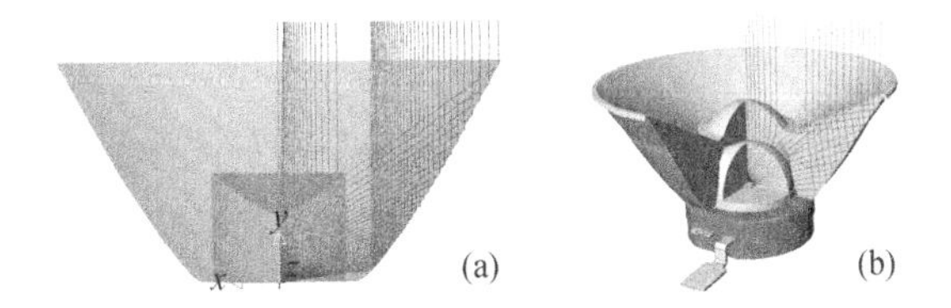

Figure 6.28 (a) PMMA LED collimation lens; and (b) UFO LED collimator lens [8]. (Reprinted with permission from O. Dross, R. Mohedano, P. Benitez *et al.*, "Review of SMS design methods and real-world applications," Nonimaging Optics and Efficient Illumination Systems, *Proceedings of SPIE Vol. 5529*, **35**, 2004. © 2004 SPIE.)

Figure 6.29 Different kinds of LED MR16 lamps. *(Color version of this figure is available online.)*

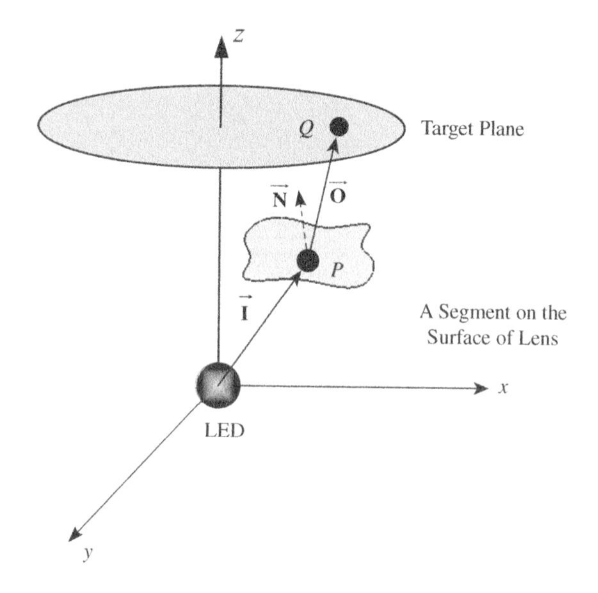

Figure 6.30 Schematic of the circularly prescribed illumination problem.

The freeform lens design method is briefly described in Figure 6.31 and it consists of three main steps as follows:

Step 1. Establishment of Light Energy Mapping Relationship

Since both the circular target and luminous intensity distribution of the light source are a central symmetry, the lens is also designed as a central symmetry. Then only the contour line C_0 of the lens' cross section needs to be calculated and the lens can be formed by rotating this contour line around the symmetry axis.

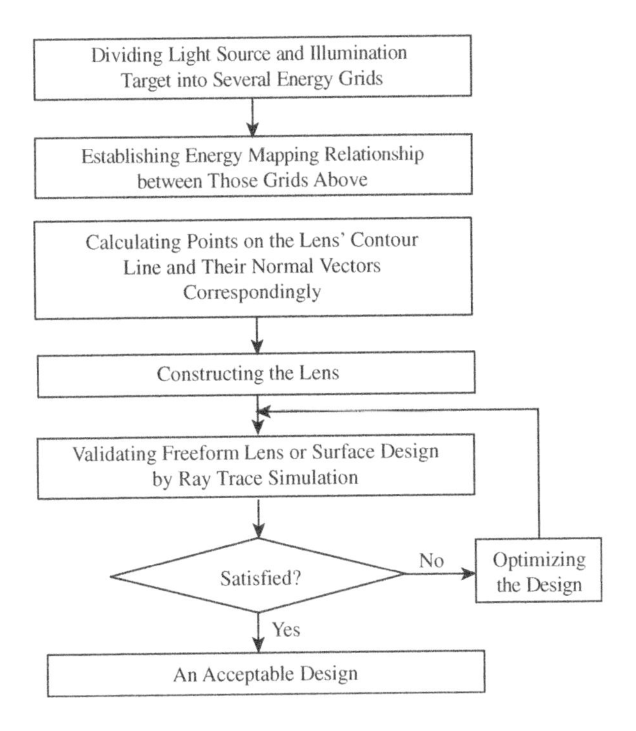

Figure 6.31 Design flowchart of the circular freeform lens design method.

In this design, both the light source and the illumination target plane are divided into M grids with equal luminous flux Φ_0 and area S_0 respectively. Then a mapping relationship is built between a pair of light source grid and target plane grid. Therefore, the average illuminance, $E_0 = \Phi_0/S_0$ of each target plane grid is the same and a uniform light pattern can be obtained when the grid is quite small comparing with the whole target plane.

Firstly, since the light source is axisymmetric, we divide the intensity space distribution of light source into M grids with equal luminous flux ϕ_0 only in the longitude direction. The intensity space distribution of light source $I(\theta)$ can be obtained from the experiment where θ is the angle between ray and the symmetry axis of the light source as shown in Figure 6.32. Suppose the total flux of light source is ϕ_{total}, according to the principle of photometry, the relationship between luminous flux ϕ and luminous intensity I is expressed as:

$$I = \frac{d\phi}{d\omega} \tag{6.3}$$

The solid angle $d\omega$ can be expressed as:

$$d\omega = \sin(\theta)d\theta d\varphi \tag{6.4}$$

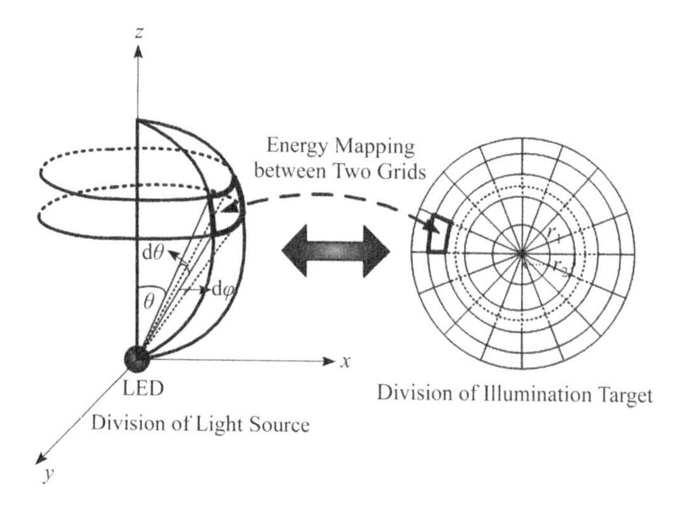

Figure 6.32 Schematic of light energy mapping between the light source and target.

Therefore, we can obtain following equations:

$$\begin{aligned} d\phi &= I(\theta)\sin(\theta)\mathrm{d}\theta\mathrm{d}\varphi \\ \phi &= \int_0^{2\pi}\int_0^{2\pi} I(\theta)\sin(\theta)\mathrm{d}\theta\mathrm{d}\varphi \end{aligned} \tag{6.5}$$

The edge angle $\theta_1, \theta_2, \ldots, \theta_i$ $(i = 1, 2, \ldots, M)$ of each part, which defines the direction of edge light of each source grid, can be calculated from Equation 6.6.

$$\begin{aligned} \phi_0 &= \frac{\phi_{\text{total}}}{M} \\ \int_0^{\theta_i} 2\pi I(\theta)\sin(\theta)\mathrm{d}\theta &= i\phi_0 \end{aligned} \tag{6.6}$$

Secondly, we divide the illumination target plane into M concentric rings with the same area S_0. Suppose the radius of the illumination target is R. Then the radius r_i $(i = 1, 2, \ldots, M)$ of each ring on the illumination target can be calculated from Equation 6.7.

$$r_i = R\sqrt{\frac{i}{M}} \tag{6.7}$$

Step 2. Construction of Lens

There are four steps to calculate the contour of lens. First of all, as shown in Figure 6.33, we fix a point as the vertex of lens which is the 1st point (P_1) on the curve C_0, and the normal vector of this point is also determined to be vertical up. The second point on the curve C_0 can be determined by the intersection of the incident ray and the tangent plane of the previous point. Secondly, the direction of the exit ray can be calculated by the point obtained in the first step and the corresponding point on the target plane. Thirdly, we calculate the present point's normal

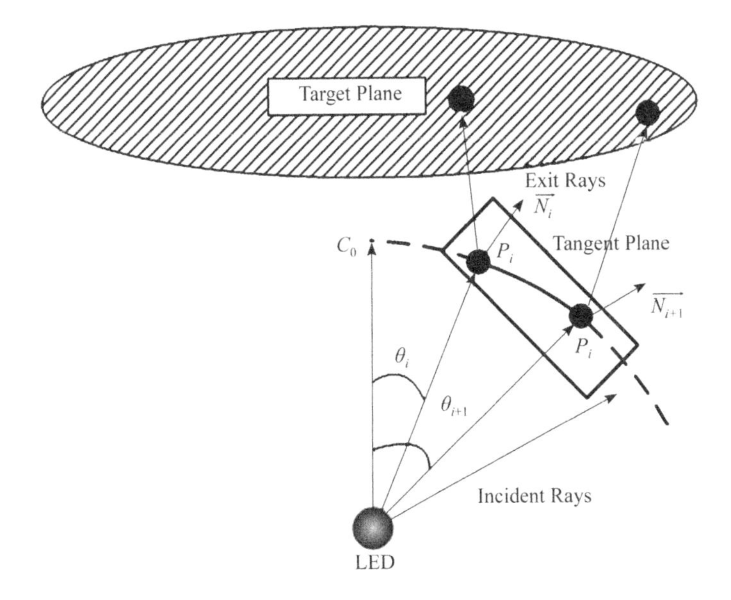

Figure 6.33 Calculations of the points on the contour line C_0 of a lens.

vector by incident ray and exit ray using the inverse procedure of Snell' law, which is expressed as follows:

$$\left[1+n^2-2n\left(\vec{O}\cdot\vec{I}\right)\right]^{\frac{1}{2}}\vec{N}=\vec{O}-n\vec{I} \tag{6.8}$$

where $\vec{\mathbf{I}}$ and $\vec{\mathbf{O}}$ are the unit vectors of incident and refracted rays; $\vec{\mathbf{N}}$ is the unit normal vector on the refracted point and n is the index of refraction in the lens. Finally we can obtain all the points and their normal vectors on the curve C_0 in this chain of calculation (Figure 6.34a). Then we fit these points to form the lens' contour and rotate this contour to obtain the entity of the lens, as shown in Figures 6.34b and 6.34c respectively. To fix to the LED, a hemisphere inner cavity

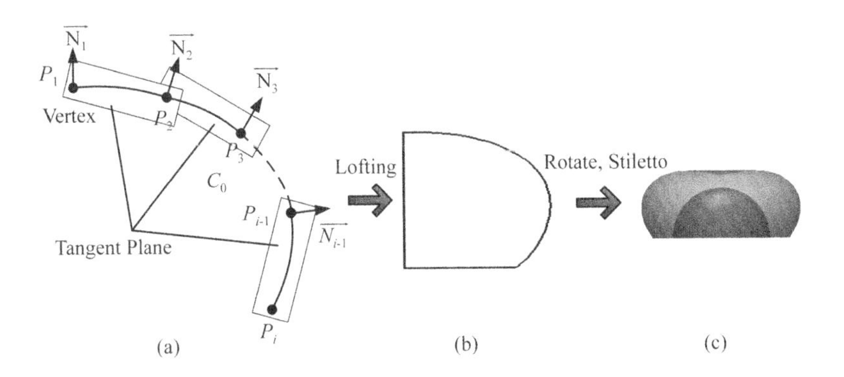

Figure 6.34 Progress of construction of the novel freeform lens.

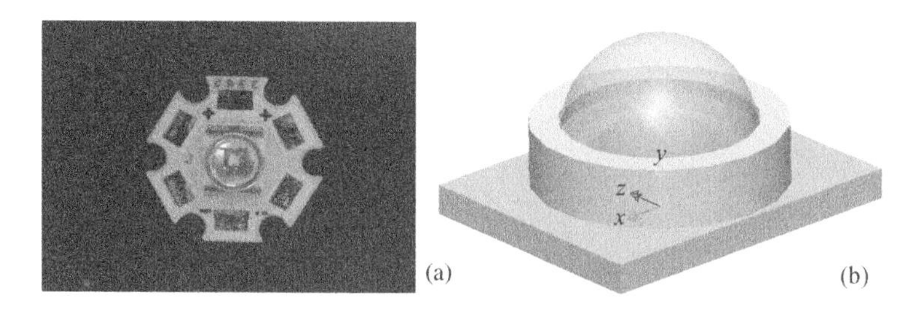

Figure 6.35 Cree® XLamp® XR-E LED: (a) material object and (b) practical optical model.

is needed at the bottom surface of the lens, which will not change the directions of lights irradiating from the light source.

Step 3. Validation of Lens Design

Since it is costly to manufacture a real circular freeform lens, numerical simulation based on Monte Carlo ray tracing method is an efficient way to validate the lens design. According to the simulation results, slight modification is needed to make the illumination performance better, such as the radius of the hemisphere inner cavity, installation position of the freeform lens, and so on.

The following is the design of MR16's freeform lens based on the Cree–XLamp LED which will give the readers a better understanding of the design method of the LED MR16's freeform lens.

Step 1. Optical Modeling for an LED

As shown in Figure 6.35, here we use the Cree–XLamp XR–E LED as the light source. A practical optical model of the Cree–XLamp XR–E LED is also to be established to evaluate the actual illumination performance of the designed lens.

First of all, we take some measurements to determine the geometrical parameters and establish a practical structural model of the LED. We obtain LIDC of this LED model by the widely used Monte Carlo ray tracing method and the LED is simulated by one million rays. Then, a precise LED optical model is established by comparing the similarity between the simulation light intensity distribution curve (LIDC) and the experimental LIDC, which is quantified by the normalized cross correlation (NCC) [10]. The NCC is written as follows:

$$NCC = \frac{\sum_{x}\sum_{y}\left(A_{xy}-\bar{A}\right)\left(B_{xy}-\bar{B}\right)}{\left[\sum_{x}\sum_{y}\left(A_{xy}-\bar{A}\right)^{2}\sum_{x}\sum_{y}\left(B_{xy}-\bar{B}\right)\right]^{1/2}} \quad (6.9)$$

where A_{xy} and B_{xy} are the intensity or irradiance of the simulation value (A) and experimental value (B); $\bar{A}\bar{A}$ ($\bar{B}$) is the mean value of A (B) across the x–y plane. As to the modeling algorithm for an LED model mentioned in Figure 6.36 [11], we adjust the scattering parameters and refraction indexes of some packaging materials used in the LED, such as phosphor, polymer, silicone, and so on, until the NCC reaches as high as 97.6%, as shown in Figure 6.37. Thus, the precise optical modeling for LED is finished.

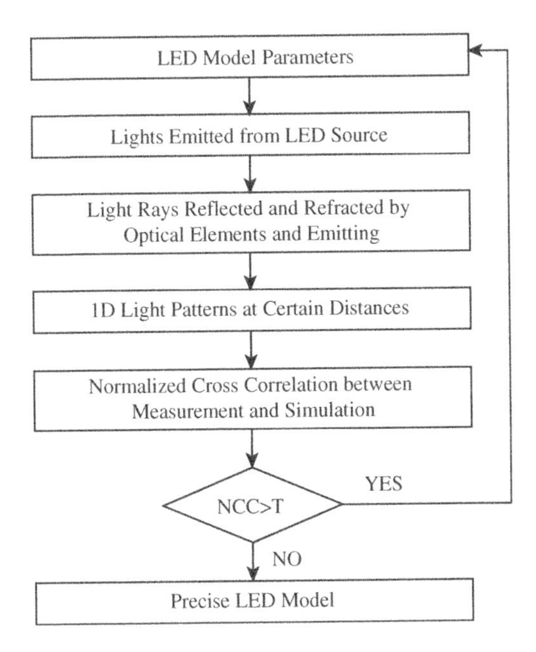

Figure 6.36 Modeling algorithm for an LED model.

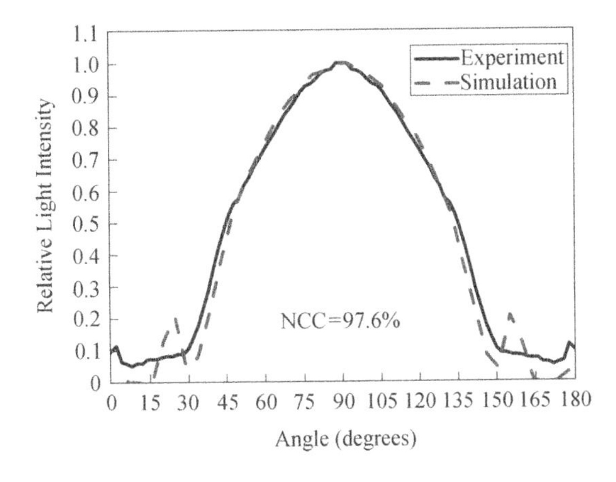

Figure 6.37 Experimental LIDC versus simulation for the Cree LED.

Step 2. Design of Freeform Lenses

Based on this method, two novel LED MR16 lamp's PMMA lenses of emerging angles of 90° and 120° respectively have been designed as examples. As the comparisons shown in Figure 6.38, the height, diameter, and volume of the 90° lens are 5.6 mm, 5.2 mm, and 105 mm^3 respectively, and are 5.6 mm, 15 mm, and 508 mm^3 respectively for the 120° lens. However the size of the traditional MR16 lens existing in the market, with the type of total

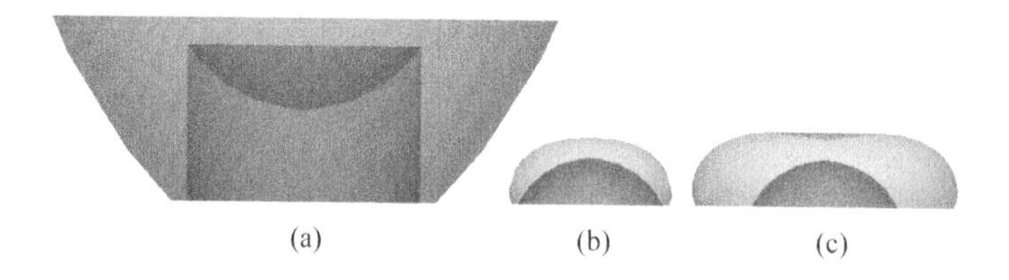

Figure 6.38 (a) A TIR lens for LED MR16 lamp existing on the market; (b) the novel 90° LED MR16 lamp lens; and (c) the novel 120° LED MR16 lamp lens.

internal reflector (TIR) lens, is much larger than these novel lenses with a height of 10 mm, diameter of 28 mm, and volume more than 3000 mm^3. The volume of these novel lenses are no more than a fifth of that of the TIR lens. Thus these novel MR16 LED lenses provide a better and effective way for those size compact illumination applications.

Step 3. Simulation Results and Analysis

We simulate the TIR LED MR16 lens and the design novel freeform lenses numerically by the widely used Monte Carlo ray tracing method. Figure 6.39 shows the simulation illuminance distribution on a test area which is 1 m away from the LED module. The light output efficiencies (LOE) of these two novel freeform lenses reachs as high as 98% while slightly less than 82% is achieved for the TIR lens. This is probably due to the fact that some lights irradiate downward after being reflected several times at the internal surface of the TIR lens and these lights are totally lost. For the 90° MR16 lens, 89% of lights exit from the lens surface into the main beam, and 90% for the 120° MR16 lens, which are much higher than that of the TIR lens with only about 60% lights exiting from the lens into the main beam.

As shown in Figures 6.39 and 6.40, the novel LED MR16 lamps have a much higher uniformity illuminance distribution across the target compared with the traditional LED MR16 lamp with a TIR lens, especially for the central illumination area. Thus the novel LED MR16 could provide a high quality and more comfortable illumination performance in applications. Moreover, the novel MR16 lamps with small distribution angles could not irradiate to the desired points on the target, by one refracted optical surface limited by the largest deviation angles of lights refracted by one optical surface.

(iii) Lens Integrated with 3D Structure

As the design of a freeform lens is very difficult, and as most LED manufacturers do not have the capability of designing a freeform lens, it is very difficult to meet the demands of LED lighting applications directly through a freeform lens, therefore, an axisymmetric lens integrated with 3D distributions is usually used to meet the demands of lighting. This optical design is simple and as to the realization process, the readers can make a reference to the 3D distributions in Section 6.1.2 and the design of circular symmetric lens in Section 6.1.3.

(iv) Non-Rotational Symmetric Freeform Lens for LED Lighting

The freeform lens is a kind of newly emerging non-imaging optics and has become the trend for LED secondary optics design due to its advantages of unique design freedom, small size, and accurate light irradiation control. The advent of multi-spindle high-precision machine tools brings forceful and totally new technology for the processing of freeform optical components, which leads to a rapid development of freeform optical components. To deal with freeform lens

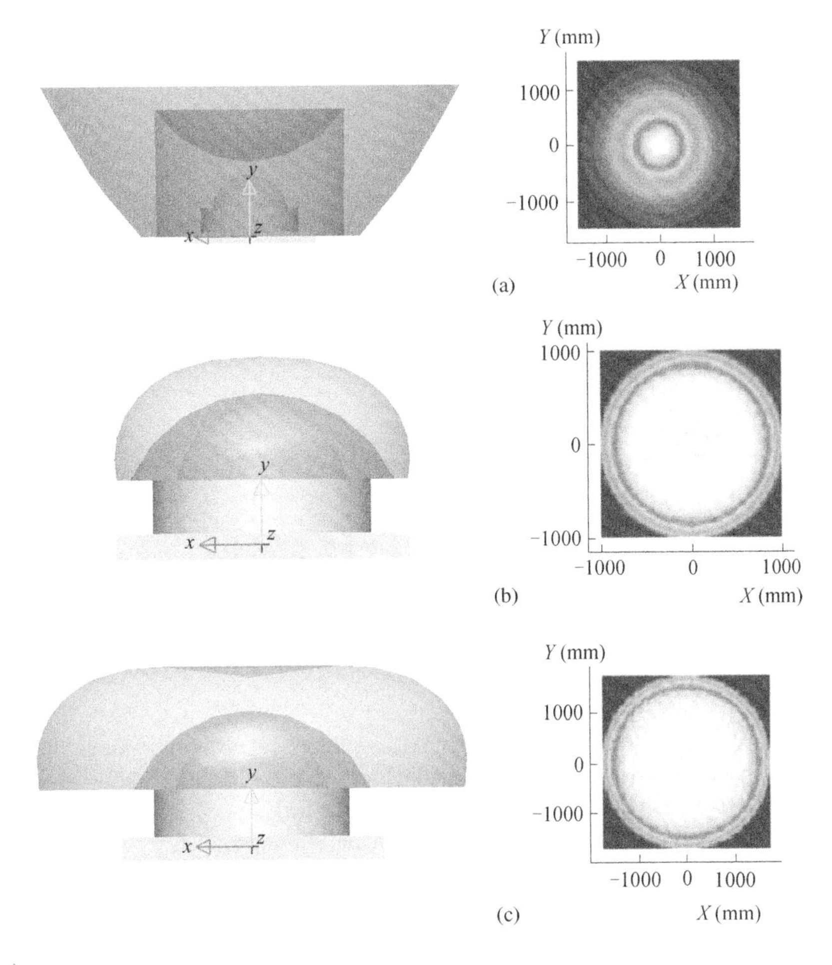

Figure 6.39 Simulation illumination performance of different LED MR16 lenses on a test plane: (a) a TIR LED MR16 lamp lens; (b) a novel 90° MR16 lens; and (c) a novel MR16 lens.

design, many different methods have been proposed, such as the simultaneous multiple surfaces (SMS) method [12–14], the tailored freeform surface method [15], the discontinuous freeform lens method [14,16,17], and so on. The following section will introduce the design methods of these kinds of freeform lens and their applications in LED lighting.

(1) Simultaneous multiple surfaces (SMS) method
The SMS 3D method is a method used in 3D freeform lens which is proposed by Benítez *et al.* [12]. Its basic strategy is as follows: firstly, according to the luminous characteristics of the light source and target surface, make two pairs of incident wavefronts W_i correspond to two pairs of exit wavefronts W_o, respectively $W_{i1} - W_{o1}$ and $W_{i2} - W_{o2}$. Then simultaneously design two freeform surfaces of an optical system, make the incident wavefronts W_{i1} and W_{i2} correspond to the exit wavefronts W_{o1} and W_{o2} after they are refracted or reflected by the optical system, as shown in Figure 6.41. The SMS method provides an

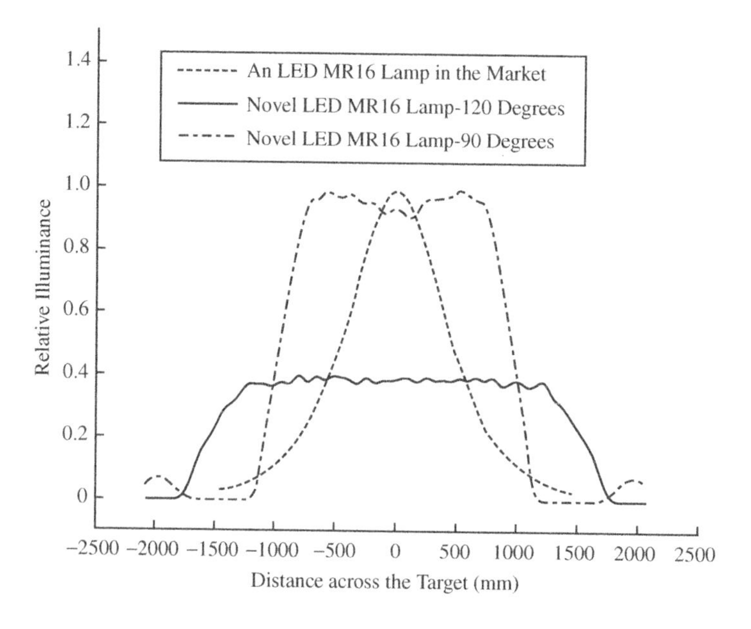

Figure 6.40 Relative illuminance distributions of different kinds of LED MR16 lenses.

optical system with two freeform surfaces that deflects the rays of the input bundles into the rays of the corresponding output bundles and vice versa. At present, only the SMS method could deal with extended light source effectively.

The advantages of the SMS method are: (1) it can simultaneously design more than one optical surface to improve the control of emerging rays; (2) it is suitable to design an optical

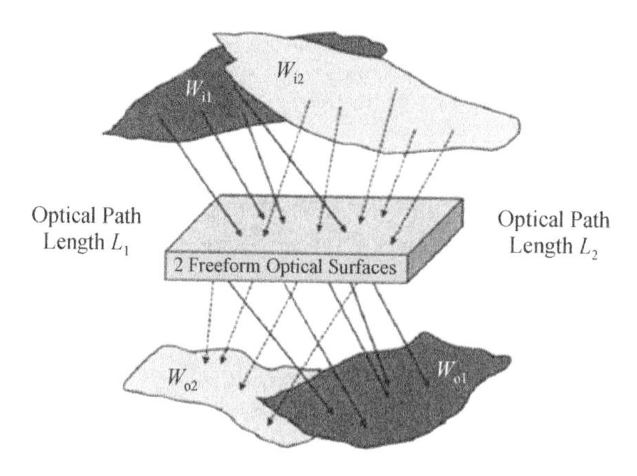

Figure 6.41 SMS method providing two surfaces for transforming two input wavefronts into two output ones [12]. (Reprinted with permission from P. Benítez, J.C. Miñano, J. Blen *et al.*, "Simultaneous multiple surface optical design method in three dimensions," *Optical Engineering*, **43**, 1489, 2004. © 2004 SPIE.)

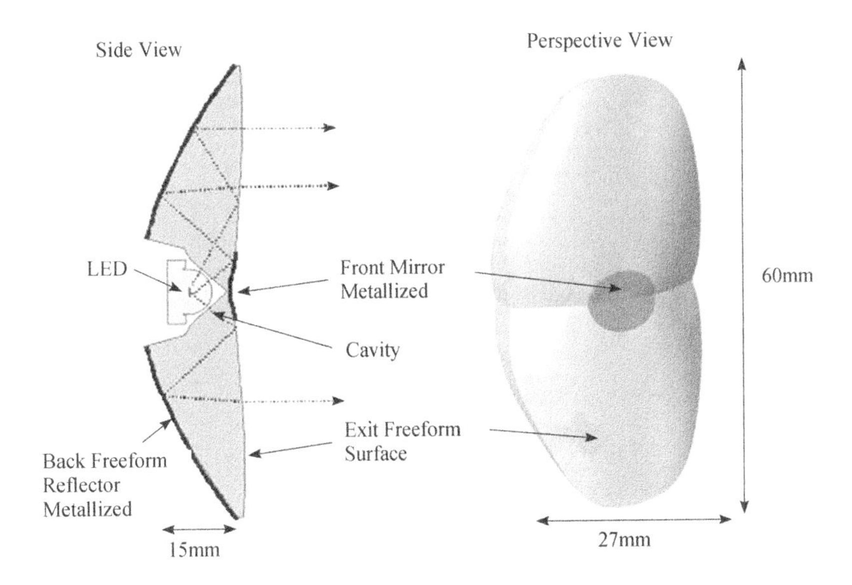

Figure 6.42 LED headlamp lens for low beam [8]. (Reprinted with permission from O. Dross, R. Mohedano, P. Benitez *et al.*, "Review of SMS design methods and real-world applications," Non-imaging Optics and Efficient Illumination Systems, *Proceedings of SPIE Vol. 5529*, **35**, 2004. © 2004 SPIE.)

system based on an extended source; (3) the rays with large emerging angles are applicable and have greater light source utilization. However, in this method the given illumination distributions need to be converted into wavefronts. This method also needs to solve some second-order nonlinear Monge–Ampere equations with a complicated and verbose calculation. Detailed description can be found in a reference [12].

The SMS 3D method provides LED lighting with an effective design method of freeform lens. Figure 6.42 presents the LED headlamp lens for low beam designed through SMS method [8].

(2) Tailored freeform surface method

The tailored freeform surface method constructs the freeform surface through a numerical solution of partial differential equations. As to a light source with a small volume, every point on the reflector or lens will distribute the radiation of the light source to a target area, then the size and distributed radiation of the target area are totally defined by the curvature and slope of the corresponding point on the optical surface. In addition, the curvature and slope are the second derivative and first derivative of this optical surface respectively, so that the optical surface can be obtained by solving the partial differential equation which consists of curvature and slope. The method which obtains the freeform surface through a numerical solution of the partial differential equation is highly efficiency and can obtain the data of the surface directly, without the need for the iterative optimization process in traditional methods. At the same time, during the solution, the continuous Gaussian curvature of the surface is adopted to ensure the local smoothness of the surface and obtain ideal illumination distribution in the area with small angles. However, it still has its limitations: it is only suitable for light sources with a small volume and it does not take into

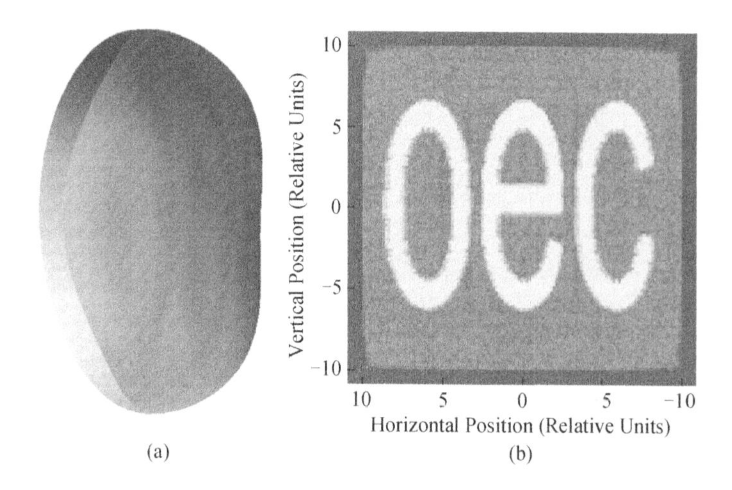

Figure 6.43 (a) A freeform lens designed based on the tailored method and (b) its numerical illumination performance [15]. (Reprinted with permission from H. Ries and J. Muschaweck, "Tailored freeform optical surfaces," *Journal of Optical Society of America A,* **19**, 3, 590–595, 2002. © 2002 Optical Society of America.)

consideration the rays with large emerging angles; the development of theory and design should be further improved. Detailed description of this method can be found in a reference [15].

Figure 6.43a is a freeform lens design based on the tailored method by OEC, adopting the beams within a half field angle of 32.8° from the light source to radiate on the lens. The simulated lighting performance is shown in Figure 6.43b. It is the three letters of the company's name (Optics & Energy Concepts) which fully indicates the accuracy and flexibility of the freeform lens to control the rays.

(3) Discontinuous freeform lens method

Both the SMS method and tailored method mentioned above are based on the coupling of input wavefronts and output wavefronts. A practical design method in 3D is to establish light energy mapping relationships between the light source and the target. The strategy of designing a freeform lens based on energy mapping is as follows. First of all, assume the light emitting from the light source all radiates on the target plane, that is, the energy of the light source is equal to that of the target plane. Then divide both the light source energy and target plane into many grids. According to the edge ray principle [18], establish the corresponding relationship between every energy grid of light source and the target plane grid by solving the energy conservation based differential/integral equation or through direct correspondence, which will be discussed in a later section. Finally, according to Snell's law and certain method of constructing curved-surfaces, calculate and obtain several points on the freeform surface, construct integrated optical surface, then validate and modify it through numerical simulation. The flow chart of this method is shown in Figure 6.44.

(a) The discontinuous freeform lens method based on solving the energy conservation based differential/integral equation

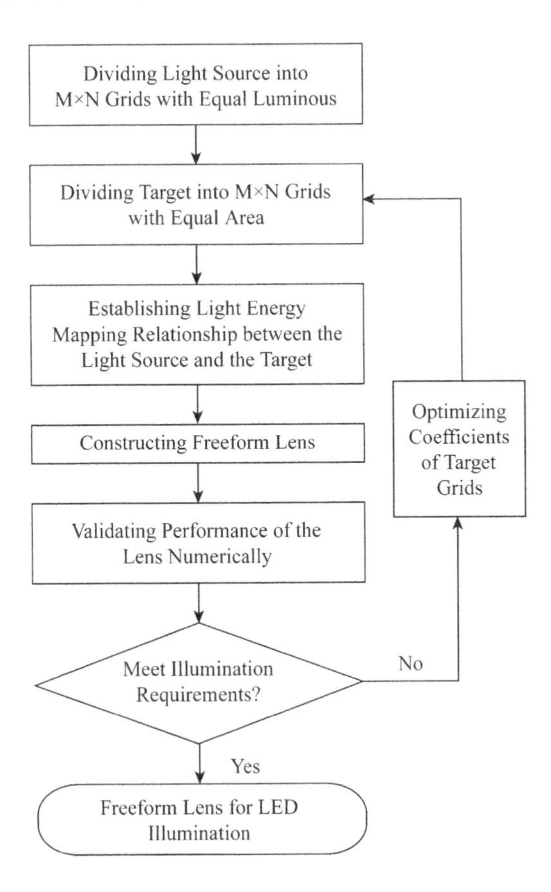

Figure 6.44 Flow chart of the freeform lens design method based on the light energy mapping.

Step 1. Establishment of Light Energy Mapping Relationship
Assuming the energy from the light source equals the energy illuminating the target plane, the energy conservation integral equation is:

$$\iint_{\Omega} I d\Omega = \iint_{\Omega} E ds \tag{6.10}$$

where I is the luminous intensity distribution of the light source and E is the illuminance distribution on the target plane. Ω defines the regions of the light directions from the point light source, and D defines the target region to be illuminated.

If we employ parametric coordinates (u, v) on the left side to represent the light direction and parametric coordinates (x, y) on the right side to represent the position on the target plane, Equation 6.10 yields:

$$\iint_{\Omega} I(u, v)|J(u, v)| fufv = \iint_{\Omega} E(x, y) ds |J(x, y)| dxdy \tag{6.11}$$

In Equation 6.11, $|J(u,v)|$ is the absolute Jacobian determinant, which represents the transformation factor using $dudv$ as the differential area, and so is $|J(x,y)|$. Transforming Equation 6.11 into differential form yields gives:

$$I(u,v)|J(u,v)|\mathrm{d}u\mathrm{d}v = E(x,y)|J(x,y)|\mathrm{d}x\mathrm{d}y \tag{6.12}$$

Then according to the initial condition and boundary condition, solve this differential equation to obtain the corresponding relationship between the light source (u, v) and the target plane (x, y): $x = f(u, v)$, $y = g(u, v)$. Next obtain the mapping relationship between the light energy distribution of the light source and the energy distribution on the target plane. For this energy mapping method, details can be found in a reference [16].

Step 2. Construction of Lens

In this section we will find out the lens which can realize the mapping between the light source and the target plane. There are four main steps to construct the outside surface of the freeform lens:

Step 2.1. Construction of the seed curve

The seed curve is the first curve to generate other lens curves and we will construct curves along a longitudinal direction. As shown in Figure 6.45, we fix a point P_0 as the vertex of the seed curve. The first subscript of a point designates the sequence number $(1, 2, \ldots, M + 1)$ of the longitude curve and the second designates the sequence number $(1, 2, \ldots, N)$ of the point on a longitude curve except the vertex. The second point $P_{1,1}$ is calculated by the intersection of incident ray $I_{1,1}$ and the tangent plane of the previous point. The direction of the refracted ray $O_{1,1}$ can also be obtained as Q_1-$P_{1,1}$, where Q_1 is the corresponding point of incident ray $I_{1,1}$ on the target plane. Then we can obtain the normal vector $N_{1,1}$ of the second point according to Snell's law. Based on this algorithm we can obtain all other points and their normal vectors on the seed curve. This method can guarantee that the tangent vectors of the seed curve are perpendicular to its calculated normal vectors.

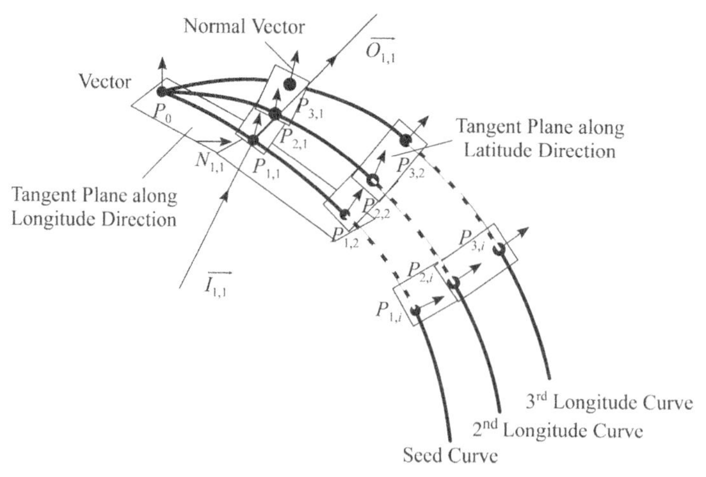

Figure 6.45 Schematic of points generation on the outside surface of the lens.

Step 2.2. Generation of other longitude curves
Since one quarter light source is divided into M parts along the latitudinal direction, there are M corresponding longitude curves to be calculated except the first seed curve. First of all we calculate the second curve whose vertex coincides with that of the seed curve. As shown in Figure 6.45, different from the seed curve algorithm, point $P_{2,i}$ on the second longitude curve is calculated by the intersection of the incident ray and the tangent plane of point $P_{1,i}$ on the previous curve. Then the following longitude curves, such as 3rd curve, 4th curve, and so on, are easy to obtain based on this algorithm.
Step 2.3. Error control
Unit normal vector (**N**) of each point is calculated based on the corresponding incident ray and exit ray at that point. However, the surface construction algorithm of lofting between curves cannot guarantee that the real unit normal vector (**N′**) of the point of the lens surface is still the same as the calculated one (**N**). Thus the direction of the exit ray will deviate from the expected one, which will result in poor illumination performance. As shown in Figure 6.46, starting from the second longitude curve, every point (for example $P_{2,i}$) on the curve only has one adjacent point (for example $P_{3,i}$) on its tangent plane in the latitude direction. Therefore deviations between the real normal vectors and calculated ones are generated on these longitude curves. The deviation can be estimated by the deviation angle θ_{d}:

$$\theta_{\mathrm{d}} = \cos^{-1}\left(\frac{NgN'}{|N||N'|}\right) \tag{6.13}$$

Since θ_{d} becomes larger with the increase of the sequence of longitude curves, a threshold θ_{dth}, for example 6°, is needed to confine the deviation. If the maximum deviation of one longitude curve is larger than θ_{dth}, we go back to the Step 1 and calculate another seed curve to replace this longitude curve.

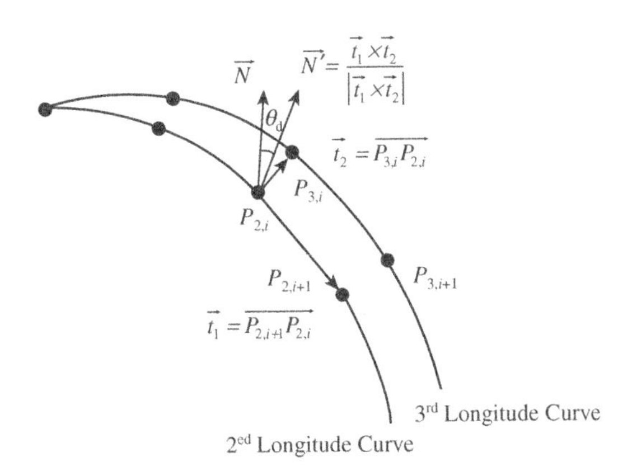

Figure 6.46 Deviation between the real unit normal vector (**N′**) and the calculated unit normal vector (**N**) of one point on the surface.

Step 2.4. Construction of surface
The lofting method is utilized to construct a smooth surface between these longitude curves. The freeform surface can be modeled using nonuniform rational B-splines (NURBS) [19], which offer a common mathematical form representing and designing freeform surfaces. Since the newly generated seed curves are discontinuous with the longitude curves before them, the surface of the lens becomes discontinuous.

Step 3. Validation of Lens Design
Due to the high cost involved in manufacturing a real discontinuous freeform lens, numerical simulation based on the Monte Carlo ray tracing method is an efficient way to validate the lens design. According to the simulation results, slight modification by trial and error is needed to make the illumination performance better. For instance, the construction of a transition surface between two sub-surfaces is not contained in this method, and the shape of the transition surface should be modified according to the simulation results and manufacturing process.

Wang *et al.* [16] adopted a similar method to design a discontinuous freeform lens, as shown in Figure 6.47a. The first seed curve is a latitude curve, not the longitude curve mentioned in the method above. Then they extended curves in the latitude direction to construct curved-surfaces. The simulation illumination performance is based on a point light source as shown in Figure 6.47b which is an 'E' light pattern. As LED is small in volume and the general emitting area is about 1 mm $\times$ 1 mm, it can be treated as a point light source when the lens size is large. Therefore, this method is suitable for LED lighting design.

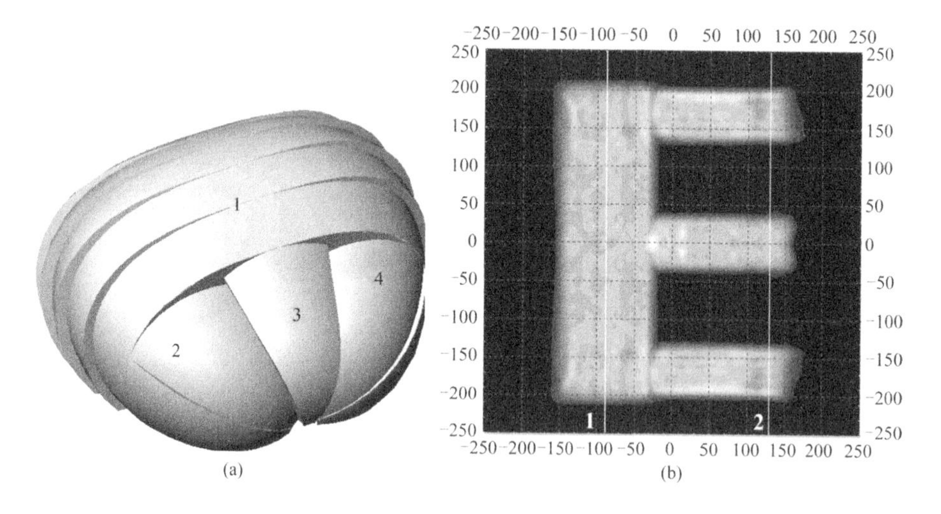

Figure 6.47 A discontinuous freeform lens to form 'E' light pattern and its simulation illumination performance [16]. (Reprinted with permission from L. Wang, K.Y. Qian and Y. Luo, "Discontinuous free-form lens design for prescribed irradiance," *Applied Optics*, **46**, 18, 3716–3723, 2007. © 2007 Optical Society of America.)

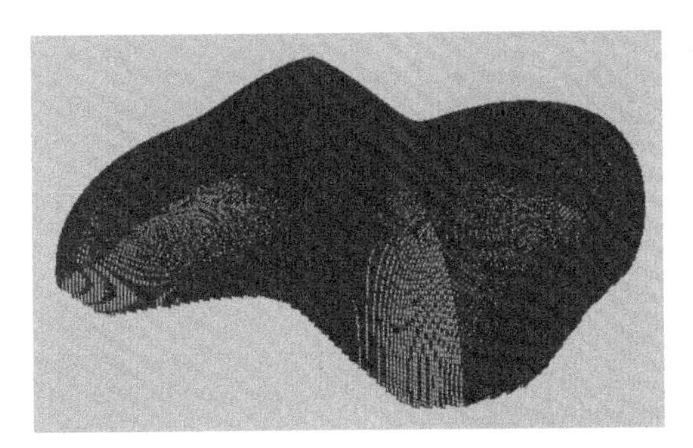

Figure 6.48 Freeform lens consisting of 450 discrete sub-surfaces [17]. (Reprinted with permission from Y. Ding, X. Liu, Z.-R. Zheng and P.-F. Gu, "Freeform LED lens for uniform illumination," *Optics Express*, **16**, 17, 12958–12966, 2008. © 2008 Optical Society of America.)

Ding *et al.* [17] also obtained the mapping relationship between the light source (θ, φ, ρ) and the target plane (x, y, z) through solving the energy conservation integral equation as follows:

$$\int_0^{2\pi} d\theta \int_0^{\phi_{\max}} I(I(\varphi))\sin\varphi d\varphi = \int E(t)dA \tag{6.14}$$

where $E(t)$ is the luminance at point t, A is the area illuminated. $I(\boldsymbol{I}(\varphi))$ is LED emitting intensity in the direction of $\boldsymbol{I}(\varphi)$. Equation 6.14 indicates the relationship between θ, φ and x, y, z, and its exact form depends on the topological mapping from the source to the target plane. More details can be found in reference [17].

Ding *et al.* [17] adopted this method to design a freeform lens to form a 4:3 rectangular light pattern. During validation, they found that if the freeform surface was smooth and modeled as a whole, the simulated result was no good. Therefore, they also adopted a discontinuous surface to solve the problem by modeling the surface with discrete sub-surfaces. As shown in Figure 6.48, the freeform lens consists of 450 pieces. Each piece is lofted from three lines which are constructed from points. It is shown in Figure 6.49 that the rectangular illumination is on the target plane, which is a 4 : 3 rectangle, whose diagonal length is about 100 mm. If the fluctuation of errors on the surface has a low frequency, the surface tolerance can be up to 2 micrometers. It is fine to design this discontinuous freeform lens numerically, but it is difficult to manufacture it in practice. Moreover, during mass production (that is, injection molding), many manufacturing factors, such as surface morphology of mold, injection molding temperature and pressure, viscosity of liquid, and so on, will affect the surface morphology of the discontinuous freeform lens and thereby affect the optical performance of the lens. Sometimes the manufacturing defects can be as large as several hundred micrometers and these defects will significantly deteriorate the illumination performance.

(b) The discontinuous freeform lens method based on light ray direct mapping

Although the discontinuous freeform lens method is based on energy conservation, and the differential/integral equation provides an accurate mathematical expression

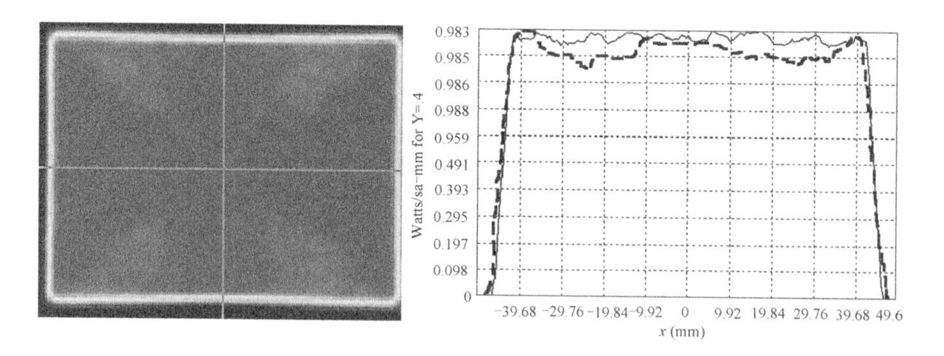

Figure 6.49 Simulation illumination performance of the discontinuous freeform lens consisting of 450 discrete sub-surfaces [17]. (Reprinted with permission from Y. Ding, X. Liu, Z.-R. Zheng and P.-F. Gu, "Freeform LED lens for uniform illumination," *Optics Express*, **16**, 17, 12958–12966, 2008. © 2008 Optical Society of America.)

between the coordinates of light source and the target, it will become complicated to handle light sources with random luminous intensity distribution curves (LIDC), non-rectangular target grids, and to find these solutions when dealing with uniform illumination problems. In this part, a discontinuous freeform lens design method in 3D based on light ray directly mapping is presented [14], with advantages of flexible energy mapping relationship, accurate light irradiation control, and easier to manufacture.

During the prescribed illumination design, in most cases, optical systems are designed based on the LIDC of light sources and the expected illuminance distributions on the target plane. As shown in Figure 6.50, the freeform lens refracts the incident ray **I**, represented by spherical coordinates (γ, θ, ρ), into the output ray **O**. Then **O** will irradiate at corresponding point $Q\ (x, y, z)$ on the target plane. According to the energy mapping relationship, edge ray principle and Snell's law, the coordinates and normal vector of point $P\ (x, y, z)$ on the surface of the freeform lens is able to be calculated.

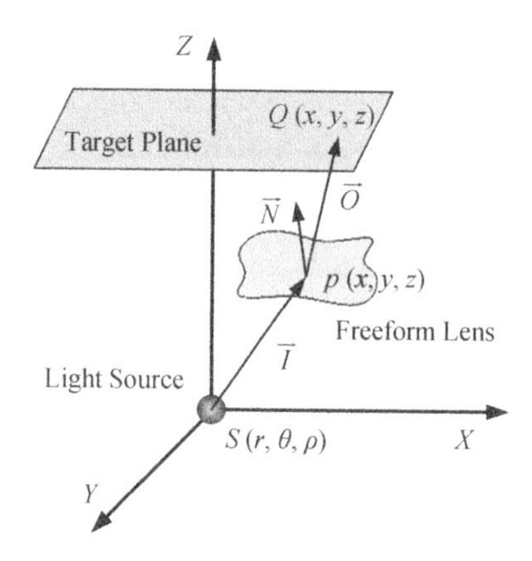

Figure 6.50 Schematic of the rectangularly prescribed illumination problem.

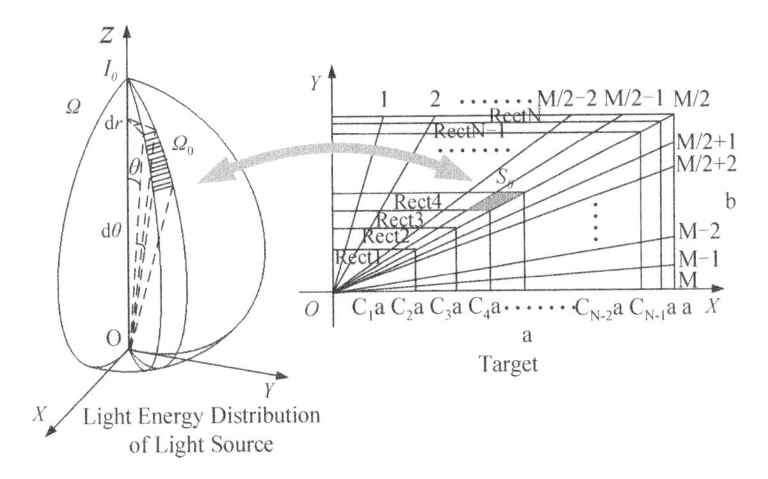

Figure 6.51 Schematic of light energy mapping between the light source and target.

In this design method, both light energy distribution of the light source and illumination of the target plane are divided into several grids with equal luminous flux ϕ_0 and area S_0 respectively. Then a mapping relationship is established between thelight source grid and target plane grid. Therefore, the average illuminance is expressed as $E_0 = \eta\phi_0/S_0$, where η is the light output efficiency of the freeform lens, each target plane grid is the same, and a uniform light pattern can be obtained when the grid is quite small compared with the whole target plane.

Since both the light source and target plane are of axial symmetry, only one quarter of the whole light source and target plane are to be considered in this discussion. First of all, the light energy distribution of the light source is divided into $M \times N$ grids with equal luminous flux. As shown in Figure 6.51, the light source's light energy distribution Ω could be regarded as being composed of a number of unit conical object Ω_0, which represents the luminous flux within the angular range with a field angle dγ in the latitudinal direction and dθ in the longitudinal direction. The luminous flux of Ω_0 can be expressed as follows:

$$\phi = \int I(\theta)\mathrm{d}\omega = \int_{\gamma_1}^{\gamma_2} \mathrm{d}\gamma \int_{\theta_1}^{\theta_2} I(\theta)\sin\theta\mathrm{d}\theta \tag{6.15}$$

where $I(\theta)$ is the luminous intensity distribution of the light source and dω is the solid angle of a one-unit conical object. Least square curve fitting in the form of polynomial is one of the most used curve fitting methods in numerical analysis. Thus, most $I(\theta)$, expressed in the form of LIDC, can be fitted by a polynomial of θ as follows:

$$I(\theta) = I_0 \sum_{i=1}^{m} a_i\theta^i \tag{6.16}$$

where I_0 is the unit luminous intensity, a is the polynomial coefficient and m is the order of this polynomial. The larger m is, the more accurate the curve fitting will be, but corresponding computational time will also increase. For most LIDCs, m should not be less than 9. Moreno *et al.* (2008) provided a better method to describe the LIDC of LED as the sum of a maximum of two or three Gaussian or cosine-power functions. By using this method, random variations of the LIDC profiles can be obtained realistically. More detailed information of this method can be found in a reference [20]. The total luminous flux of this one quarter light source can be expressed as follows:

$$\phi_{\text{total}} = \int_0^{\pi/2} \mathrm{d}\gamma \int_0^{\pi/2} I(\theta)\sin\theta \mathrm{d}\theta = \frac{\pi}{2}\int_0^{\pi/2} I(\theta)\sin\theta \mathrm{d}\theta \tag{6.17}$$

We divide the Ω into M fan-shaped plates along the latitudinal direction with an equal angle of $\Delta\gamma = \pi/2M$ and equal luminous flux of ϕ_{total}/M, as shown in Figure 6.51. Then each fan-shaped plate is divided into N parts equally along the longitudinal direction. The field angle $\Delta\theta_{i+1}$, which is different for every j, of each separate conical object along longitudinal direction can be obtained by an iterative calculation as follows:

$$\frac{\pi}{2M}\int_{\theta_j}^{\theta_{j+1}} I(\theta)\sin\theta \mathrm{d}\theta = \frac{\phi_{\text{total}}}{MN} \quad (j = 0, 1, 2, \ldots, N-1, \theta_0 = 0) \tag{6.18}$$

$$\Delta\theta_{j+1} = \theta_{j+1} - \theta j \quad (j = 0, 1, 2, \ldots, N-1) \tag{6.19}$$

Thus the light source has been divided into $M \times N$ sub-sources with equal luminous flux. The directions of rays, which define the boundary of one sub-source, also have been defined.

Secondly, to establish the mapping relationship with the light source, the one quarter target plane is also divided into $M \times N$ grids with equal area. A warped polar grid topology is appropriate to fit a rectangle. Since the target plane is perpendicular to the central axis of the light source, z is a constant Z_0 for all points on the target. As shown in Figure 6.51, the length and width of the one quarter rectangle target plane is a and b respectively and the area is $S=ab$. Firstly, the target plane is divided into N parts equally by sub-rectangles Rect_1, Rect_2, ..., Rect_N, which have the same length-width ratios as the whole plane. The relationship of area $S_{\text{Rec}k}$ of each sub-rectangle is:

$$S_{\text{Rect}_k} = kS_{\text{Rect}_1} \quad (k = 1, 2, \ldots, N) \tag{6.20}$$

where $S_{\text{Rect}_k} = S$. Therefore the rectangular target plane has been divided into N parts with the same area of S/N. The coefficient C_q of length of C_q a of each sub-rectangle can be obtained as follows:

$$C_q = \left(\frac{q}{N}\right)^{1/2} \quad (q = 1, 2, \ldots, N) \tag{6.21}$$

Then the plane is divided into M parts equally along the central radiation directions by M–1 radial lines. The endpoints of these lines equally divide the edges of the target plane, which ensures that the plane has been divided into M parts with an equal area S/M. The edges of the sub-rectangles and radial lines construct the warped polar grids. Therefore the plane has been divided into $M \times N$ grids S_0 with an equal area of S/MN.

According to edge ray principle, rays from the edge of the source should strike the edge of the target. This principle is true in 2D, and in 3D the skew invariant will lead to light loss, but which could be partly recovered by increasing the number of grids. Therefore if we desire to map the light energy in Ω_0 onto the target grid S_0, we should ensure that four rays, which construct the Ω_0 as the boundary, irradiate at the four corresponding end points of the target grid S_0 after being refracted by the freeform lens. Since the light source and target plane are both divided into $M \times N$ grids, each ray from the light source could have only one corresponding irradiate point on the target plane. Thus the light energy mapping relationship between the light source and target plane has been established. When dealing with non-rectangular illumination problems, it only needs to change the sizes and shapes of the grids on the target plane, which makes it easier and more flexible to re-establish the light energy mapping relationships comparing with the methods based on solving the energy conservation based differential/integral equation.

The methods of construction of the lens and validation of the lens design are the same as the method mentioned above and readers can refer to those design processes during design.

Let's consider some work we did before. Wang *et al.* (2009) designed a PMMA discontinuous freeform lens for road lighting based on this method [21]. Figure 6.52 depicts the schematic of the discontinuous freeform lens and its simulation illumination

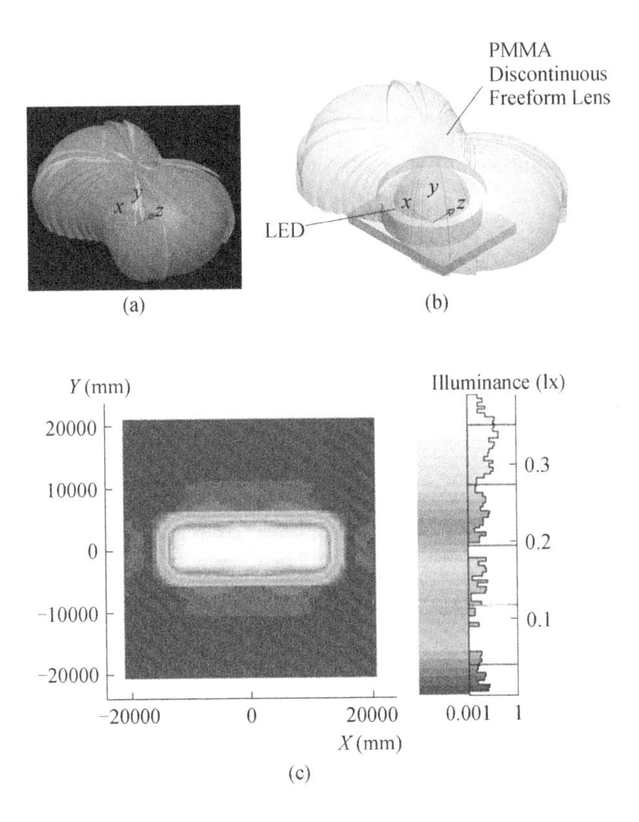

Figure 6.52 (a) Schematic of the discontinuous freeform lens for road lighting; (b) LED module with this lens; and (c) simulation illumination performance of the lens [21]. *(Color version of this figure is available online.)*

performance at 8 m away. The whole discontinuous freeform lens was constructed by only 40 sub-surfaces, which made it easier to manufacture and few manufacture defects would be induced comparing with 450 sub-surfaces used by [17]. From the simulation results we found that more than 95% light energy was uniformly distributed in the central area of about 28 m long and 10 m wide, which is suitable for road lighting. This discontinuous freeform lens was also manufactured by an injection molding method. Figures 6.53a and 6.53b show the practical freeform lens and the LED testing module. The total luminous flux of the LED module with and without the freeform lens were measured by UV–VIS–near IR spectrum photo colorimeter measurement and integrating sphere. The light output efficiency of the PMMA lens reached as high as 90.5%, which was quite close to the simulation results. Figure 6.53c depicted the light pattern at 70 centimeters away from LED. We found that most of the light energy was distributed in a nearly rectangle area with the length of 240 centimeters and width of 83 centimeters, and the light pattern would enlarge to 27.5 m long and 9.5 m wide at the height of 8 m according to the light rectilinear propagation principle, which was also in agreement with the expected shape.

Section 6.1.2 introduced a method for adopting the reflector and 3D arrangement to design the LED road light. The main limitation in that method is that the light pattern of a single LED integrated with one reflector cannot meet the requirement of road lighting; only through superimposing more than one light pattern can the sub-rectangular light pattern needed in road lighting be realized. The advantages of that method are obvious in terms of its simple design and flexibility; the obvious disadvantages exist in its complicated structure, large volume, heavy weight, and inconvenience in assembly and installation.

Now, an LED circular light pattern can be turned into a rectangular light pattern only through a single lens, thus the optical design of the LED road light can be realized through

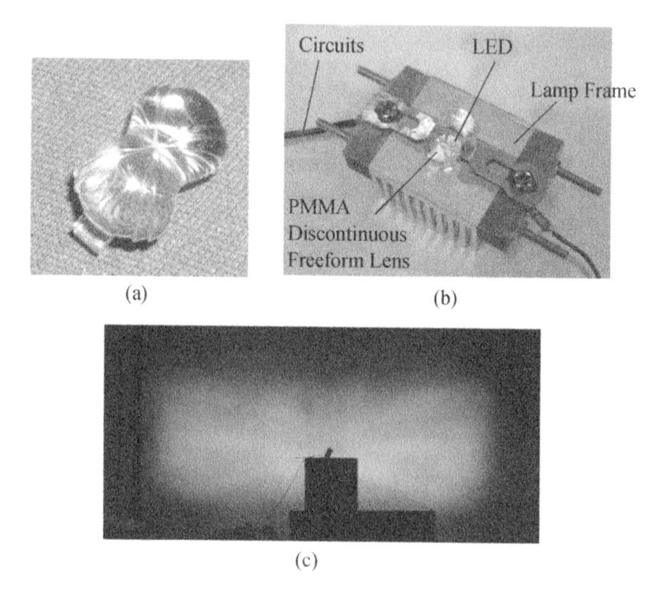

Figure 6.53 (a) Photo of the discontinuous freeform lens for road lighting; (b) LED testing module; and (c) light pattern of the lens [21].

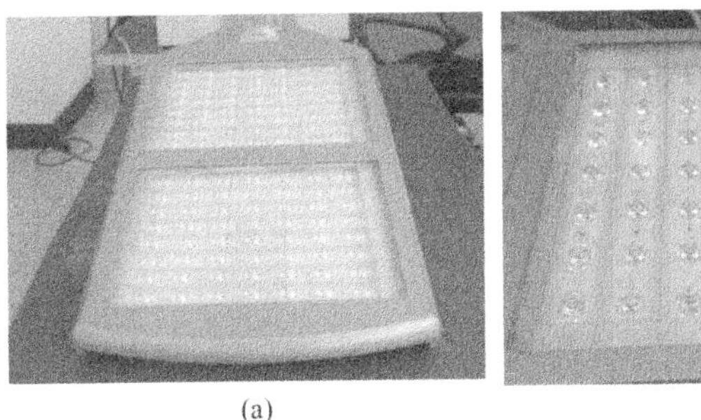

(a)

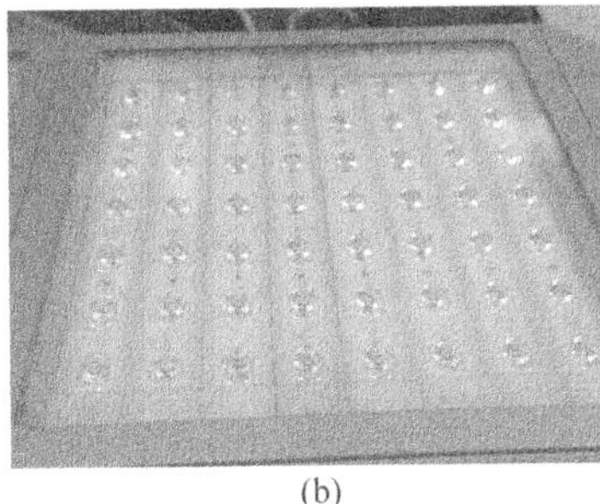

(b)

Figure 6.54 (a) The 112W LED road light based on the discontinuous lenses; and (b) enlarged view of the lenses. (Reproduced with permission from www.gd-realfaith.com, Guangdong Real Faith Enterprises Group Co., Ltd., accessed April 12, 2011.)

the planar array of freeform lenses. Because every freeform lens integrated LED module can generate a uniform rectangular light pattern, the planar array of more than one module can also generate a uniform rectangular light pattern, which is suitable for road lighting. At the same time, the superimposition of more than one rectangular light pattern eliminates the light pattern's dark stripes in single module lighting caused by manufacturing defects, achieving more uniform light pattern.

An 112 W LED road light based on the discontinuous lenses array is shown in Figure 6.54a, which consists of two 56 W LED road light source modules. This is because those LED road lights with various powers are required to meet the demands of urban road lighting. Adopt the design method of modules: employ the LED road light with a single module or double modules on the urban collector road and branch road and employ the LED road light with three modules on the major road, realizing a module based power scalable LED road light, thus reducing the design and manufacturing cost of a road light. Every 56 W LED module consists of eight rows and seven columns of LEDs (56 in total), each LED integrated with a discontinuous freeform lenses, as shown in Figure 6.54b. This LED road light enjoys such advantages as simple structure, beautiful appearance, small volume, low height, light weight, and more convenient assembly and installation.

Figure 6.55 presents the actual illumination performance of the road light: it is bright with good uniformity; the average illumination is 20.2lx and the illumination uniformity reaches 0.62. Figure 6.56 presents a comparison of illumination performance for the112 W LED road light with a 250 W traditional high pressure sodium road light. It is easy for us to find that the illumination performance is much better than that of the traditional high pressure sodium lamp.

With the same design method, Liu *et al.* also designed and manufactured a PMMA discontinuous freeform lens used in tunnel lighting. As shown in Figure 6.57a, the whole discontinuous freeform lens was constructed by only 32 sub-surfaces, even fewer than that of the discontinuous freeform lens for road lighting. The largest value of length, width, and height of the lens were 10.6 mm, 9.7 mm and 5.1 mm respectively. The illumination performance of this LED tunnel freeform lens at 80 cm away is shown in Figure 6.57b. The light pattern is a uniform rectangle with its size being 119 cm $\times$ 138 cm and then the size of the light pattern when it is 5.5 m high is 8.2 m $\times$ 9.5 m, which fully meets the demands of tunnel lighting.

Figure 6.55 Actual illumination performance of the 112W LED road lights at Meishi Street, Nanhai District, Foshan City, Guangdong Province, China. (Reproduced with permission from www.gd-realfaith.com, Guangdong Real Faith Enterprises Group Co., Ltd., accessed April 12, 2011.) *(Color version of this figure is available online.)*

Figure 6.56 Illumination performance comparison of LED road light (white light) with traditional high pressure sodium road light (yellow light). (Reproduced with permission from www.gd-realfaith.com, Guangdong Real Faith Enterprises Group Co., Ltd., accessed April 12, 2011.) *(Color version of this figure is available online.)*

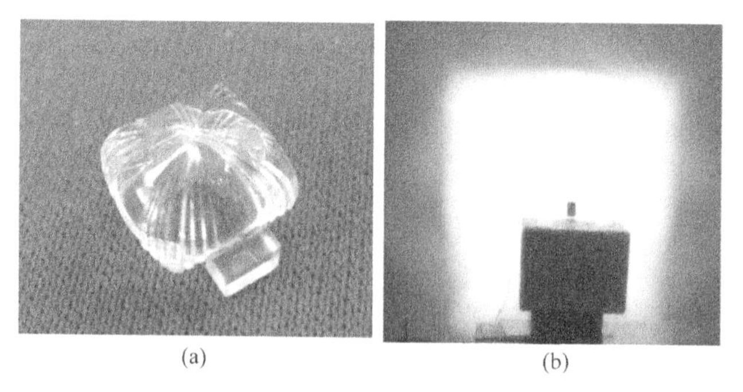

Figure 6.57 (a) Photo of a discontinuous freeform lens for tunnel lighting; and (b) its illumination performance at 80 cm away.

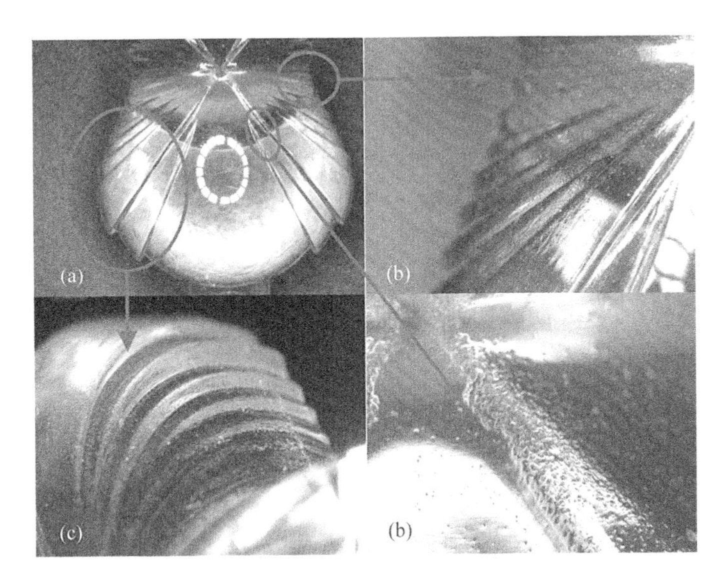

Figure 6.58 Micrographs of different parts of the PMMA discontinuous freeform lens for road lighting [21].

Manufacturing defects are difficult to avoid for the freeform lens, especially for the discontinuous freeform lens constructed by several discrete sub-surfaces. Surface roughness is one of the most common manufacturing defects existing in discontinuous freeform lenses [21]. As shown in Figure 6.58, we can clearly find that there are a lot of micron-sized particles distributed on the surface of the discontinuous freeform road light lens, especially on the transition surfaces, which are supposed to be smooth. Furthermore, some parts of this discontinuous freeform lens are totally composed of a number of discontinuous particles with the size of tens of micrometers and the surface morphology of these parts is quite different from that expected, which will result in severe scattering. These particles may be produced during the manufacturing processes and could mainly be caused by the unpolished surface of mould. Moreover, unsuitable injection molding temperature, pressure, and viscosity of liquid also could result in these particles.

Although the shape of the light pattern in Figure 6.53 agrees quite well with the design target, obvious dark stripes exist on the light pattern, especially in the middle-upper part of the pattern, which will decrease the uniformity of the pattern and the performance of illumination. Moreover, we also can find that the relative positions of these dark stripes on the light pattern are the same as the relative positions of transition surfaces on the lens surface.

A roughed surface of the lens could increase the chance of light scattering at the interface of lens and air. By comparing Figure 6.59a with Figure 6.59b, we can find that the roughed optical surface scatters lights randomly and the direction of exit lights are quite different from the designed ones. The scattered lights generated at the roughed transition surface will deviate from the expected irradiation directions and probably will overlap with other lights, which will result in dark stripes appearing in the direction of transition surfaces on

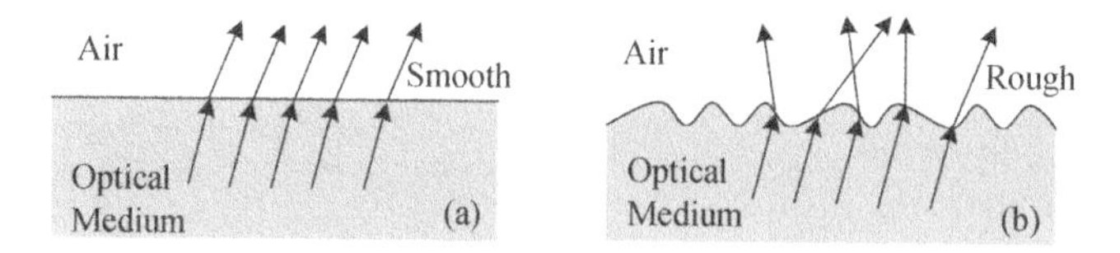

Figure 6.59 (a) Schematic of lights propagation at smooth and (b) rough optical surface.

the light pattern. Fortunately, the area of transition surfaces accounts for only less than 10% of the whole lens surface area, therefore the dark stripes also have a small effect on the shape of light pattern. However, the uniformity of the light pattern on the illumination target can be reduced significantly [21]. The effects of manufacturing defects on the illumination performance of the freeform lens could be reduced by designing a continuous freeform lens, which will be introduced in the next section.

(4) Continuous freeform lens method

For the discontinuous freeform lens method mentioned above, discontinuous seed curves are introduced to guarantee the deviations between the real normal vectors and calculated ones lower than the threshold θ_{dth}. Otherwise, the direction of exit rays will deviate from the expected ones significantly. However, if we adopt another light energy mapping relationship between the light source and the target plane, it is possible that the deviations of exit rays will not significantly deteriorate the illumination performance and it is still acceptable for applications. Therefore, it is possible to design a continuous freeform lens for high quality illumination in this situation.

As shown in Figure 6.60, the light source's intensity distribution Ω is specified by coordinates (u, v), where u is the angle between the light ray and X axis, and v is the angle between the Z axis and the plane containing light rays and X axis. The volume Ω represents the total luminous flux ϕ_{total} of this one quarter light source and the luminous flux of unit

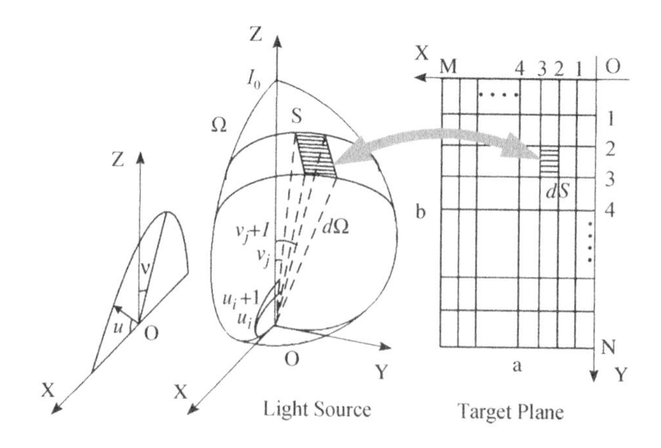

Figure 6.60 Schematic of light energy mapping between the light source and target.

object dΩ could be expressed as follows:

$$\phi(u,v) = \int_{v_j}^{v_{j+1}} \int_{u_i}^{u_{i+1}} I(u,v)\sin u\, du\, dv \tag{6.22}$$

where $I(u, v)$ is the luminous intensity distribution of light source. We divide Ω into M parts along u direction and N parts along v direction equally, and then each division point $S(u_i, v_j)$ on the surface of Ω could be obtained through solving Equations 6.23 and 6.24 as follows:

$$\int_0^{\pi/2} \int_0^{u_i} I(u,v)\sin u\, du\, dv = \frac{i\phi_{\text{total}}}{M} \qquad (i = 0, 1, \ldots, M) \tag{6.23}$$

$$\int_0^{v_j} \int_0^{\pi/2} I(u,v)\sin u\, du\, dv = \frac{j\phi_{\text{total}}}{N} \qquad (j = 0, 1, \ldots, N) \tag{6.24}$$

Rectangular grids are adopted on the target plane. The width (W_{grid}) and length (L_{grid}) of each grid on the target plane can be expressed as follows:

$$W_{\text{grid},i} = \frac{a}{M} \qquad (i = 0, 1, \ldots, M) \tag{6.25}$$

$$L_{\text{grid},j} = \frac{b}{N} \qquad (j = 0, 1, \ldots, N) \tag{6.26}$$

Then, according to the edge ray principle, we can establish a light energy mapping relationship between the unit object $d\Omega$ of the light source and the rectangular grid of the target plane. The following steps of the construction of the lens and validation of the design are the same as discussed above.

As shown in Figure 6.61, a continuous freeform lens is designed according to the new light energy mapping relationship. Compared to the discontinuous freeform lens designed in Figure 6.53, the surface of this lens is quite smooth. The lens forms a rectangular light pattern with a uniform illuminance distribution that is shown in Figure 6.61. It is found that the quality of the light pattern is better than that shown in Figure 6.53 for a discontinuous

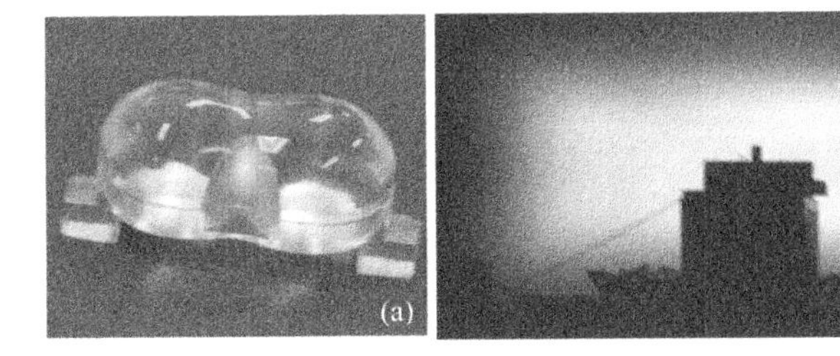

Figure 6.61 (a) Continuous freeform lens for LED road lighting; and (b) its illumination performance in laboratory.

freeform lens. These results demonstrate that the continuous freeform method without error control is an effective way to design a freeform lens for LED lighting.

Road illuminance is an important index in evaluating the road illumination performance, but it is only a reflection of the amount of received luminous flux per unit area on the road, not a true reflection of the light and dark situation of the road surface that the human eye can see. The visual object on the motorway is the driver and what he or she sees as the luminance on the road. Therefore luminance is an objective physical quantity which is close as to the practical situation that the human eye sees. At present, illuminance is being phased out of evaluation in North America, Europe, Japan and and so on. As to the luminance uniformity, freeform lenses are also of great usefulness.

Figure 6.62 shows the road illumination performance of a type of 250 W high pressure sodium lamp simulated with DIALux software and the simulated result evaluation is shown in Figure 6.63 in which we can see that the glare threshold increment of the road TI can meet the national demands but the overall uniformity of the road luminance U_0 and longitudinal uniformity of the luminance U_I are extreme low, far below the national standards. The low luminance uniformity generally generates an internally bright and dark zebra crossing, which leads to visual fatigue of the drivers and may well result in traffic accidents. Figure 6.64a is a type of 112 W LED road light with uniform luminance and a free form lens, and its simulation result is shown in Figures 6.64c and 6.65. The LED road light's overall uniformity of road luminance U_0 and longitudinal uniformity of luminance U_I are respectively 0.4 and 0.7, which satisfy the national standards and will not lead to drivers' visual fatigue. The glare threshold increment TI is seven, which is much lower than the national standard, thus ensuring that the driver will avoid the disturbance of glare and improve visual comfort and safety in night driving.

Freeform lens have become the development trend of the optical design of LED road lights due to their advantages of small volume, high light extraction efficiency, accurate beams control, and so on Figure 6.66 shows a few LED road lights based on freeform lenses.

(5) Integrated freeform lens for LED backlighting

Backlighting, especially for large-scale display equipment, is one of areas with the most rapid development in LED application. In recent years, the market for flat panel display (FPD) has become much larger with the development of consumer electronics. LCD flat panel display accounts for nearly 90% of the flat panel display shares. In the field of large-screen flat-panel TVs, which shows rapid development, compared with plasma TV, the LCD TV has many advantages in the market.

According to the optical structure, LED backlit modules are generally divided into edge lighting and direct lighting. The light sources of the edge lighting are placed at the edge of the panel and the rays are distributed onto the panel through a light guide plate, as shown in Figure 6.67. The light sources of the direct lighting are placed under the panel and the rays distributed onto the panel through a reflective cavity, shown in Figure 6.68. Generally, edge-lit backlighting is used in display equipment of medium and small sizes, while direct-lit backlighting is mostly used in backlighting equipment with a large size and a super large size.

A light guide plate is adopted in the edge lighting. When the panel area exceeds a limit, the light guide plate cannot distribute the rays on the panel uniformly. Compared with edge lighting, direct lighting does not adopt a light guide plate, so that its optical performance is

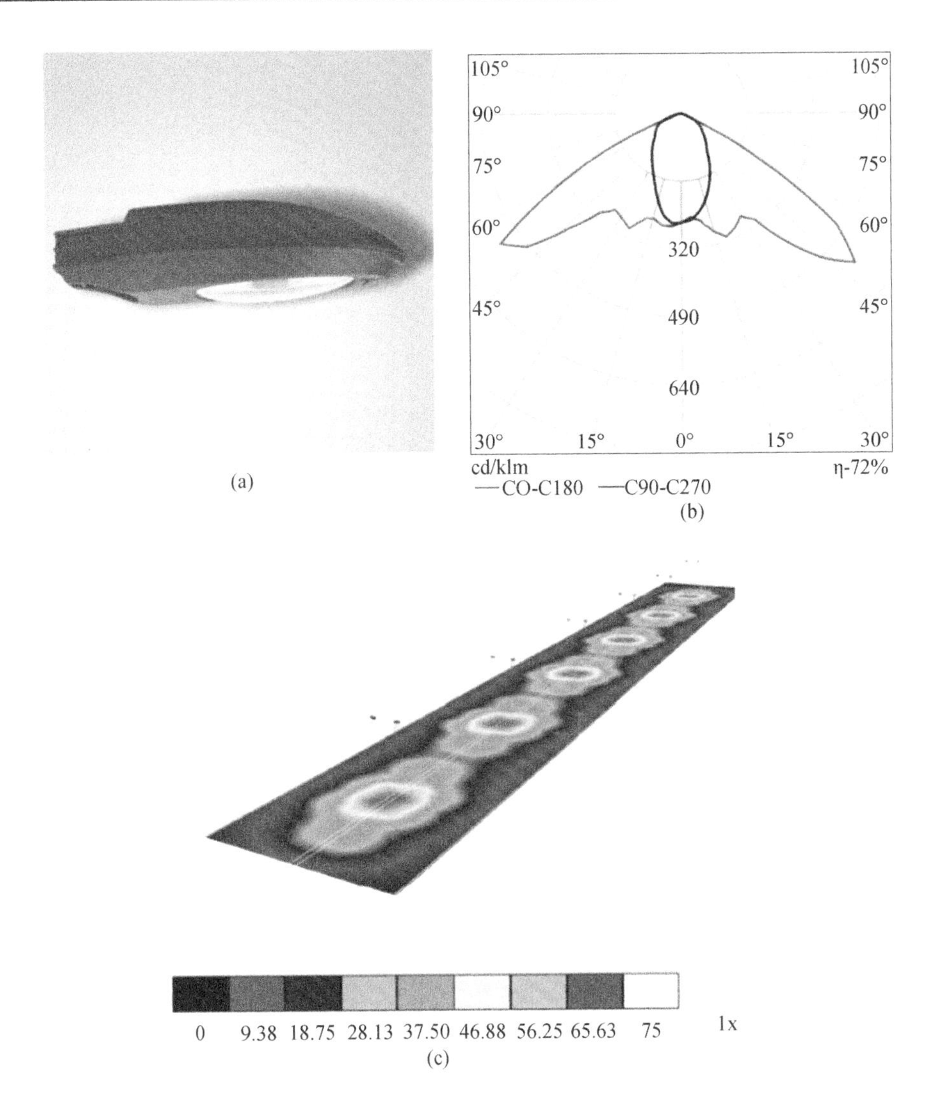

Figure 6.62 (a) A 250W high pressure sodium lamp; (b) its luminous intensity distribution curve; and (c) its simulation illumination performance. *(Color version of this figure is available online.)*

	Average Luminance [cd/m^2]	UO	UI	TI[%]
Simulation Results	1.24	0.1	0.3	2
Requirements (ME4a)	≥0.75	≥0.4	≥0.6	≤15
Satisfaction	✓	✗	✗	✓

Figure 6.63 Simulation results of a 250 W high pressure sodium lamp.

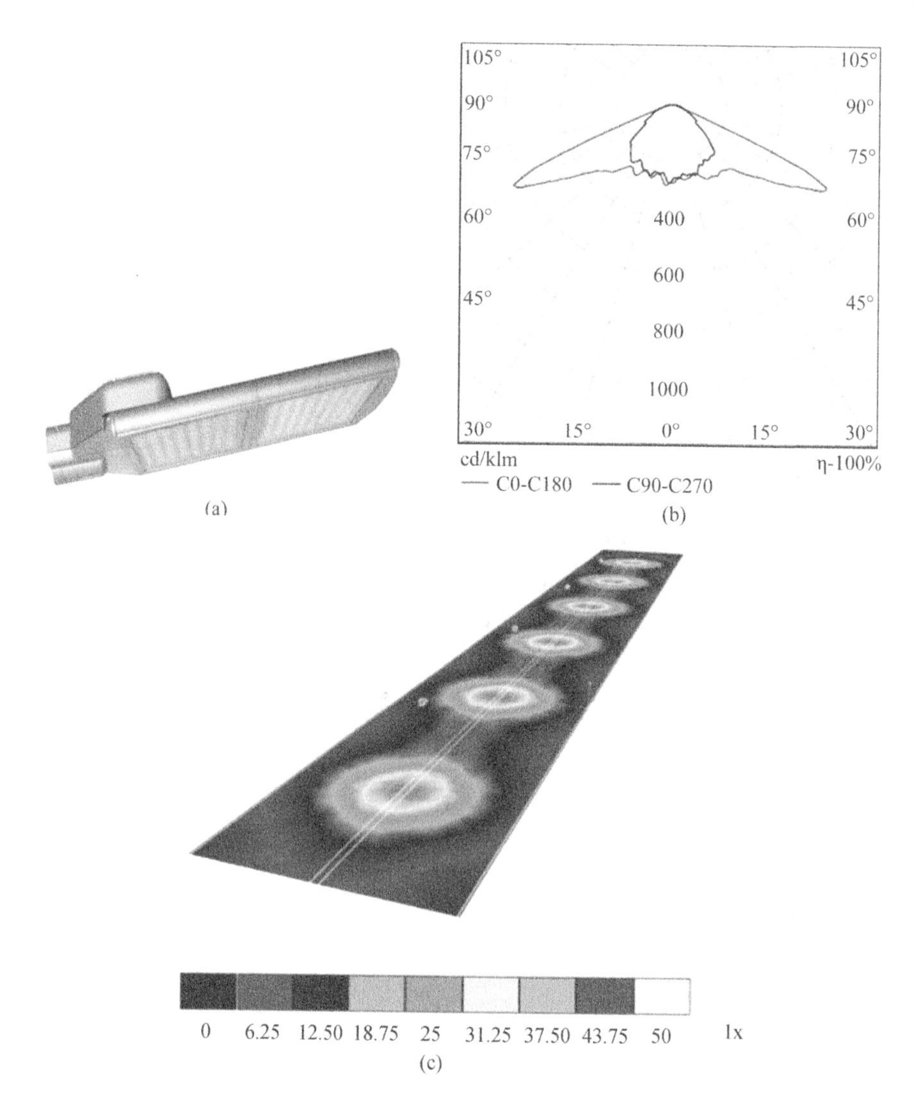

Figure 6.64 (a) A 112 W LED road light with continuous freeform lenses; (b) its luminous intensity distribution curve; and (c) its simulation illumination performance. (Reproduced with permission from www.gd-realfaith.com, Guangdong Real Faith Enterprises Group Co., Ltd., accessed April 12, 2011.) *(Color version of this figure is available online.)*

	Average Luminance [cd/m²]	UO	UI	TI[%]
Simulation Results	0.87	0.4	0.7	7
Requirements (ME4a)	≥0.75	≥0.4	≥0.6	≤15
Satisfaction	√	√	√	√

Figure 6.65 Simulation results of the 112 W LED road light. (Reproduced with permission from www.gd-realfaith.com, Guangdong Real Faith Enterprises Group Co., Ltd., accessed April 12, 2011.)

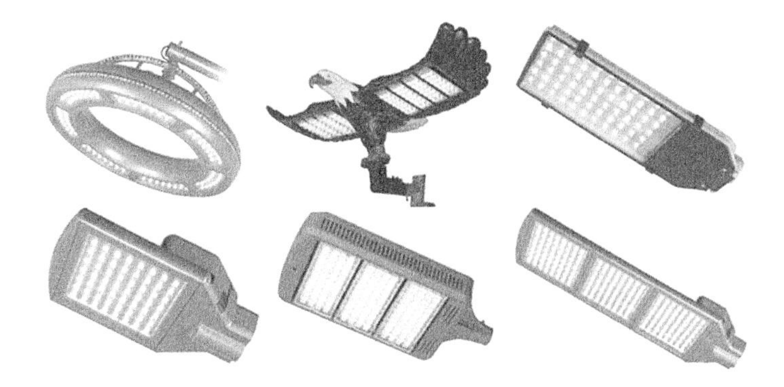

Figure 6.66 LED road lights based on freeform lenses on the market. (Reproduced with permission from www.gd-realfaith.com, Guangdong Real Faith Enterprises Group Co., Ltd., accessed April 12, 2011.)

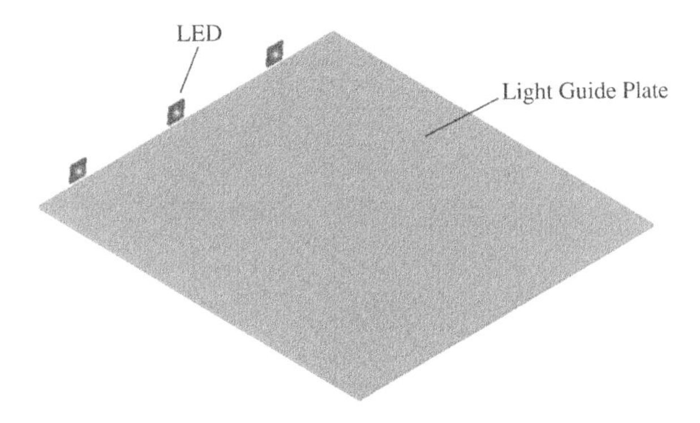

Figure 6.67 Schematic of edge lighting backlighting.

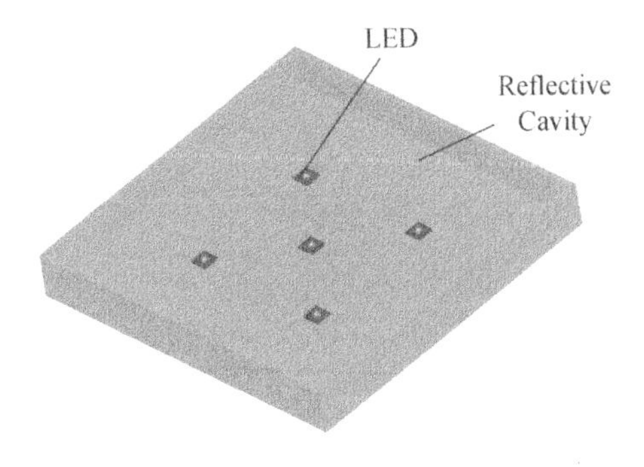

Figure 6.68 Schematic of direct lighting backlighting.

not affected by the size of the panel and good uniformity can be achieved on the large size backlighting. In addition, as in LED direct lighting, there is a certain relationship between every LED and the panel area, thus in a totally dark area, the corresponding LED can be turned off, which can effectively solve the problem of dynamic contrast reduction caused by the light leakage of the liquid crystal in a closed state. The following section will introduce the application of freeform lenses in direct lighting.

At present, the side emitter LED represented by Lumileds is the main technology of high power LED backlighting based on direct lighting. The application specific LED package (ASLP) adopting a compact freeform lens provides another viable technology for LED backlighting. The following section will introduce these two methods.

(a) Side emitter LED

Side emitter direct LED backlighting technology mainly adopts s a kind of side emitter LED advocated firstly by Lumileds [22]. A practical module is shown in Figure 6.69 and schematic of light extraction is shown in Figure 6.70. Most of the light energy is concentrated within ±20° in the horizontal direction to emit. The luminous intensity

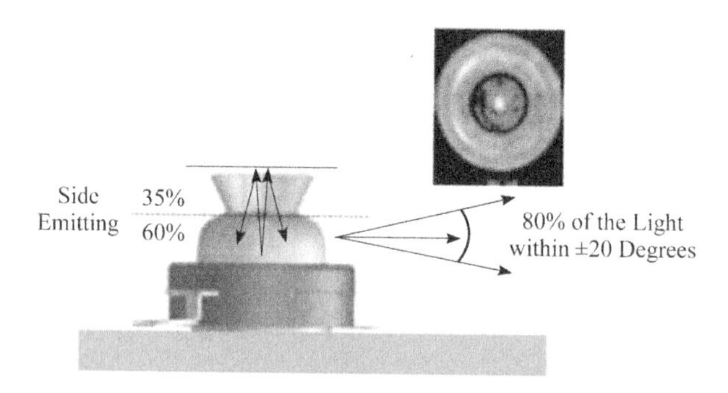

Figure 6.69 Schematic of Lumileds side emitter LED [22]. (This figure was reproduced from "High brightness direct LED backlight for LCD-TV," R.S. West, H. Kobijn, W. Sillevis-Smitt *et al.*, published in the *SID International Symposium Digest of Technical Papers*, **43**, 4, 1262–1265, 2003, with permission by The Society for Information Display.)

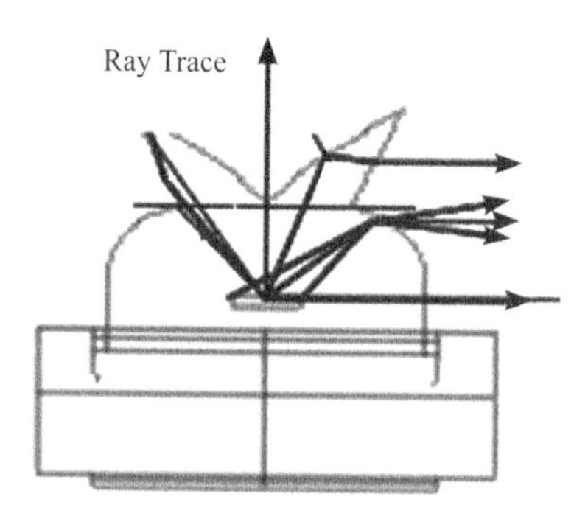

Figure 6.70 Light extraction schematic of side emitter LED [22].

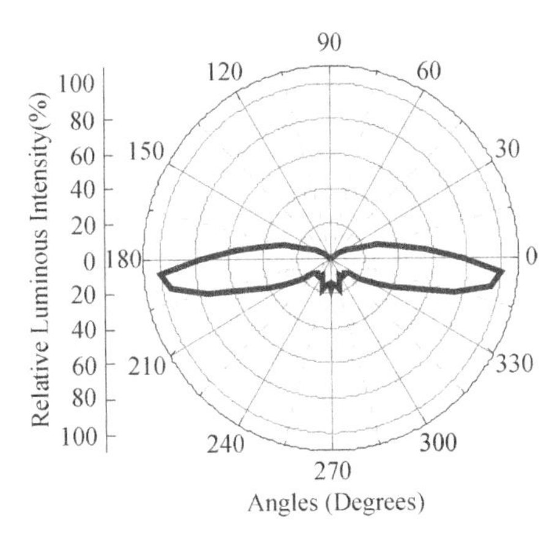

Figure 6.71 Luminous intensity distribution curve of the single side emitter LED.

distribution curve (LIDC) of a single side emitter LED is shown in Figure 6.71. Above the LED is placed a reflector plate with high reflectivity (~95%), which further reduces the light extraction of the LED in the vertical direction.

Every backlighting module (BLU) consists of four LEDs (1 red, 1 blue, and 2 green), as shown in Figure 6.72. 24 BLUs form two LED arrays and every array consists of 48 LEDs. The distance between two arrays is 94 mm, the height of the diffuser plate away from the light source is 50 mm, and the panel area is 503 mm × 282 mm. The interior wall of the reflective cavity which supports the light source and panel is made from materials with high reflectivity (~90%), as shown in Figure 6.73.

The principle of the direct lighting backlighting system is that most rays emitting from LEDs do not radiate toward the panel directly but reach the panel after the continuous reflection through the bottom surface as well as the side wall, and this kind of structure is provided with a very high illumination (luminance) uniformity. Moreover, during the long optical distance, three kinds of rays with different wavelengths are fully mixed, so this structure is also provided with a very high uniformity of chromaticity.

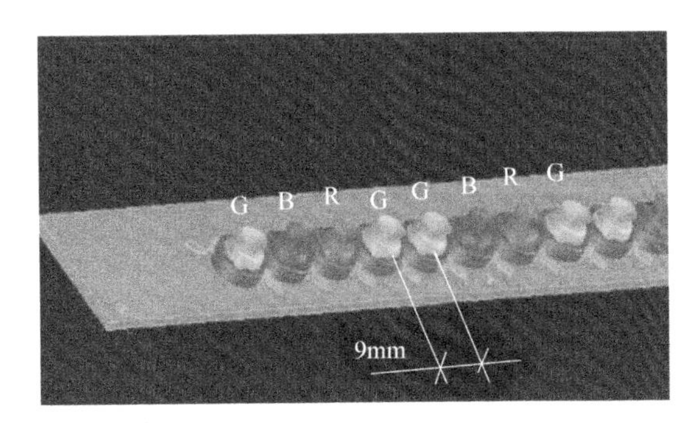

Figure 6.72 Schematic of BLU consists of four side emitter LEDs [22].

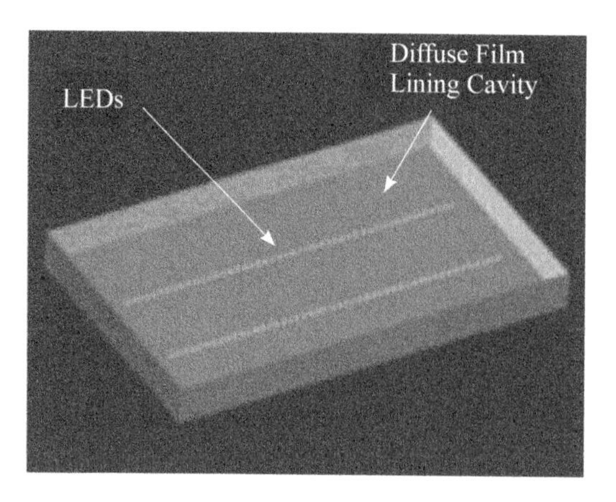

Figure 6.73 Backlighting system made of RGB side emitter LEDs and reflective cavity [22].

The simulated illumination light pattern on the panel (diffuser plate) and the chromaticity CIE coordinates are shown in Figures 6.74 and 6.75 and both of them can meet the demands of display backlighting after going through the diffuser and BEF (Brightness Enhancement Film). However, every time the rays are reflected by the internal wall, there must be a certain amount of absorption loss. As in this structure, the rays are reflected by the internal wall many times, the total loss is great. The simulation results show that the efficiency of the system is 65.7% (excluding the efficiency of the LED primary lens). Considering that the primary lens's light extraction rate of side emitter LED is not high, half of the light energy from the LED chip cannot be effectively utilized in this system.

(b) ASLP

According to the customers' requirements of light energy distribution, ASLP can design the primary freeform lens which can meet the demands directly [23]. When ASLP is applied in backlighting, the primary principle is that the light emitted from every LED corresponds to a light pattern with certain area and certain uniformity on the panel. A rectangular light pattern with large size and high uniformity, which is demanded in the backlighting system, can be obtained through the combination and superimposition of single light patterns.

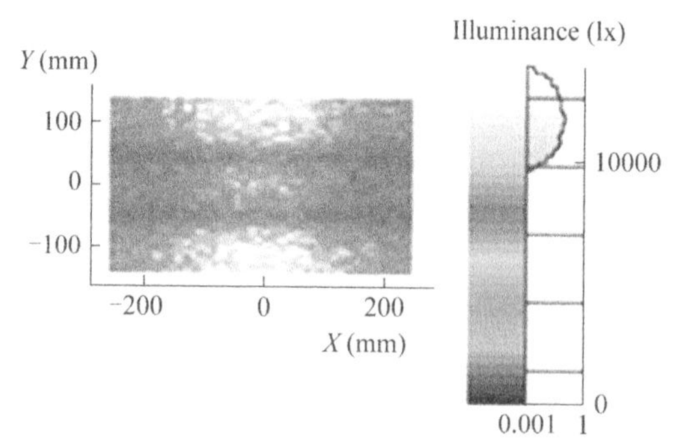

Figure 6.74 Illuminance distribution on the panel of the side emitter LED direct backlighting system (50 mm high). *(Color version of this figure is available online.)*

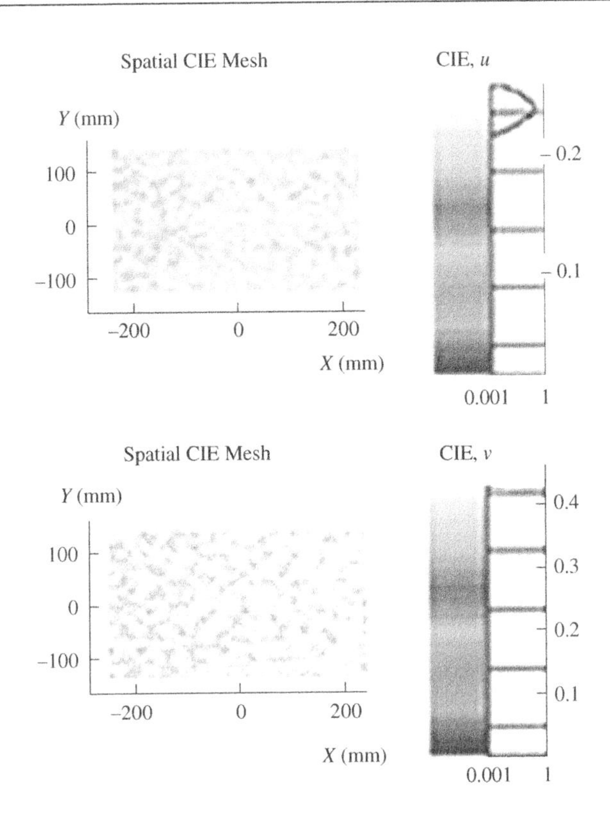

Figure 6.75 Chromaticity CIE coordinates distribution of the side emitter LED direct backlighting system (50 mm high). *(Color version of this figure is available online.)*

The major difference between ASLP LED backlighting and other direct LED backlighting is that ASLP can accurately control the light pattern formed by a single LED. The single light pattern can be sharp-edged uniform rectangle, or sub-rectangle with graded edge, or circular pattern, and so on. The actual design should be decided according to the size and height of the panel, the power of LED, the structure of heat dissipation, and so on Moreover, no matter what the corresponding light pattern of the single LED is, most of the lights emitting from the light source in ASLP LED backlighting reach the panel (diffuser plate) directly. Furthermore, the light extraction rate of the primary lens in ASLP is almost 100%, and the efficiency of the whole backlighting system is very high (~90–95%).

In order to show the advantages of ASLP LED backlighting when compared with the side emitter LED's system, the direct backlighting with size of 22 inches (503 mm × 282 mm) is redesigned. Supposing that the thickness of the backlighting panel is 40 mm, a type of freeform lens which forms a 200 mm × 200 mm rectangular light pattern with graded edge at the height of 40 mm is designed. The light pattern is shown in Figure 6.76. Figure 6.77 shows the BLU is made of four ASLPs (1 red, 1blue, and 2 green).

20 BLUs are distributed at the bottom of the backlighting panel with the array of 4 × 5, shown in Figure 6.78 and the distance between BLUs has been optimized. The illuminance

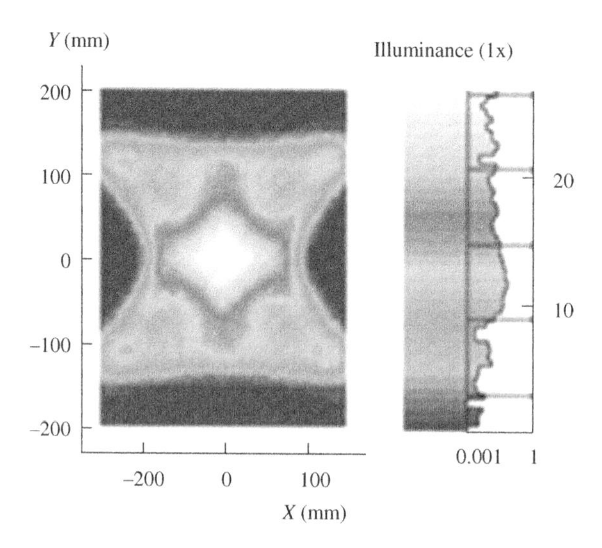

Figure 6.76 Light pattern of a single ASLP for direct backlighting at 40 mm height. *(Color version of this figure is available online.)*

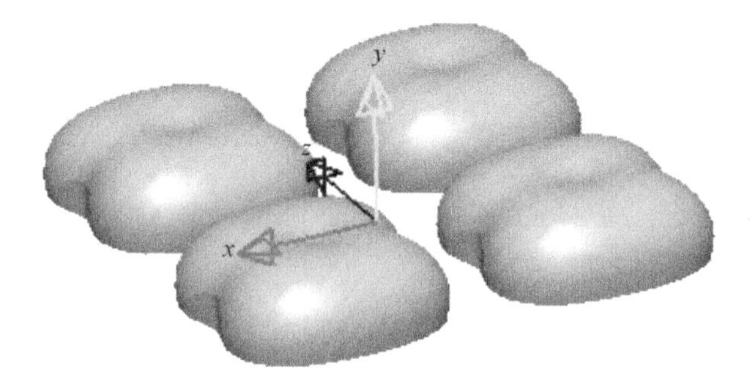

Figure 6.77 Schematic of a group of BLU with 4 ASLPs for direct backlighting.

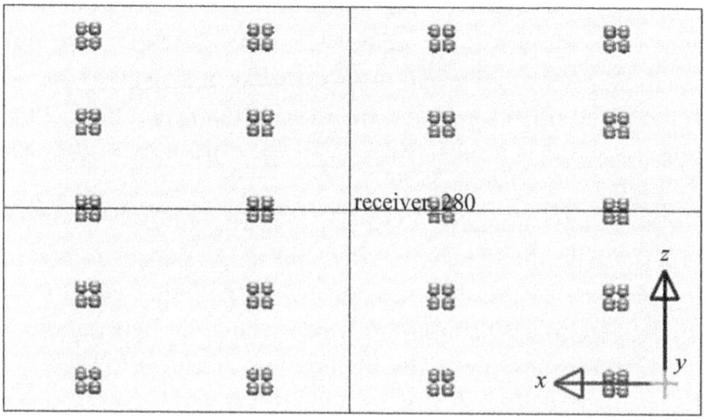

Figure 6.78 Schematic of the ASLP direct LED backlighting system.

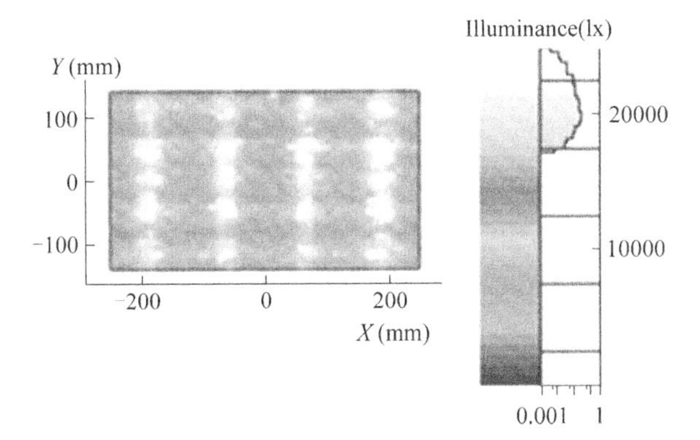

Figure 6.79 Illuminance distribution on the panel of the ASLP LED direct backlighting system. *(Color version of this figure is available online.)*

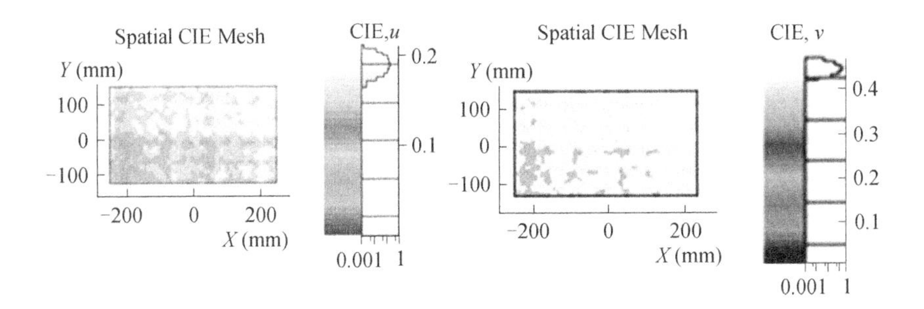

Figure 6.80 Chromaticity CIE coordinates distribution of the ASLP LED direct backlighting system. *(Color version of this figure is available online.)*

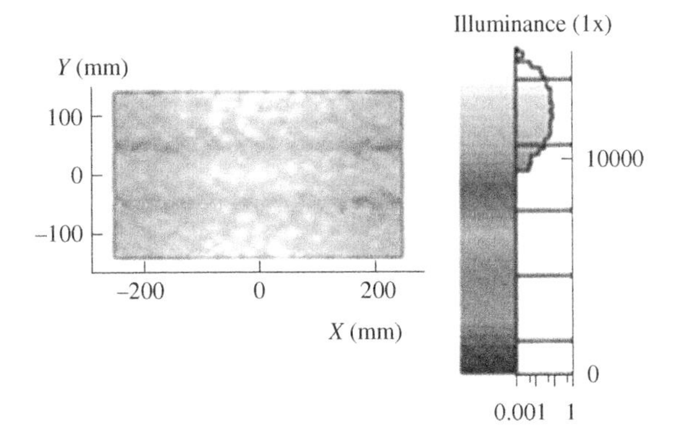

Figure 6.81 Illuminance distribution on the panel of the side emitter LED direct backlighting system (40 mm high). *(Color version of this figure is available online.)*

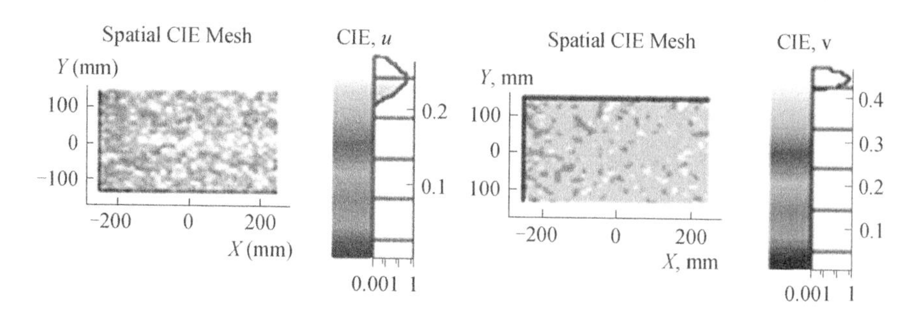

Figure 6.82 Chromaticity CIE coordinates distribution of the side emitter LED direct backlighting system (40 mm). *(Color version of this figure is available online.)*

Table 6.2 Comparison between ASLP LED backlighting and side emitter LED backlighting (without diffuse pate and BEF)

	ASLP	Side Emitter
LED Numbers	80	96
Light Efficiency	96.0%	<67.3%
Reflectivity of the Bottom Surface and the Side Wall	90%	90%
Illuminance Uniformity	0.828	0.769
Chromaticity Uniformity ($\Delta u'v'$)	0.031	0.042

distribution on the panel and the chromaticity CIE coordinates distribution of the ASLP LED direct backlighting system are shown in Figures 6.79 and 6.80 respectively.

The side emitter LED backlighting system which is 40 mm high has also been modeled for the comparison. The illuminance distribution on the panel and the chromaticity CIE coordinates distribution are shown in Figures 6.81 and 6.82. The comparison between these two backlighting systems is shown in Table 6.2.

From Table 6.2, we can see that compared with side emitter LED direct backlighting, ASLP LED direct backlighting not only enjoys great advantages in light efficiency, but also in illuminance uniformity which can be improved through methods such as structure optimization, embedded diffuser plate, and so on. Compared with the side emitter LED, ASLP enjoys the incomparably congenital advantage in its high efficient light extraction. Furthermore, each single ASLP is in one-to-one correspondence with the panel area which is more beneficial to the realization of dynamic contrast technology.

6.1.4 Diffuser

(i) Diffuser for LED Backlighting

An LCD backlighting system, is mainly composed of a light source, light guide panel, and reflective film (direct backlighting does not need a light guide), diffuser, and BEF, as shown in Figures 6.83 and 6.84. The light guide transfers linear light source (for example, CCFL) or point light source (for example, LED) into surface light source vertically emitted through its internal optical structure, but the light emitted by the light guide has problems such as low uniformity, nonuniform color mixing,

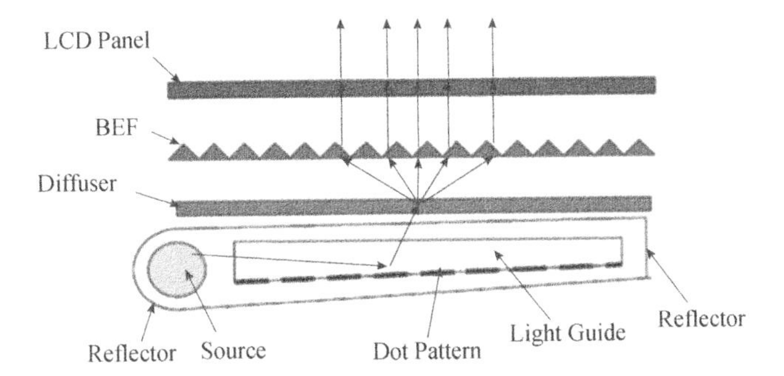

Figure 6.83 Schematic of structure of LCD backlighting panel (side-down).

and so on. If the light is supplied directly from the light guide for the brightness enhancement film, it is difficult to meet the requirements of LCD backlighting. The role of the diffuser is to provide a uniform surface light source for the LCD screen by making light emitted from the light guide more uniform through its own scattering effect.

The optical property of the diffuser is characterized by diffuse transmittance, diffuse reflectance, and the degree of light atomization. The diffuser should also have properties of relatively low thermal deformation and moisture deformation to prevent the fluctuation of the optical properties caused by a change of external dimensions. In addition, the diffuser should have a good UV absorption property.

In recent years, in order to reduce the thickness of the backlighting panel and the cost of purchasing the diffuser, there have been manufacturers integrating the diffuser into the light guide panel, realizing the diffusion effect by the use of the top surface of the light guide panel. This method still has the same mechanism for realizing light diffusion without using the diffuser.

There are many ways for the diffuser to realize diffusion and there are various patent technologies in this field, among which common ones are: (1) adding chemosynthetic light diffusing impurity into the transparent material; (2) fabricating optical micro-structure; (3) printing special ink; and (4) using polymer materials of graded refractive index, and so on. The following section will introduce the first two methods.

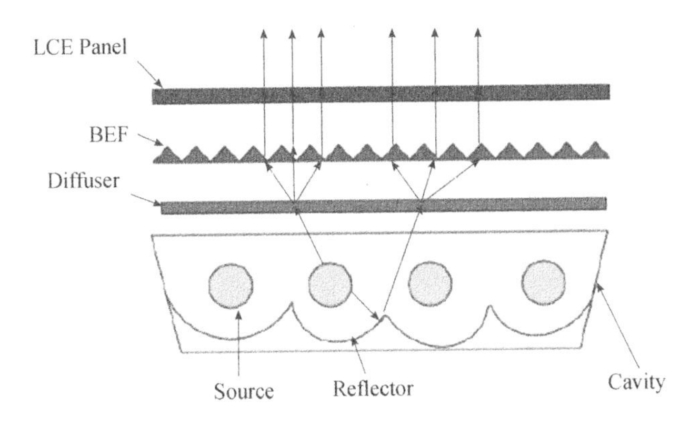

Figure 6.84 Schematic of structure of LCD backlighting panel (direct).

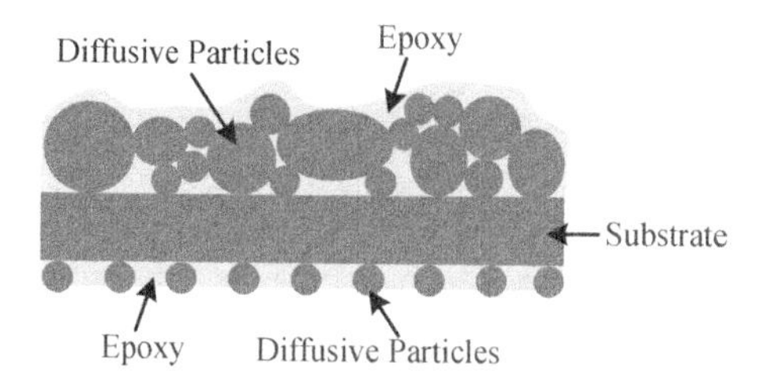

Figure 6.85 Structure of the backlighting diffuser.

(1) The structure of a diffuser using light diffuse impurity includes substrate and light diffuse film, as shown in Figure 6.85. The materials of the substrate are mostly PMMA, PC, or its derivative copolymers of small thermal deformation, small moisture absorption, and high light transmittance. The diffuse impurities used by the diffuser include silicone polymer, methyl methacrylate-styrene (MS) copolymers, and so on. The chemical particles added into the diffusion film material disperse in resin layers as scattering particles. As shown in Figure 6.85, light will continuously pass through two mediums with different refractive indices when passing by a diffusion layer, at the same time a lot of light refraction, reflection, and scattering will happen, which results in light diffusion. In some new diffuse structures, the diffuse impurity can be added into substrate to enhance the atomization ability of the diffuser. In addition, the ultraviolet absorbent (for example, benzotriazole) is added according to specific situations.
(2) Another common diffuser structure is the array optical micro-structure with scattering function. For example, Chang *et al.* introduced a diffuser using a conical curved surface microlens array (MLA) [24]. Its profile structure diagram is shown in Figure 6.86.

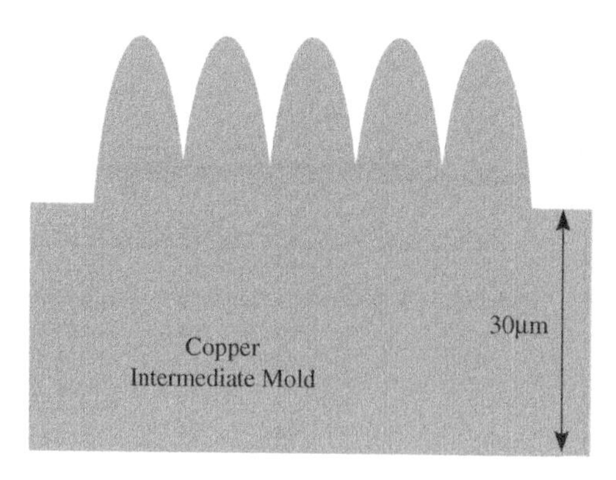

Figure 6.86 Schematic of a diffuser using conical curved surface lens array [24]. (Reprinted with permission from S.I. Chang and J.B. Yoon, "Microlens array diffuser for a light-emitting diode backlight system," *Optics Letters*, **31**, 20, 3016–3018, 2006. © 2006 Optical Society of America.)

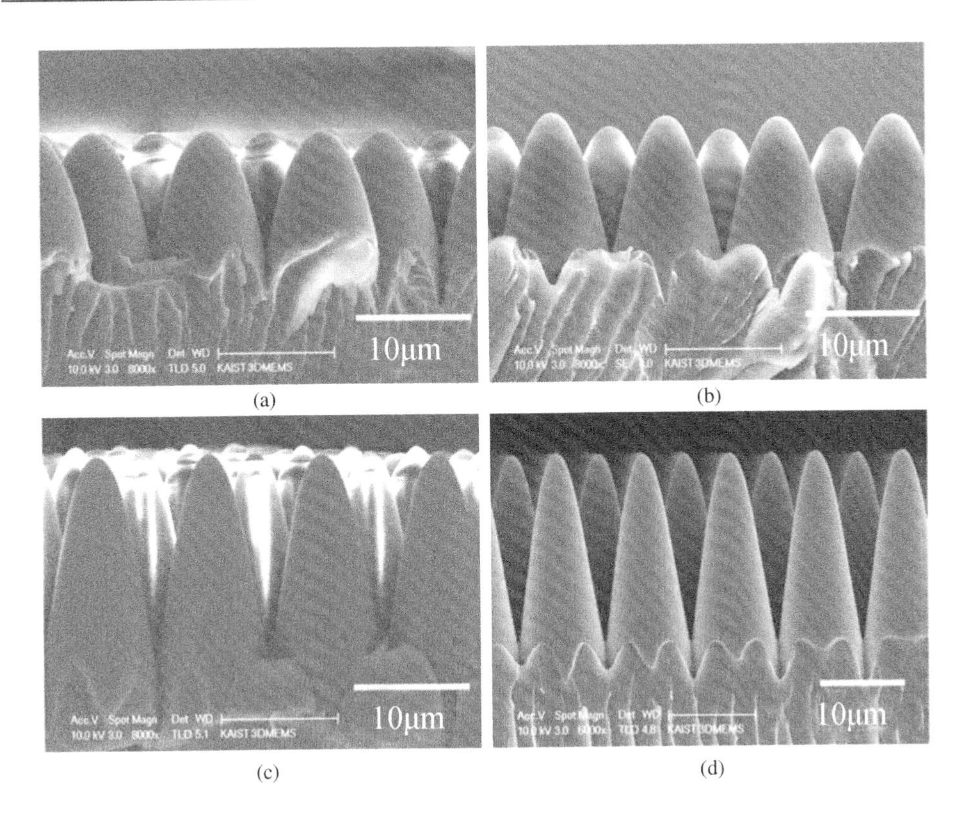

Figure 6.87 Scanning electron microscopy photographs of the fabricated conical curved surface lenses arrays [24]. (Reprinted with permission from S.I. Chang and J.B. Yoon, "Microlens array diffuser for a light-emitting diode backlight system," *Optics Letters*, **31**, 20, 3016–3018, 2006. © 2006 Optical Society of America.)

The scanning electron microscopy photographs are shown in Figure 6.87. The profile of the microlens can be expressed as:

$$h(r) = \frac{1}{R}\frac{r^2}{\sqrt{1+[1-(K+1)r^2/R^2]}} \tag{6.27}$$

Where h is the height of the microlens, $r = (x^2 + y^2)^{1/2}$ is the distance from the surface of the curvature to the optical axis, R is the radius of the curvature at the vertex, and K is the aspherical constant.

The spatial distribution change of light energy before and after using this MLA diffuser is shown in Figure 6.88. We can find that this MLA diffuser improves the light which concentrates energy in a vertical direction to approximate a batwing type.

(3) If fabricating an optical structure on the surface of the light guide panel, the light guide panel and diffuser will be integrated together. A structure of scattering netted dots fabricated on the surface of the light guide is shown in Figure 6.89 [25]. The light emitted from the light source is changed into a surface light source in the light guide and scattering is realized at the same time.

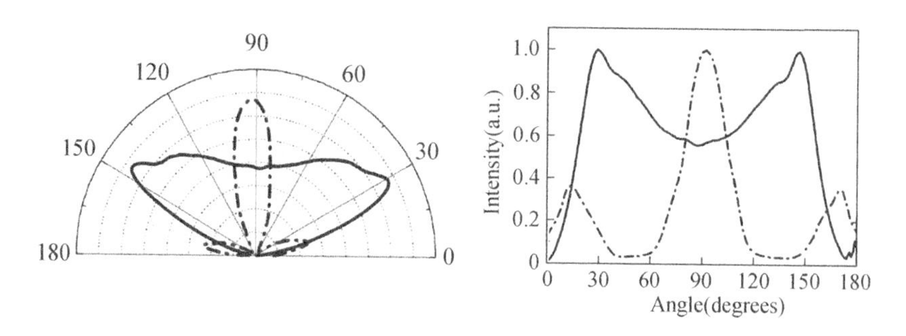

Figure 6.88 Light energy spatial distribution with (solid line) and without (dashed line) the MLA diffuser [24]. (Reprinted with permission from S.I. Chang and J.B. Yoon, "Microlens array diffuser for a light-emitting diode backlight system," *Optics Letters*, **31**, 20, 3016–3018, 2006. © 2006 Optical Society of America.)

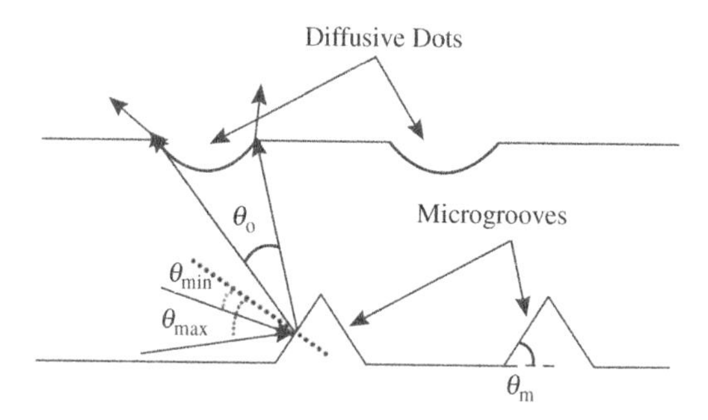

Figure 6.89 An optical structure integrating light guide with diffuser [25].

(ii) Diffuser for LED Daylight Lamp

LED luminaires need to provide a soft and comfortable lighting environment for indoor applications. However, because the LED is a point light source, it is prone to glare and a good lighting environment cannot be provided if directly applying LED daylight lamps to lighting. The diffuser/diffusion film can effectively diffuse light and uniformly mixes light before it is emitted from the lamp to make the light emitting soft and comfortable, becoming the first choice of high-quality LED indoor lamps, as shown in Figure 6.90.

The structure of the diffuser of the LED daylight lamp is similar to that used in backlighting which is composed of a substrate and diffusion layer, as shown in Figure 6.91. The substrate mostly uses PMMA, PC, or PTFE, and so on. The diffusion layer is composed of the glass beads, PMMA beads, ethyl polyamines beads, and so on, playing the role of scattering light. The adhesive layer is used to bond the substrate and diffusion layer and some diffusers do not use bonding but spraying, printing, and so on.

The LED daylight lamp is composed of several LED arrays, but the quality of lighting needs to be improved through a diffuser. Most manufacturers have installed a diffusion layer inside the lamp tubes, making the light emitting soft and uniform, as shown in Figure 6.92. Some also

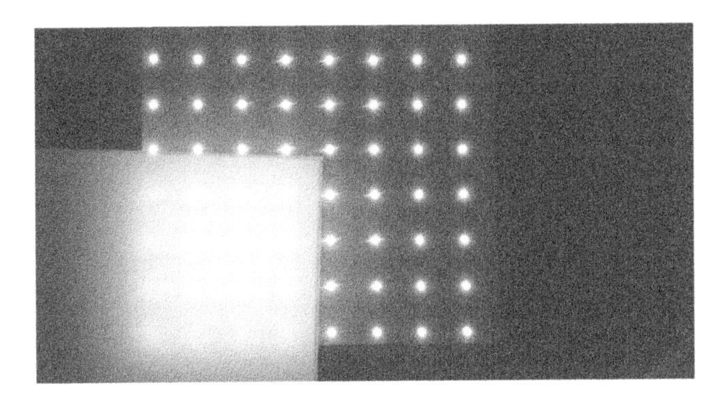

Figure 6.90 LED diffuser.

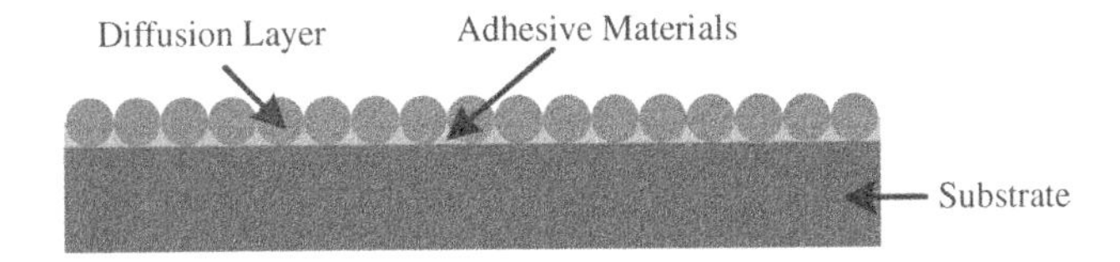

Figure 6.91 Schematic of LED daylight lamp diffuser.

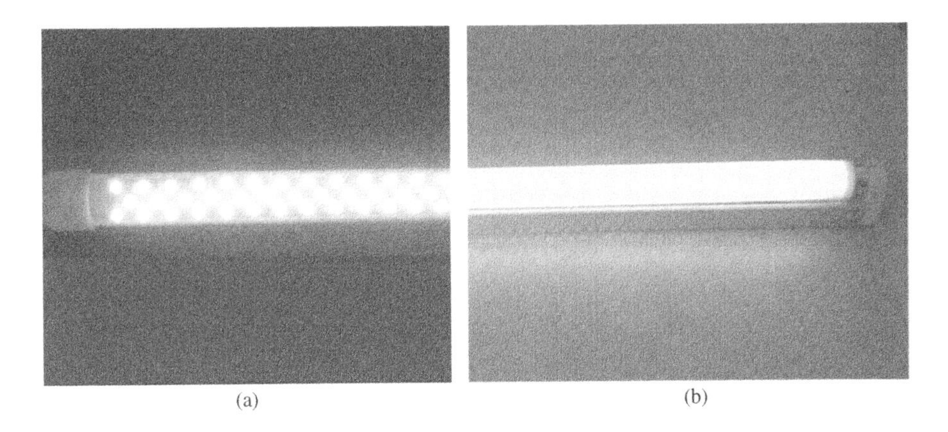

Figure 6.92 (a) An ordinary LED daylight lamp and (b) an LED daylight lamp with diffuser.

set a diffuser in the light extraction port of the grid lamp to play the role of mixing light, as shown in Figure 6.93.

6.1.5 Color Design and Control in LED Applications

From the invention of the first LED, the LED has include bands from the infrared to the ultraviolet band, especially the nitride blue LED essential in white lighting developed by

Figure 6.93 An LED grid lamp with diffusers.

Japanese scientist Shuji Nakamura in 1993. Blue and green LEDs using InGaN as main materials are realized in mass production which marks the entering of the era of white LED. The use of the LED develops from monochromatic indicator lighting to white-based general lighting.

Color design and control is not the primary problem of the monochromatic LED lighting, but in the era of white-light lighting, the color temperature of the light source, color rendering, and color stability directly influence user's comfort, and the importance of color design and control is equal to that of light efficiency.

(i) Color Design in LED Applications

At present, there are three methods of realizing white LED lighting:

(1) *Blue LED chip plus yellow phosphor*: a GaN-based LED chip emits blue light to stimulate the YAG phosphor to emit yellow light. Yellow and blue light are mixed to produce white light, the color temperature of white light can be controlled by adjusting the amount of phosphor. The white LED technology is simple and mature and has a high luminous efficiency. It is the main method of producing white LED lighting. Due to the lack of red light, however, it is very difficult to achieve low color temperature. Using CRI, it is hard to achieve a high level of more than 90 and a great difference exit for individual LEDs in terms of color uniformity. It is suitable for road lighting, lighting in public places, to name a few.
(2) *UV–LED chip plus RGB trichromatic phosphors*: obtain tricolor light through exciting multicolor phosphor by near-ultraviolet emitted by the LED chip and mix them to form white light. The color rendering and color uniformity of white light obtained by this method has greatly improved as compared to method (1), but the biggest problem is that currently the wavelength conversion efficiency of multi-color phosphor to ultraviolet is not high. The overall luminous efficiency for the UV–LED chip is low, not being able to demonstrate the energy-saving advantages of LED. Moreover general plastic packaging

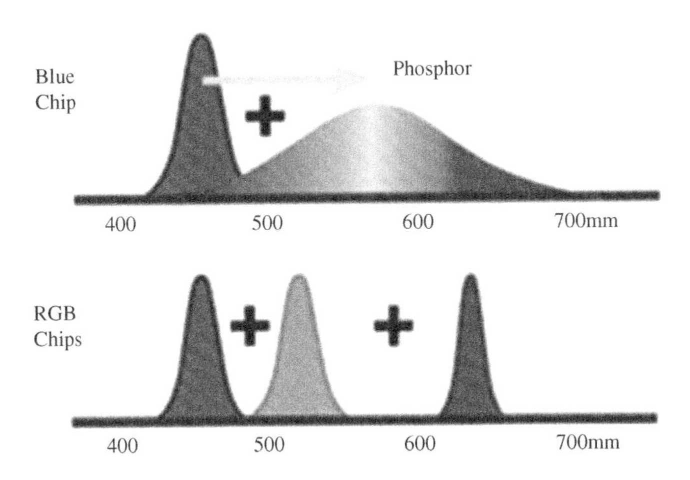

Figure 6.94 Comparison of spectrum of white light through two different approaches. *(Color version of this figure is available online.)*

gel is UV-sensitive and plastic packaging degeneration can cause a considerable degree of luminous decay. The application of this approach is rarely used currently and significant efforts are needed in research.

(3) *RGB trichromatic LEDs*: the lights emitted by LED chips of three different wavelengths (red, blue, green) are mixed to form white light. It can be seen from the principle of additive color mixing that any color within the scope of visible light can be obtained by mixing RGB trichromatic with different proportions. Color temperature can be precisely controlled by independently adjusting different driving luminous fluxes of LEDs of different wavelengths, and the white light obtained by this method has a relatively wide spectrum and high color-rendering index. The comparison between the white light spectrum of trichromatic LEDs and the white light spectrum based on blue LEDs and phosphor LEDs is shown in Figure 6.94.

The color design of LEDs is determined by using either a monochromatic or white LED according to different application occasions. If using a white LED, the way of determining color mixing is in accordance with the requirements of color temperature and color-rendering index. One example was to carry out the design of LED lighting for office use [26] where white light needs to be used. The office considered had a certain requirement for color rendering of the light source. An LED designed by RGB trichromatic, was chosen as it was believed to be superior to a design using a blue LED and yellow phosphor. Color temperature of the light source needed to be determined, because a cool light source was believed to help increase work efficiency, and color temperature was therefore set to be 6500 K. The CIE chromaticity coordinates (0.3135, 0.3237) were obtained according to the color temperature. The chromaticity coordinate was changed into a ratio of the RGB trichromatic luminous flux, $\phi_r : \phi_g : \phi_b = 0.3005 : 0.5879 : 0.1116$ by a color equation (Equation 6.28). A technical manual produced by LED manufacturers was required for the quantitative determination of the proportion and driving current of an LED of different wavelengths.

$$\begin{pmatrix} \frac{X_w}{Y_w} \\ \frac{1}{Y_w} \\ 1 \end{pmatrix} = \begin{pmatrix} \frac{X_r}{Y_r} & \frac{X_g}{Y_g} & \frac{X_b}{Y_b} \\ \frac{1}{Y_r} & \frac{1}{Y_g} & \frac{1}{Y_b} \\ 1 & 1 & 1 \end{pmatrix} \begin{pmatrix} \phi_r \\ \phi_g \\ \phi_b \end{pmatrix} \tag{6.28}$$

where (x_r, y_r), (x_g, y_g), (x_b, y_b) are chromaticity coordinates of RGB triomatic chromaticity, and (x_w, y_w) is chromaticity coordinates of white light.

Here is an introduction of color design in some typical LED application fields.

(1) LED display
The LED full color display screen as shown in Figure 6.95 is widely used in stations, banks, securities, hospitals, stadiums, municipal squares, stages and airports, to name a few. With the LED full color display screen, there are mainly three problems that should be considered: (5) how to improve color uniformity of LED display screen; (6) how to raise its degree of color reduction; and (7) how to expand its color gamut in order to save more natural color.

(2) LED interior lighting and decoration
LED interior lighting like that shown in Figure 6.96 mainly refers to or depends on interior decoration luminaires. As to interior decoration, color LEDs are widely used in public building places and develop rapidly due to the special need in lighting in public areas. The LED serves as a special illuminating tool for architects and designers. Nearly all colors can be produced by using an LED with various colors such as red, orange, yellow, green, blue, and white. Apart from this, a variety of flexible forms can also be made, such as points,

Figure 6.95 LED displays. *(Color version of this figure is available online.)*

Figure 6.96 LED interior lighting. *(Color version of this figure is available online.)*

lines, circles, and three-dimensional shapes, and even some special forms that can be shaped according to different customers' requirements. It is small, light, and thin in appearance and can be easily controlled. The color LED is also popular in stages, ground decoration, and for step indication in theatres, to name a few. The special illuminating effects are produced by using an LED, which can be proven from its applications in entertainment places, building facades, and some landscapes. In a word, the development in the use of the LED is rapid in interior lighting. There are no limits for the lighting performances that can be achieved and the only limit is the imagination of the designers. Furthermore, the LED will be widely accepted in the architecture community if there is a further lowering of the cost, a raising of the light output, and a guarantee of the reliability.

The main features of LED interior lighting are as follows: (1) good color rendering: the LED luminaires with high color rendering index (generally up to 90) are usable in kitchens, counters, and surgery operation tables and so on; (2) intelligence: presently, it is reflected in LED lighting control and circuit design and mainly concentrates on the fact that the color and the gray scale are adjustable. Guided by this function, the actual brightness and color needed for the LED luminaires can be adjusted for practical situations. In this way, the glare is lowered and energy is further saved, and what's more, the artistic awareness of the space is intensified. In addition, in certain situational lighting, the shift of light is controlled by music or adjusted by the atmosphere one needs to create. The design of the LED lighting is humanistic.

(3) LED traffic signal lighting

Having devoted many years to replacement work, the main cities in China have nearly replaced all the traditional traffic lights with LED ones as shown in Figure 6.97. Now the traffic lights signaling red, yellow, and green on the urban road are almost all LED ones.

Figure 6.97 LED traffic signal lights. *(Color version of this figure is available online.)*

(4) LED outdoor landscape lighting
LED outdoor landscape lighting like that shown in Figure 6.98 has a distinct advantage in decorating the contours of the architecture. It is able to present the carvings and the portrayals on the architecture in more detail and more realistically. Meanwhile, as the need for urban nightscape lighting becomes more urgent, the LED will have a promising market. Urban landscape lighting is so huge an engineering project that it not only needs the help of high strength gas discharge lamps to set the keynote of the urban lighting, but also needs some variable colors adorned by LED lighting. Thus, the LED will demonstrate its power by its energy saving, reliability, richness in lights, colors, and forms.

(5) LED automobile lighting
Applications of automobile lighting mainly include:

(a) Vehicle interior light source: The LED was first used as an indicator lamp on vehicle dashboards and many automobile manufacturers have already used the LED as indicator lamps of dashboards and control units. Recently, because of the reduction in price, the LED has been further applied to all vehicle lamps including interior reading lamps and footwell lamps.

(b) Signal light: It has been 20 years since the LED has been used as a high-mount brake lamp. So far, the global market share of LED high-mount brake lamp is about 40%. Manufacturers have applied the LED to the vehicle rear lamp group such as brake lamps and steering lamps.

(c) Headlamp: It is a relatively demanding part among the automobile light sources. Although there exist some disadvantages of the LED, such as deficiency in brightness

Figure 6.98 LED outdoor landscape lighting. *(Color version of this figure is available online.)*

per unit and a high price, and there exists great competition from the HID lamp in the market, the LED has been used in some future concept vehicles and a few high end vehicles. However, judging from the past development track of HID as automobile headlamps and the future technical progress of the LED, it is predicted that LED headlamps will be commercialized in the next few years.

(ii) Color Control in LED Applications

Besides its richness in colors and having a good color rendering, the LED has another advantage in that it can be easily controlled dynamically. Generally speaking, it is difficult for traditional luminaires to adjust their colors and color temperatures. Thus in order to achieve dynamic illumination in different situations, lots of luminaires should be used. But as for the LED, especially for the tricolor white light LED, it is easy to change between different colors and also between different color temperatures by only adjusting the input current. The LED has the following light mixing methods: (1) analog; (2) pulse width modulation (PWM); and (3) frequency modulation. For example, in a Boeing 787 airliner, a technique that "simulates sky light" is used in the LED illumination luminaires in the passenger compartment. The LED light source can simulate the dawn–dust change in a day by changing its color temperature and brightness in order to improve the sleep quality of passengers on a long trip.

As for the most prospective tricolor white light LED lighting, how to keep the stability of color has become another important issue. The light extraction properties of the LED with different wavelengths vary with the junction temperature, and their light degradation and failure properties are also different. After being in use for a period of time, the rate of the main wavelength, half-width of spectrum, and peak power of the LED with different wavelength will change. The change will cause the color temperature to drift in mixed light. It is reported that the extent of this color temperature drift can reach over 1000 K, which affects the use severely. In order to characterize the offset of color, color difference $\Delta u'v'$ in the RGB system (u, v) based on CIE1960 is used:

$$\Delta u'v' = \sqrt{(u - u_0)^2 + (v - v_0)^2} \tag{6.29}$$

where (u_0, v_0) represents chromaticity coordinates of the needed colors, and (u, v) represents chromaticity coordinates of the real color of light source. $\Delta u'v'$ is generally required to be under 0.003.

Recently, the solution to color temperature drift is achieved through feedback control. Generally speaking, feedback control can be mainly divided into the following classes:

(1) Temperature feedback control
The luminescence properties of the power-type LED are very sensitive to junction temperature, the changing of which will cause a change in light efficiency and wavelength. But the junction temperature is hard to measure directly. The general method used recently is to measure the temperature of the heat sink and to adjust the input current dynamically according to the relationship between the temperature and the luminous flux and wavelength. The disadvantage of this method is that it cannot reflect the color temperature drift caused by different light degradation for LEDs with different wavelength.
(2) Luminous flux feedback control
Measure luminous flux of every kind of LED independently by sensors, and then compare them to their initial luminous fluxes, and finally adjust the input current dynamically to ensure the stability of the chroma of the mixed white light. This method can compensate for the defects of color temperature drift caused by different light degradation and failure in the LED with different wavelengths, but it is incapable of solving color temperature drift caused by the drift of main wavelength.
(3) Temperature and luminous flux feedback control
This method combines the advantages of both temperature feedback control and luminous flux feedback control. It can greatly compensate for the defects caused by different light degradation and failure, as well as main wavelength drift. However, it also has the weakness of being a rather complicated system.
(4) Chroma feedback control
This method can measure the spectrum of mixed white light by adding a color filter to the photoelectric sensor. Match the spectrum with the color matching function to get the chromaticity coordinates of white light. Then adjust the input current of different wavelength LEDs by a control system. This method is immune to temperature and time which influence the stability of color, and its adjustment accuracy is high, but its cost is also high.

(iii) LED Color Temperature Selection in Road Lighting (Figure 6.99)

At present, because light extraction efficiency (l m/W) of high power LED with neutral white light (3500–4500 K) or cool white light (4500–8000 K) is generally higher than that with warm white light (2600–3500 K), most LED road lights adopt an LED with neutral white light or cool white light as their light sources. However, citizens in some cities report that after replacing the traditional high pressure sodium lamps (HPSL) with LED lamps, the road illuminance improves but it is dusky as a whole and makes people feel oppressive and lonely. That is

Figure 6.99 Road illumination performance of LED lamps with different color temperatures at Meishi Street, Nanhai District, Foshan City, Guangdong Province, China. *(Color version of this figure is available online.)*

because the human eye is most sensitive to light with a 555 nm wavelength. Due to the fact that there is relatively more long wavelength light existing in HPSLs and LED lamps with warm white light, things tend to look brighter than those in neutral white light and cool white light when the light power is the same. What's more, warm white light looks more comfortable and softer and it gives people a feeling of being warm and safe. Moreover, if considered from the point of view of security, yellow light or warm white light are also more suitable than neutral white light or cool white light for road illumination because of their claimed good performance in penetrating fog. Therefore, LED lamps with warm white light are more suitable for road illumination.

(iv) Colorful LED Applications in Water Cube of Beijing 2008 Olympic Games

The color of the Beijing Olympic National Aquatics Centre, or the Water Cube, is controlled by colorful LEDs. It is an architecture with blue, symbolizing water, as its essential tone. However, if blue is added to red which symbolizes fire, the beauty of water will be enhanced by combining the contradictory unity of water and fire. Four hundred and ninety-six thousand of high power LEDs were used in the Water Cube. Controlled by the automation system, three colors, red,

Figure 6.100 Colorful LED applications in the Water Cube. (Reproduced by permission of photographer Mr. Tiehan Duan, Shenyang Yuanda Aluminum Industry Engineering Co. Ltd.). *(Color version of this figure is available online.)*

green, and blue (RGB), were made to create 256 × 256 × 256 colors and varied colors were combined to achieve rich and varied dynamic changing effects and to form different kinds of pictures which provided broad areas for scene design. The illumination performance of the Water Cube (Figure 6.100) presented the design intent perfectly, built a space which takes water as life theme, and shows a glittering and colorful Water Cube to people all over the world. The LED landscape lighting of the Water Cube achieved great success both during the Olympic Games and in operation after the Olympic Games, earned international attention as well as praise, and makes the Water Cube one of the most famous symbols of the Olympic Games.

(v) Colorful LED Applications in Buddhist Temple

Next, we are provided with another LED visual feast. The LED lamps and its light control system in theWuxi Lingshan Buddhist Temple (from Figure 6.101 to Figure 6.106) are an important part of the Lingshan Phase III Project, a key tourism project in Jiangsu province [27]. And they were also essential elements in constructing the "site of world Buddhism forum". This project used the LED lamp and control system to sketch the gorgeous interior architecture of the Buddhist temple in detail, to exert a strong visual impact on the tourists and to make the spectators melt into Buddhist culture not only physiologically but also psychologically.

There are five main kinds of LED lamps used in Wuxi Lingshan Buddhist Temple: point light source, projecting luminaires, outline luminaires, buried luminaires, and tube lamps. And LED lamps are mainly of 3 W@700 mA. As for the controller, there are two types: one is a DMX (Digital Multiple X) controller which is based on the RS485 communication; the other is a serial control system with a serial translocation store. Run by many such kinds of luminaires and systems, the color and rhythm of the LED, designed according to the designers' imagination, are fully presented in the Buddhist Temple. As to the light distribution of projected light on the Altar, when performing the full color light distribution, designers overcame constraints of tradition,

Figure 6.101 Wuxi Lingshan Buddhist temple [27]. *(Color version of this figure is available online.)*

and designed an RGB combination type in order to highlight a religious and artistic atmosphere. What's more, a white LED point light source is twinkling against the dark blue background, assuming an illusive starring scene in the palace of 75 m high. There is a famous saying in China regarding this project: now that the Buddha has used LEDs, why not humans?

6.2 Thermal Management

6.2.1 Analysis of System Thermal Resistance [28]

For LED lighting products, a low system thermal resistance of the product is highly appreciated for achieving good reliability and optical performance. Usually, the LED products need a multi-chip array packaging or packaging with many individual discrete chips; for such a system, heat generated by the LED chip flows through interface materials between the LED chip and the board and then spreads into the heat sink, and finally transfers into the environment. Therefore, the system thermal resistance of the LED product consists of four parts of thermal resistances. The first part is thermal resistance of the LED chip, R_c, which is related to the chip packaging technology. The second part is thermal resistance of interface

Figure 6.102 Culture and art gallery of the Buddhist temple—the hall [27]. *(Color version of this figure is available online.)*

Figure 6.103 Tower hall of Buddhist temple (starring scene) [27]. *(Color version of this figure is available online.)*

Figure 6.104 Pure land of Vairocana of Buddhist temple (colored glaze backlighting) [27]. *(Color version of this figure is available online.)*

Figure 6.105 Top of the altar of Buddhist temple (being changeable in color) [27]. *(Color version of this figure is available online.)*

materials between LEDs and heat sink, R_{int}, which is mainly determined by the bonding and adhesive materials and their thicknesses. The third part is the spreading thermal resistance as heat dissipates from the interface material into the heat sink, R_{sp}, which is affected by many geometrical sizes such as chip size and heat sink size, and the material of heat sink, and so on. The last part is the thermal resistance between the heat sink and the environment, R_{env}, mainly determined by way of the heat dissipation to the environment. A system thermal resistance network of LEDs light source is presented in Figure 6.107.

(i) Interface Thermal Resistance

Interface thermal resistance usually includes two parts, one part is bulk material thermal resistance, the other part is contacting thermal resistance. For the bulk material thermal

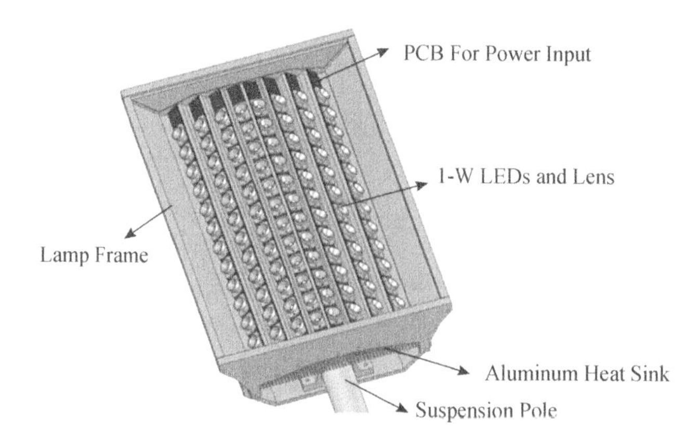

Figure 6.106 An LED road lamp.

resistance, it mainly depends on the thermal conductivity of the interface material and its thickness. For contacting thermal resistance, the three materials' chemical properties are very important. The two materials which will be connected by the thermal interface material should have good compatibility with each other and also with the thermal interface material itself so that the contacting thermal resistance will be low.

(ii) Spreading Thermal Resistance

Spreading thermal resistance exists when heat conducts from a component with a small contacting area into a larger one. Spreading thermal resistance may be the major part of the total thermal resistance of the heat transfer process. In the four thermal resistances shown in Figure 6.107, thermal spreading resistance is complex to be calculated. For the LED application products, many LEDs are distributed on the heat sink and each LED creates heat, therefore, it can be considered that many heat sources are distributed on the heat sink. The heat

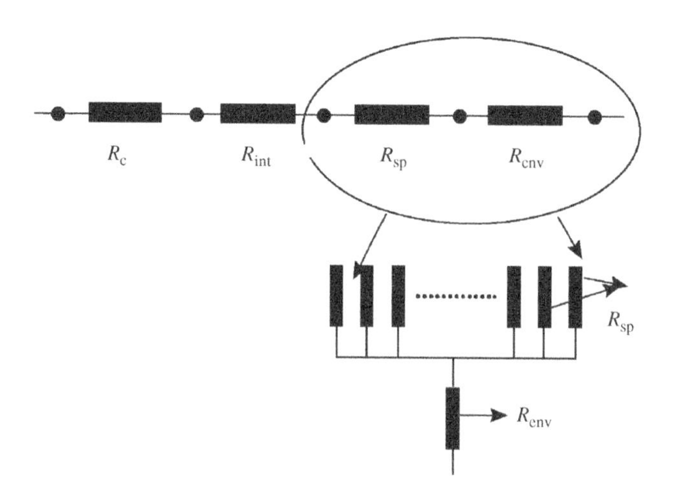

Figure 6.107 System thermal resistance network of LEDs module.

produced by each LED module transfers to the heat sink and finally dissipates into the environment. During the heat transfer process from each LED to the heat sink, one thermal spreading resistance for an LED module exists because heat transfers from the small area of the LED chip to the heat sink with a larger area. Therefore, there are many thermal spreading resistances for each LED chip in the LEDs light source. These thermal spreading resistances, R_{Ls}, are connected in parallel because of a matrix array of LEDs for heat sink. For typical LEDs light source with LED module array, since the size of each module is the same and their sizes are very small compared with the heat sink size, although the position for every heat source is different in the base plate, the thermal spreading resistances for each LED still can be regarded as the same. As a result, it is very easy to simplify and calculate the final thermal spreading resistance R_{sp} of the LEDs module array, which can be expressed as:

$$R_{sp} = \frac{R_{Ls}}{n} \tag{6.30}$$

where n is the LED number of the LEDs module.

For the thermal spreading resistance of single LED R_{Ls}, its maximum value is defined as:

$$R_{LS\_max} = \frac{T_{max} - \bar{T}_b}{Q} \tag{6.31}$$

where T_{max} is the maximum temperature in the contacting area between the LED and the heat sink, $\bar{T}_b$ is the average temperatures over the LED and base area of heat sink, and Q is the heat transfer rate of each LED.

A closed-form equation for calculating spreading thermal resistance has been proposed by Lee *et al.* [29,30]. Based on these references:

$$R_{Ls\_max} = \frac{\psi_{max}}{k \cdot a \cdot \sqrt{\pi}} \tag{6.32}$$

where a is the equivalent radius of the LED, and k is the thermal conductivity of the heat sink material.

$$\psi_{max} = \frac{\varepsilon \cdot \tau}{\sqrt{\pi}} + \frac{1}{\sqrt{\pi}} \cdot (1 - \varepsilon) \cdot \Phi_c \tag{6.33}$$

In Equation 6.33:

$$\Phi_c = \frac{\tan h(\lambda_c \cdot \tau) + \frac{\lambda_c}{Bi}}{1 + \frac{\lambda_c}{Bi} \cdot \tanh(\lambda_c \cdot \tau)} \tag{6.34}$$

Where:

$$\lambda_c = \pi + \frac{1}{\sqrt{\pi} \cdot \varepsilon} \tag{6.35}$$

In the equations presented above, b is the equivalent radius of the heat sink, t is averaging thickness of the heat sink. R_{env} is thermal resistance between heat sink and environment. ε is dimensionless heat source radius, it can be expressed as:

$$\varepsilon = \frac{a}{b} \tag{6.36}$$

τ is dimensionless heat sink thickness, which is defined as:

$$\tau = \frac{t}{b} \tag{6.37}$$

Bi is effective Biot Number, it is given by:

$$Bi = \frac{h_{eff} \cdot b}{k} \tag{6.38}$$

where h_{eff} is the equivalent heat transfer coefficient for the heat sink of the LEDs module.

Substituting Equations 6.32–6.38 into Equation 6.31, the maximum thermal spreading resistance of the single LED can be calculated. As a result, spreading thermal resistance of the LEDs module can be obtained based on Equation 6.30.

(iii) Environment Thermal Resistance

For the LED light source, environment thermal resistance exists in the process by which heat transfers from the heat sink to the environment. The environment thermal resistance is mainly dependent on the way heat is transfered.

As mentioned in Chapter 4, Newton's Law of Cooling is a classic method for calculating heat exchange in convection. There is a parameter h, heat transfer coefficient, used for denoting the heat transfer intensity of heat convection in the expression. According to Newton's Law of Cooling, and the definition of thermal resistance, environment thermal resistance based on convection is given by:

$$R_{env} = \frac{1}{hA} \tag{6.39}$$

where A is the heat exchange area. For thermal radiation or other heat transfer modes that happen around a heat source, an equivalent heat transfer coefficient, h_{equ}, which is used to equate the actual heat dissipation between the heat sink and ambient, is usually found for calculating the heat exchange rate and environment thermal resistance. Therefore, no matter what the heat transfer mode is, the environment thermal resistance can be obtained as:

$$R_{env} = \frac{1}{h_{equ}A} \tag{6.40}$$

6.2.2 Types of Heat Dissipation to Environment

Types of heat dissipation to the environment can be divided into passive and active cooling. Passive cooling is the heat transfer process without any artificially imposed force and extra energy consumption, such as free convection. On the other hand, active heat dissipation needs imposed forces or input power. At present, forced air and liquid cooling, semiconductor refrigeration, microjet cooling, ultrasonic heat dissipation and superconducting cooling, and so on, are the underlying technologies used for active cooling. The follow section will introduce typical passive and active cooling solutions.

(i) Passive Heat Dissipation Methods

For the LED light source, a plate fin heat sink is the most commonly used cooling solution. It is a passive heat dissipation method because only natural convection existing in the surfaces of the heat sink. A plate fin heat sink has many advantages, such as low cost, simple structure, and high reliability. However, as it dissipates heat into the environment mainly by natural convection, only a small amount of heat can be transferred. Therefore, compared with other cooling solutions, if the provided cooling space is the same, it is more suitable for a low power LED light source. According to Newton's Law of Cooling, heat transfer coefficient h and heat exchange area A are the key factors affecting the heat transfer intensity of natural convection. As a result, engineers usually improve the heat dissipation performance of plate fin heat sinks by increasing the fin number for enlarging heat exchange area A or designing the space between the fins and the height of the fins to enhance heat transfer coefficient h. By optimization the design of the plate fin heat sink, it is applied to general LED lighting products quite well.

With the exception of the fin heat sink, the heat pipe with fin heat sink is also a typical passive cooling solution. This technology uses a conductive component called a heat pipe, which was invented by Grover G.M. from American National Lab Los Alamos in 1963. Inside a heat pipe, at the hot interface a fluid turns to vapor and the gas naturally flows and condenses at the cold interface. The liquid falls and is moved back by capillary action to the hot interface to evaporate again. A heat pipe can transport a large amount of heat with a very small temperature difference between the hot and cold ends since there is phase transition in the heat pipe. Therefore, using heat pipes in air cooling will greatly improve the performance of heat dissipation. Figure 6.108 shows a heat sink with a heat pipe.

(ii) Active Heat Dissipation Methods

There are many kinds of active cooling methods, some typical types are discussed here.

(1) Forced air convection

As presented above, free convection from the plate fin heat sink is limited and is not feasibly used to dissipate heat for future special high-power LED light source. To meet the heat dissipation requirement of the high-power LED light source which also requires a small volume, forced convection is applied. Forced convection is driven by artificial force or power, such as a fan or pump, which will accelerate the air flow rate and greatly increase heat transfer coefficient h. Therefore, forced convection can improve the heat exchange rate significantly compared with free convection.

However, forced convection by fan will bring some problems. If the LED is used in indoor illumination, silence is the main demand besides illumination for customers.

Figure 6.108 A commercial heat pipe (copper) coupled with fin heat sink (aluminum).

Though the fan or pump has a good cooling effect, it will make noise and that is not tolerable for customers. Moreover, the reliability of the LED device will be reduced because of the moving parts in the fan. All these disadvantages restrict the application of a fan in high-power LED cooling. However, we should have some trade-off between performance and reliability when choosing the fan heat sink to cool the LED light source.

(2) Semiconductor refrigeration

Semiconductor refrigeration is also called thermoelectric refrigeration or thermoelectric cooling. The theoretical basis of semiconductor refrigeration is the thermoelectric effect of solid. In the environment without external magnetic field, the thermoelectric effect consists of five parts: heat conduction, loss of joule heat, Seebeck effect, Peltire effect, and Thomson effect. The advantages of semiconductor refrigeration are: high cooling density, compatibility with IC technology, no moving parts and abrasion, compact construction, and the possibility to increase the integration.

What Figure 6.109 shows is a thermoelectric couple connected by a p-type and a n-type semiconductor elements. After DC power is on, a temperature difference and heat transfer will occur at the junctions, and it will cause a temperature decrease and heat absorption at the top junction contacted with chip. While at the junction below which is called the hot end, the temperature is rising and the heat is released.

In order to further improve the efficiency of semiconductor refrigeration, the application of multistage semiconductor refrigeration is put forward, and it is required to integrate some heat sinks to enhance the heat exchange with the ambient, as shown in Figure 6.110.

(3) Active liquid cooling

Active liquid cooling depends on the pump-driven fluid flow to take away heat to achieve the heat dissipation effect. A liquid cooling system is usually composed of a pump, heat exchanger, cooling plate, and so on. The liquid flow absorbs heat generated by the chip

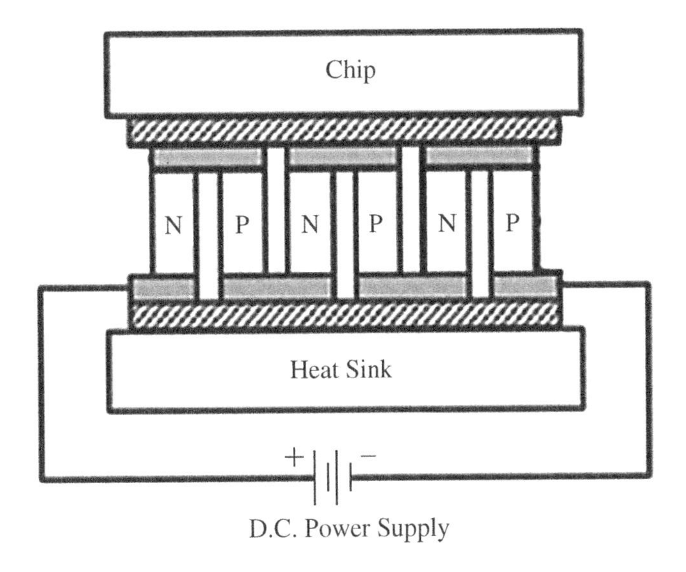

Figure 6.109 Structure of semiconductor refrigeration.

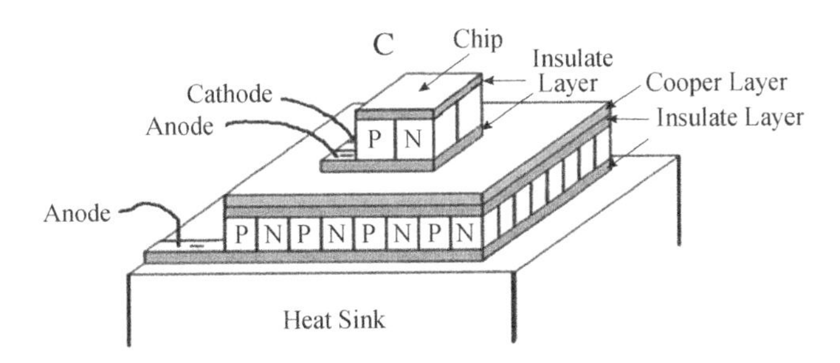

Figure 6.110 Integration of double-stage semiconductor refrigeration with heat sink.

through the cooling plate and transfers the heat to the heat exchanger which dissipates the heat to the ambient. The advantage of such a cooling system is its high cooling performance and that it can cool a very high heat flux device. Its disadvantages are that it is difficult for customers to understand the technicalities involved, that the cost is relatively high and that the moving parts such as pumps make the system's reliability doubtful. All these disadvantages do not make it a good choice for normal customers. Since the heat dissipation ability of liquid cooling is very good, it is suitable for superpower LED light sources.

Microjet and microchannel are two important active liquid cooling methods. The two methods are supposed to be the most potential cooling ones with powerful cooling capacity.

Since the invention of the microchannel in the 1980s, it has been widely applied. In a microchannel cooling system, the external force usually provided by pumps is used to press the working medium into microchannels to enhance the cooling effect of convection. Recently, the microchannel has been widely used to cool electronic devices and components.

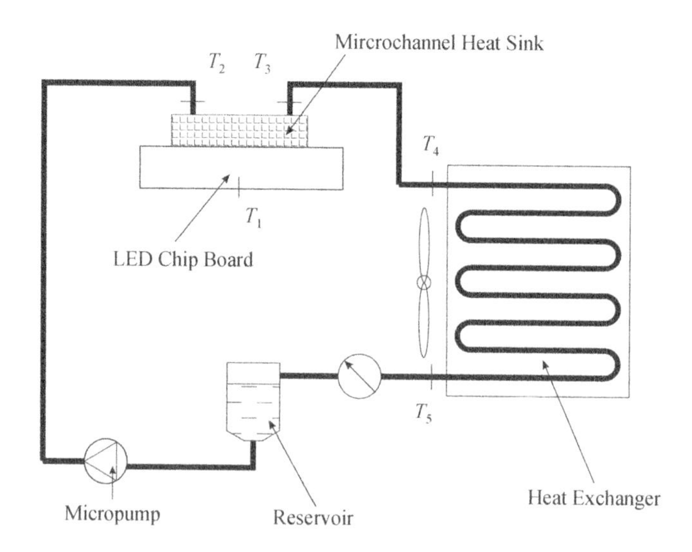

Figure 6.111 Schematic diagram of a microchannel cooling system.

Figure 6.111 demonstrates a closed-loop microchannel cooling system for an LED light source. It composes of four parts: a microchannel heatsink, a micropump, a reservoir, and a small heat exchanger with a fan. When a light source needs to be cooled, the system starts to work. Water or other fluid in the closed system is driven into the heatsink device through an inlet by a micropump. The inlet and outlet of the heatsink are located in the top side of the heat transfer device, from which the fluid is directly pushed onto the bottom plate attached with the LED chip board. Since the microchannel heatsink has a very large heat transfer coefficient, the heat created by LED chips is easily removed by the recycling fluid in the system. The fluid is heated and its temperature increases after flowing out of the heat sink device, then the heated fluid enters into the heat exchanger with fins and fans. The heat exchanger will cool the fluid and the heat will be dissipated into the external environment. The cooled fluid will be delivered into the reservoir to ensure that the fluid entering the micropump is in a liquid state to keep the micropump work properly. From the low outlet of the reservoir, the cooled fluid is then pumped back into the heat sink, thus forming a closed-loop flow system. It should be noted here that the real size of the system can be designed as one small packaging according to application requirements.

Figure 6.112 shows a typical microchannel heat sink. In the device, hollow silicon channels are connected with chips and the fluid flows through the channels.

Microjet is another active liquid cooling technology. Figure 6.113 illustrates a closed-loop LED microjet array cooling system. It is composed of three parts: a microjet array device, a micropump, and a mini fluid container with a heat sink. When the LED needs to be cooled, the system is activated. Water or other fluids or gases in the closed-loop system are driven into the microjet array device through an inlet by a micropump. Many microjets will form inside the jet device, which are directly impinged onto the bottom plate of the LED array. Since the impinging jet has a very high heat transfer coefficient, the heat created

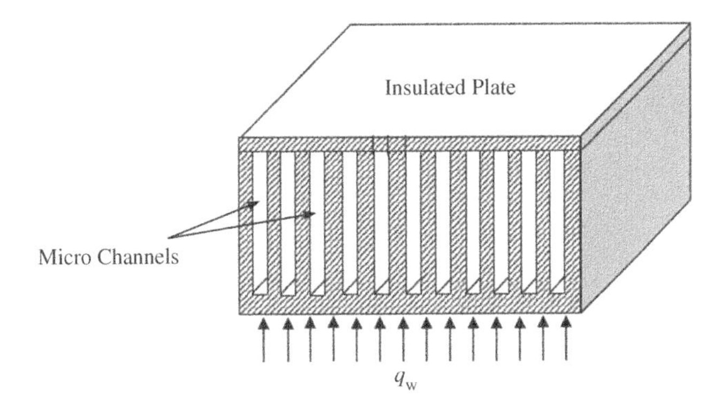

Figure 6.112 Microchannel water-cooling device.

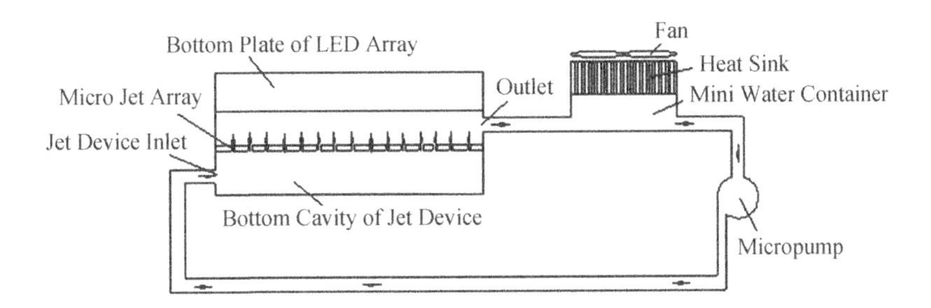

Figure 6.113 Closed-loop microjet array cooling system.

by the LEDs is easily removed by the recycling fluid in the system. The fluid is heated and its temperature increases after flowing out of the jet device, and then the heated fluid enters into the mini fluid container. The heat sink, which has a fan on the fluid container, will cool the fluid and the heat will dissipate into the external environment. The cooled fluid is delivered into the jet device to cool the LED array, again driven by the force of the micropump in the system. The above processes constitute one operation cycle of the total system.

Figure 6.114 shows the structure of the jet device of Figure 6.113 in detail. It consists of several layers, which are (from top to bottom) the chip array layer, the top plate of the jet cavity, the impinging jet cavity, the microjet array layer and the bottom cavity. Cooled fluid enters into the device through the inlet, which is open at one side of the bottom cavity layer. The fluid flows through the microjet array and forms many microjets, as shown in Figure 6.114. With sufficient driving force, the jets will impinge onto the top plate of the jet cavity which is bonded with the chip substrate of the LEDs. The heat conducted into the top plate of the jet cavity through the LED chips will dissipate into the cooled fluid quickly due to the high heat transfer efficiency of the impinging jet. The fluid temperature increases and the heated fluid flows out from the jet array outlet, which is open at one side of the top jet cavity layer. Through this process, the heat from the LED chips will be transferred into the fluid efficiently.

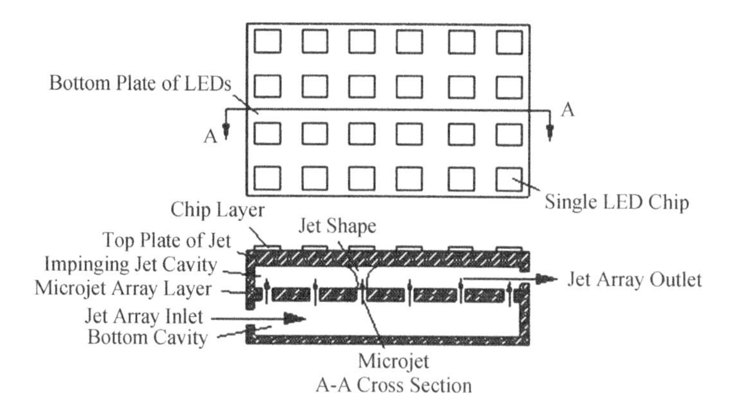

Figure 6.114 Microjet array device.

6.2.3 Design and Optimization of Fin Heat Sink

Although there are many cooling solutions for LED heat dissipation, as discussed before, fin heat sink is still the mainstream cooling solution for current LED light sources because of its obvious advantages in reliability and cost. Therefore, in this section, we will concentrate on fin heat sink design.

(i) Fin Optimization Model [31]

(1) Design model

The heat sink of the LED road lamp is usually designed as a horizontally-located plate fin heat sink, as shown in Figure 6.115. Although the heat transfer coefficient is comparatively low in natural convection (usually less than 10 W/(K.m^2)), the plate-fin natural convection heat sinks offer distinctive advantages in cost and reliability.

In the design and optimization of the horizontally-located plate fin heat sinks, heat transfer coefficient is a key factor. However, the averaging heat transfer coefficient is associated with the fin dimensions, which are the optimization factors and they are

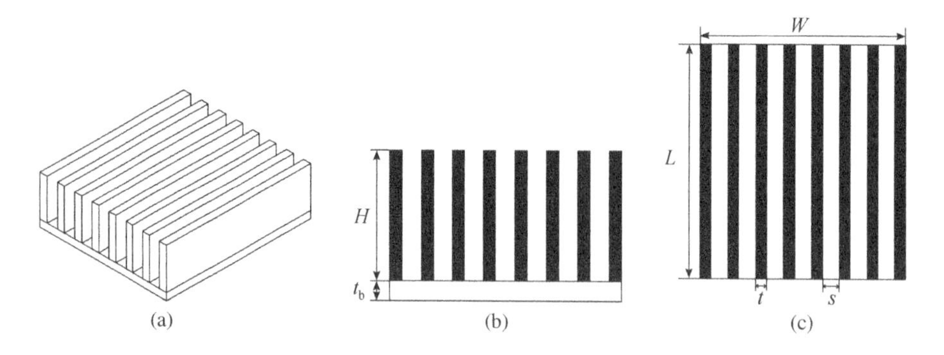

Figure 6.115 Horizontally-located rectangular plate fin heat sink, (a) horizontal configuration; (b) side elevation; and (c) top view.

strongly coupled. In addition, due to the overlapping of the boundary layers between the adjacent fins, it is difficult to solve boundary layer equations, correspondingly, it is very difficult to calculate the fin heat transfer coefficient during computation process.

For the application of a horizontally-located plate fin heat sink in the LED, heat is generated by the LED chips and then conducted through an aluminum alloy base board and finally dissipated to the surroundings by convection. To simplify the heat transfer and optimize the heat sink of the high power LED road lamp, there are certain assumptions: (1) the material is isotropic; and (2) the spreading and contact resistance is ignored in the heat sink body itself.

(2) Design and optimization method

Figure 6.115 shows the heat sink dimensions and their indications. The total heat Q_{hs} that a heat sink can dissipate is expressed by the equation:

$$Q_{hs} = Q_{bp} + n \cdot Q_{fin} \tag{6.41}$$

where Q_{bp} is the heat that is dissipated by the exposed base of the heat sink and is defined by Equation 6.42, Q_{fin} is the heat that is dissipated by the fins of the heat sink and is defined by Equation 6.44:

$$Q_{bp} = h_{bp} \cdot (n-1) \cdot \theta_{bp} \cdot A_{bp} \tag{6.42}$$

where h_{bp} is the average heat transfer coefficient of the exposed base area, n is the number of plate-fins of the array, θ_{bp} is the excess temperature from the heat sink base to the ambient temperature, and A_{bp} is the surface area of the exposed base and could be defined as:

$$A_{bp} = s \cdot L \tag{6.43}$$

where s is the fin spacing, and L is the fin thickness.

$$Q_{fin} = h_{fin} \cdot A_{fin} \cdot \theta_{bp} \tag{6.44}$$

where h_{fin} is the averaging heat transfer coefficient of the heat sink, and A_{fin} is the surface area of a single fin and could be defined as:

$$A_{fin} = 2(H \cdot t + L \cdot H + L \cdot t/2) \tag{6.45}$$

where H is the fin height and t is the fin thickness.

For most of the applications, especially for the heat sink design of the LED road lamp, the heat transfer and the approximate temperature difference between the heat sink and the environment are given, the heat dissipation are designed to meet the amount of heat dissipation and the heat sink mass. Since most plate fin heat sinks are produced with extrusion aluminum alloy, with considerations of manufacturability and strength, the ranges of the fin parameters should be as follows; (1) fin thickness is between 1 mm and 3 mm; (2) fin spacing is between 1 mm and 15 mm; and (3) fin height is between 25 mm and 50 mm.

In order to obtain the averaged heat transfer coefficient of the heat sink, the total heat dissipation area can be divided into two parts. One is from the exposed base area and the other is from the fin array.

(a) Heat dissipation from exposed base area
(1) When the ratio of fin spacing to fin height is less than 0.28, the flow inside the fins is enclosed space natural convection. In this case, the characteristic dimension is the height of enclosed space. If the value in the square brackets of Equation 6.48 is negative, the Nusselt number of base plate Nu_{bp} is replaced with 1. Equation 6.46 to Equation 6.47 are available when the Raleigh number of base plate Rabp $<4 \times 10^6$ Grashof number of base plate can be defined as:

$$Gr_{bp} = g \cdot B \cdot \theta_{bp} H^3 / v_{bp}^2 \tag{6.46}$$

where v_{bp} is the mean kinematic viscosity of air which is around the base plate.
And then:

$$Ra_{bp} = Gr_{bp} \cdot Pr_{bp} \tag{6.47}$$

where Pr_{bp} is Prandtl number of air which is around the base plate.
The Nusselt number of base plate Nu_{bp} can be given by: [32]

$$Nu_{bp} = 1 + 1.44 \cdot \left[1 - 1708/Ra_{bp}\right] + \left[\left(Ra_{bp}/5830\right)^{1/3} - 1\right] \tag{6.48}$$

(2) When the ratio of fin spacing to fin height is more than 0.28, large space natural convection is assumed [33]. In this case, characteristic dimension is $(s + L)/2$. Then:

$$Gr_{bp} = g \cdot \beta \cdot \theta_{bp} \cdot ((s + L)/2)^3 / v_{bp}^2 \tag{6.49}$$

where g is the acceleration of gravity, β is the thermal coefficient of expansion.
Then Ra_{bp} is obtained by the Equation 6.47
If $Ra_{bp} < 2 \times 10^4$, then:

$$Nu_{bp} = 1 \tag{6.50}$$

If $2 \times 10^4 < Ra_{bp} < 8 \times 10^6$, then:

$$Nu_{bp} = 0.54 \cdot Ra_{bp}^{1/4} \tag{6.51}$$

If $8 \times 10^6 < Ra_{bp} < 10^{11}$, then:

$$Nu_{bp} = 0.15 \cdot Ra_{bp}^{1/3} \tag{6.52}$$

And we can obtain different Nu_{bp} values according to the different value of Ra_{bp}. Then the averaged heat transfer coefficient of base plate h_{bp} could be written as:

$$h_{bp} = Nu_{bp} \cdot k_{bp}/l_{bp} \tag{6.53}$$

where l_{bp} is the characteristic dimension of the base plate.

(b) Heat dissipation from fin array

(1) For enclosed space nature convection, the problem is to determine the characteristic dimension. Owing to the different temperature of the fin surface, it is necessary to replace excess temperature with the heat flux to calculate the Grashof number of the fin Gr_{fin}, so the place with the lowest temperature should be the center of the space between the two fins. In other words, the characteristic dimension is half of the fin space s/2. Then:

$$Gr_{fin} = g \cdot \beta \cdot (Q_{fin}/(2 \cdot H \cdot L)) \cdot (s/2)^4/(k_f \cdot v_f^2) \tag{6.54}$$

where k_f is the thermal conductivity of air around the fin, v_f is the mean kinematic viscosity of air around the fin.

Therefore, the heat dissipation from the single fin is given by:

$$Q_{fin} = k_{fin} \cdot A_c \cdot \theta_{bp} \cdot m \cdot \tanh(m \cdot H) \tag{6.55}$$

where kfin is the thermal conductivity of fin, Ac is the cross section area of the fin and can be defined as $A_c = H \cdot t$, m is the fin parameter and can be written as $m = (h_{fin} \cdot P/k_{fin} \cdot A_c)^{1/2}$, hfin is the averaged heat transfer coefficient of the fin, P is the cross section circumference of the fin which can be written as $P = 2(H + t)$, then Raleigh number of the fin Ra_{fin} can be defined as:

$$Ra_{fin} = Gr_{fin} \cdot Pr_{fin} \tag{6.56}$$

where Pr_{fin} is the Prandtl number of the air which is around the fin.

If $Ra_{fin} < 10^4$, the heat transfer in the vertical enclosed space is pure conduction, then the Nusselt number of the fin Nu_{fin} can be defined as:

$$Nu_{fin} = 1 \tag{6.57}$$

If $10^4 < Ra_{fin} < 10^7$, then:

$$Nu_{fin} = 0.42 \cdot Ra_{fin}^{1/4} \cdot Pr_{fin}^{0.012} \cdot (H/(s/2))^{-0.3} \tag{6.58}$$

If $10^7 < Ra_{fin} < 10^9$, then:

$$Nu_{fin} = 0.46 \cdot Ra_{fin}^{1/3}(13) \tag{6.59}$$

(2) When the ratio of fin spacing to fin height is more than 0.28, it is large space natural convection. The characteristic dimension is the fin height H. Then:

$$Gr_{\text{fin}} = g \cdot \beta \cdot (Q_{\text{fin}}/(2 \cdot H \cdot L)) \cdot H^4/(k_f \cdot v_{\text{f}}^2) \tag{6.60}$$

and Ra_{fin} is obtained by the Equation 6.56

Then:

$$Nu_{\text{fin}} = 0.6 \cdot Ra_{\text{fin}}^{1/5} \tag{6.61}$$

We can obtain a different Nu_{fin} value according to the different value of Ra_{fin}. Then the averaged heat transfer coefficient of the fin h_{fin} could be written as:

$$h_{\text{fin}} = Nu_{\text{fin}} \cdot k_{\text{fin}}/l_{\text{fin}} \tag{6.62}$$

where l_{fin} is the characteristic dimension of the fin.

The fin surface is not isothermal, but when heat sink is in steady state, in other words, the temperature distribution of fin surface never changes, it means that heat through each fin surface is constant. Therefore, the heat dissipation from the single fin is a function of the averaged fin heat transfer coefficient, their relation is provided by:

$$Q_{\text{fin}} = f(h_{\text{fin}}) \tag{6.63}$$

where its inverse function is:

$$h_{\text{fin}} = f - 1(Q_{\text{fin}}) \tag{6.64}$$

Then:

$$g(h) = h - h_{\text{fin}} \tag{6.65}$$

where h is obtained through Newton iteration:

$$\varphi(x) = x - f(x)/f'(x) \tag{6.66}$$

Then iteration convergence is used to obtain the value of h_{fin}.

(ii) Fin Optimization Code

When other parameters have been determined, the geometry of the heat sink needs to be optimized to include fin height H, fin thickness t, and fin spacing s. Because the function with the heat transfer coefficient contains t and h, the relation turns into a transcendental equation. After the partial difference on H, t, and s, it is rather difficult to seek the best solution using the Lagrange multiplier due to complex expression. Therefore, it is necessary to dispose the points discontinuously, and establish a matrix to store the values of these points and the heat transfer coefficient by an iteration structure of the program. We choose an optimization fin geometry

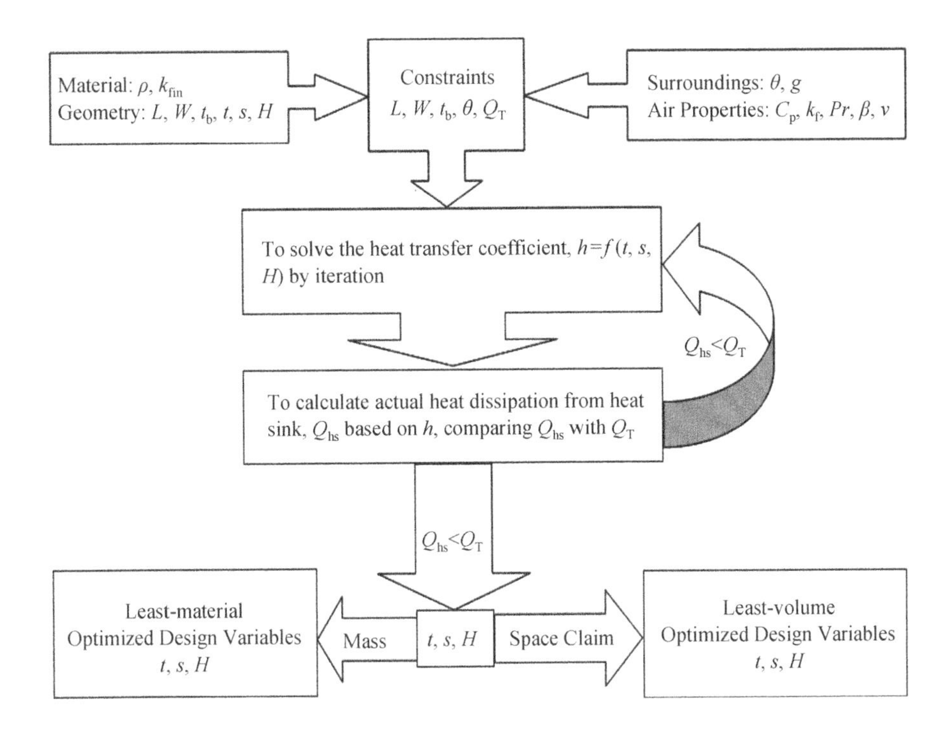

Figure 6.116 Plate Fin Optimization Flowchart.

according to least-material and least-volume criteria based on the data given by the MATLAB® program. A flowchart for plate-fin optimization is shown in Figure 6.116.

6.2.4 Design Examples of Thermal Management of Typical LED Lighting Systems

(i) LED Road Lamps [34]

A 112 W LED road lamp was used as a design example. For the heat sink of this lamp, parameters to be evaluated are given below. The temperature difference between the surface of the base plate and the ambient is 18 °C. The base plate dimensions are as follows, the thickness is 3 mm, the length and width are 530 mm and 350 mm respectively. The material density of the base plate is 2700 kg/m^3, and its thermal conductivity is 160 W/m-K. The gravitation acceleration is 9.8 m/s^2.

By using the aforementioned design and optimization code, the final 112 W LED road lamp heat sink was designed and shown in Figure 6.117. Here the fin height H is 17 mm, fin thickness t is 2 mm, and fin spacing s is 5 mm.

For the 112 watts LED road lamp shown in Figure 6.117, one hundred and twelve high power LED modules are bonded onto the heat sink. They are distributed on the heat sink base in seven rows. All the LED modules are the same, their input powers are 1 watt and the total input power for this lamp is 112 watts. When the electronic power is supplied, LEDs generate light and heat. The heat is dissipated out into the environment through the aluminum base and fins on the base.

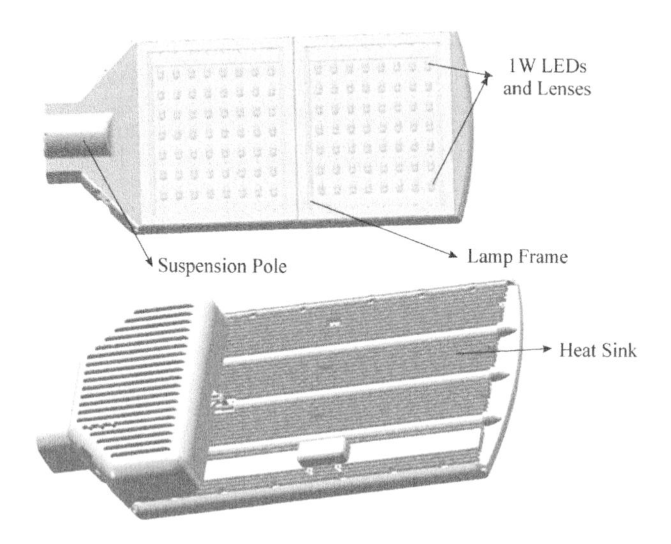

Figure 6.117 Schematic diagram of 112 watts LED road lamp.

(ii) Ultra High Power LED Illumination Systems

The ultra high power LED illumination system shown is a 220 W LED white light source [35–38]. An active cooling solution, microjet based cooling system was used for its thermal management because of high power density.

Figure 6.118 shows the 220 W LED demonstration system. A microjet array cooling system for the LED light source was designed as shown in Figure 6.119. A voltmeter and a current meter were used to indicate the input power. There were three buttons on the control panel to control the total power input, cooling system, and LED startup. A knob adjusts the input power magnitude and resultant luminance of the LED light source. The designed power input from this lamp was approximately 220 W. For such an application, the power consumption of the fan and micropump were about 3.6 W and 2.2 W respectively. The 220 W LED lighting fixture consisted of 64 high power LED chips which were packaged on a 4 cm by 4 cm metal substrate.

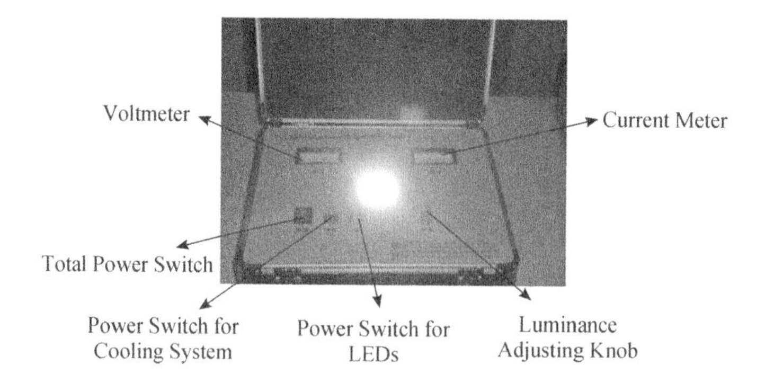

Figure 6.118 Demonstration system of 220 W LED lighting fixture. *(Color version of this figure is available online.)*

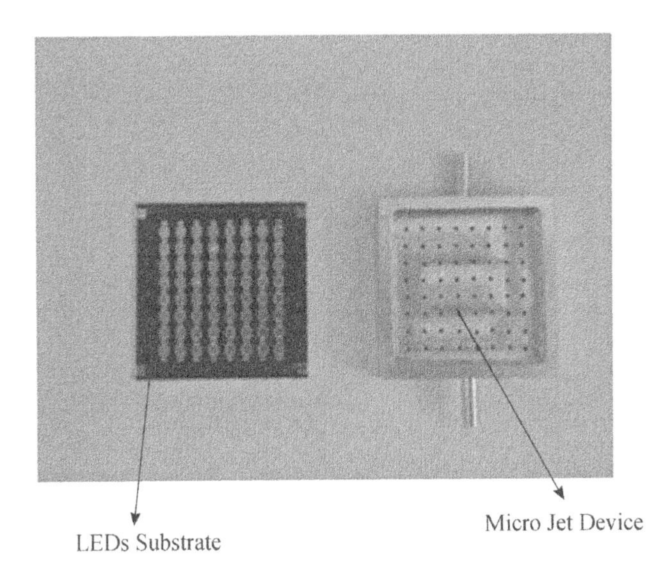

Figure 6.119 LED chip substrate and microjet array device. *(Color version of this figure is available online.)*

The microjet device and LED chip substrate inside the light source are shown in Figure 6.122. 64 microjets were uniformly distributed in a 3.6 cm by 3.6 cm cavity. The diameter of the microjet was 1 mm. The flow rate of the micropump in the system was about 18.5 mL/s.

(iii) LED Bulb

Besides the LED road light, the LED bulb is another emerging product in the market. LED bulbs will replace incandescent bulbs in the near future for both reducing environmental concern and energy consumption. However, since a bulb is usually small in size, it is therefore difficult to remove the heat generated by LEDs from the inner part of the bulb to the ambient. The thermal problem is prominent in an LED bulb.

Appearance of the analyzed 4 W LED bulb is shown in Figure 6.120. Besides electric devices, the bulb consists of three main components: glass lampshade, heat dissipation

Figure 6.120 A 4 W LED bulb analyzed. *(Color version of this figure is available online.)*

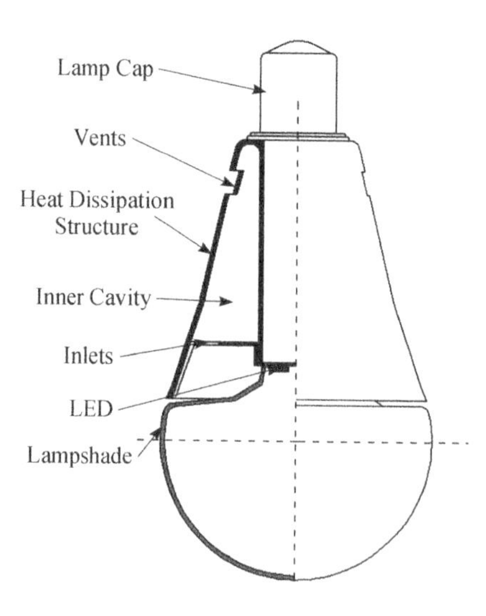

Figure 6.121 Structural diagram of the LED bulb.

structure with inner cavity made of aluminum or aluminum alloy, and lamp cap. Figure 6.121 shows the bulb structure in details. There are six vents on the top of the heat dissipation structure. With inlets at the bottom of the heat dissipation structure and inner cavity, cool air can flow into the bulb and carry the bulb generated heat to the ambient, which will enhance thermal performance of the heat dissipation structure.

To obtain data needed for the afterward thermal analysis, two experiments were conducted. In the two experiments, surface temperatures of the bulb and lamp holder were measured by thermocouples bonded on those surfaces. In experiment one the bulb worked in normal condition while in experiment two vents on the top of the bulb were sealed by aluminum foil. Both experiments were conducted at room temperature and the bulb connected with a lamp holder and placed vertically. The temperatures of those surfaces were recorded in the steady state and are represented in Table 6.3.

As shown in Figure 6.122, heat generated by LEDs transfers to the ambient mainly by three paths: dissipating to surrounding through lampshade, heat dissipation structure, and lamp holder. Both natural convection and radiation heat transfer occur on the surfaces of the bulb and lamp holder. Since emissivity of the materials for the lampshade and the lamp holder, and the temperature of those surfaces are relatively low, the radiation heat transfer that occurs on those surfaces can be ignored. Heat transfer process occurring in the heat dissipation structure is

Table 6.3 Surface temperature of the LED bulb and lamp holder

Experiment	Temperature of Surface (°C)			Room Temperature (°C)
	Lampshade	Heat Dissipation Structure	Lamp Holder	
1	32.6	52.6	37.5	25.9
2	36.5	55.7	40.9	25.9

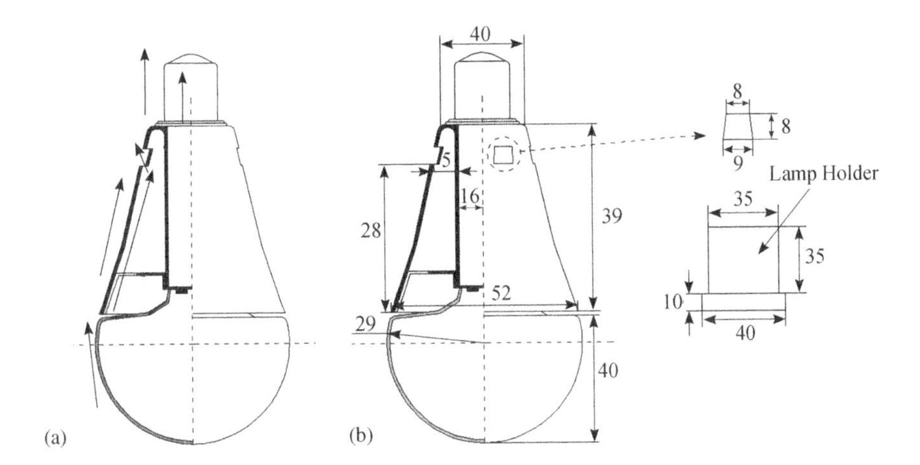

Figure 6.122 (a) Heat dissipation paths in the LED bulb; (b) dimensions of the LED bulb and lamp holder (unit: mm).

more complex. Seen from Figure 6.122, heat coming out of LEDs is conducted to the metal structure and then transfers to its surroundings by natural convection from the surface of the inner cavity and the outer surface of the metal structure, and radiation heat transfer of the outer surface.

Newton's Law of Cooling can be used to estimate the amount of heat dissipated to the ambient by natural convection from the outer surfaces of the bulb and lamp holder and it is given by:

$$Q = Ah(T_w - T_\infty) \tag{6.67}$$

where A is the surface area, h is the heat transfer coefficient of the surface, Tw is the temperature of the surface, and $T\infty$ is the ambient temperature. h is calculated by:

$$h = \frac{Nu \cdot \lambda}{l} \tag{6.68}$$

where Nu is a dimensionless Nusselt number, λ is the thermal conductivity of air which is determined by Tw and $T\infty$, and l is characteristic length for the thermal calculation. To simplify the thermal analysis, all of the outer surfaces of the bulb and lamp holder are treated as extended vertical plate. As a result, Nu is given as: [39]

$$\begin{aligned} Nu &= 0.59(Gr \cdot \mathrm{Pr})^{1/4} \quad 1 \times 10^4 < Gr \cdot \mathrm{Pr} < 1 \times 10^9 \\ Nu &= 0.021(Gr \cdot \mathrm{Pr})^{2/3} \quad 1 \times 10^9 < Gr \cdot \mathrm{Pr} < 1 \times 10^{13} \end{aligned} \tag{6.69}$$

With:

$$Gr = \frac{g\beta(T_W - T_\infty)l^3}{\nu^2} \tag{6.70}$$

where g is the acceleration of gravity, $\beta = \frac{1}{(T_w + T_\infty)/2}$, v is the kinematic viscosity of air, determined by T_w and T_∞. Gr and Pr are dimensionless numbers named Grashof number and Prandtl number determined by T_w and T_∞, respectively.

Radiation heat transfer between the outer surface of the heat dissipation structure and its surroundings can be analyzed by Stefan–Boltzmann law of thermal radiation. Since the heat dissipation structure is completely enclosed by the surrounding air, radiant exchange of the radiation heat transfer is expressed as:

$$Q_r = \varepsilon \sigma A (T_w^4 - T_\infty^4) \tag{6.71}$$

where ε is the emissivity of the heat dissipation structure, σ is the Stefan–Boltzmann constant with the value of $5.669 \times 10^{-8}\,\mathrm{W/m^2{\cdot}K^4}$.

With vents and inlets, the inner cavity can be modeled as a space between two parallel, vertical plates extended from curved surface of the heat dissipation structure. For natural convection from inner surfaces of parallel vertical plates, Nu is given as: [40,41]

$$Nu = \left[\frac{576}{(Ra')^2} + \frac{2.873}{\sqrt{Ra'}} \right]^{-1/2} \tag{6.72}$$

With:

$$Ra' = Gr \cdot Pr \cdot \frac{l}{L} \tag{6.73}$$

where L is the height of the plate. Gr can be calculated by Equation 6.70.

Figure 6.122b presents dimensions of the LED bulb and lamp holder. Thermal analysis results based on the data obtained from experiments one and two are represented in Table 6.4 and Table 6.5 The outer surface area of the heat dissipation structure in experiment two is larger than the one in experiment one because of the aluminum foil sealed on. The heat dissipation structure is made up of an aluminum or aluminum alloy and its outer surface may be anodized. Therefore, the emissivity of the outer surface of the heat dissipation structure is set to be 0.9. Half of the average pitch between the outer surface and the inner cylinder surface of the heat dissipation structure is used as the characteristic length l for Equation 6.73. L in Equation 6.73 is 32 mm. The temperature of the heat dissipation structure is considered to be uniform. The heat dissipation ratio is defined as:

$$r = \frac{Q}{Q_t} \times 100\% \tag{6.74}$$

The total heat dissipation rates of the LED bulb obtained by the two thermal analyses are nearly the same and they are 3.401 W and 3.284 W, respectively. In the first analysis it is found that the metal heat dissipation structure could dissipate 83.8% of the total heat, of which 32.4% is dissipated by free convection from its outer surface, another 25.5% by free convection from its inner cavity, and the remaining 25.9% by radiation from the outer surface. These results show that most of the heat is dissipated to the ambient by the metal heat dissipation structure

Table 6.4 Thermal analysis results based on experiment 1

	Natural convection from surfaces				Radiation Heat Transfer	
Structure	Lampshade	Lamp Holder	Outer Surface of Heat Dissipation Structure	Inner Cavity		
Characteristic Length (mm)	40	45	39	3.75	Area (mm^2)	5294.3
Area (mm^2)	7288.3	4003.5	5294.3	6282.9		
Gr	55665.0	132077.0	175701.0	156.0	Ambient Temperature (°C)	25.9
$Gr^{*}Pr$(Or Ra')	39044	92573	122868	12.8	Surface Temperature (°C)	52.6
Nu	8.29	10.29	11.01	0.48	Emissivity	0.9
h(W/(m^2.K))	5.49	6.11	7.79	3.53	Q(W)	0.881
Q(W)	0.268	0.283	1.101	0.868		
Heat Dissipation Ratio (%)	7.9	8.3	32.4	25.5	25.9	
Total Heat Dissipation Rate Qt (W)			3.401			

Table 6.5 Thermal analysis results based on experiment 2

Natural Convection from Surfaces					Radiation Heat Transfer	
Structure	Lampshade	Lamp Holder	Outer Surface of Heat Dissipation Structure	Inner Cavity		
Characteristic Length (mm)	40	45	39	–	Area (mm^2)	5702.3
Area (mm^2)	7288.3	4003.5	5702.3	–		
Gr	84116.0	163943.0	195133.0	–	Ambient Temperature (°C)	25.9
$Gr^{*}Pr$(Or Ra')	58948.0	114810	136456	–	Surface Temperature (°C)	55.7
Nu	9.19	10.86	11.34	–	Emissivity	0.9
h(W/(m^2.K))	5.93	6.52	8.00	–	Q(W)	1.076
Q(W)	0.458	0.391	1.359	0		
Heat Dissipation Ratio (%)	13.9	11.9	41.4	0.0	32.8	
Total Heat Dissipation Rate Qt (W)			3.284			

and the bulk of heat transfers to the surrounding by the inner cavity. Moreover, radiation heat transfer plays an important role in heat dissipation of the metal heat radiator. Therefore, the thermal design of the LED bulb should focus on the heat dissipation structure and the cavity of the radiator should be reserved in the structure or other ways to increase heat dissipation surfaces should be considered.

6.3 Drive Circuit and Intelligent Control Design

The high-power LED road lamps used at present are either simple and direct lighting, with a lack of intelligent control, or some of them may have a little intelligent control, but it can neither automatically examine the lighting situation of the road lamp nor can it conveniently carry out manual regulation and control. On the other hand, cables are used to control lighting and regulate the LED road lamp by the existing management system of lighting; the communication protocol is complex and the construction and operating costs are relatively high. An LED wireless intelligent control system is a good solution to these problems above.

6.3.1 Typical LED Wireless Intelligent Control System

A typical high-power LED intelligent control system (Figure 6.123) sets the central controller MCU and constant-current driving circuit as the core, and includes other items such as optical detectors, temperature sensors, possible accelerometer, memory, wireless communication module, and so on.

In this wireless intelligent control system, a constant-current driving circuit provides constant current for high-power LED arrays. At the same time, the driving circuit has a control terminal which can accept the PWM signal from the central controller, thereby adjusting the output current. The optical detector and temperature sensor can detect the brightness of LED lamps and the chip's temperature of LED lamps, and then send the detected data to the central controller. The central controller intelligently control the output current of the driving circuit by analyzing the detected luminous intensity and temperature data, thereby adjusting the brightness of LED lamps and at the same time storing the data into memory; The system also has a wireless communication module, which can send the working information of

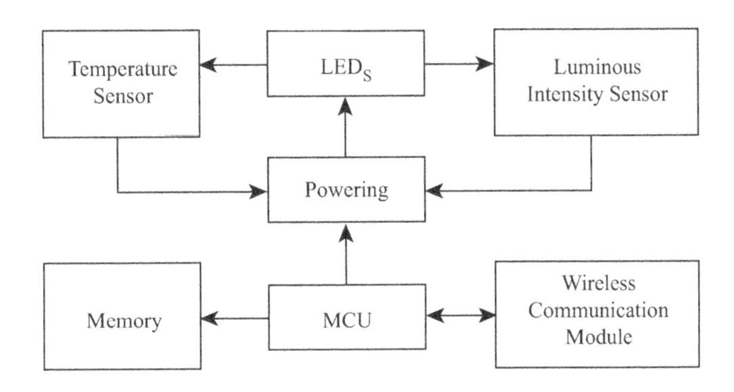

Figure 6.123 LED intelligent control system diagram.

the lamps stored in memory to the malfunction detection vehicles and can also receive control instructions which are given by the Control Center.

6.3.2 *Working Principles of Wireless Intelligent Control System*

The function of the intelligent control system is to control the brightness of LED lamps. The optical sensor installed in the part of the lamp's light source detects the lamp's brightness, illuminance, or luminous signal, converts it into a digital signal after the A/D conversion, and inputs it into the MCU controller through I2C bus. The temperature sensor installed in the internal lamps detects the lamp's temperature, converts it into digital signals after the A/D conversion, and inputs it into the MCU controller through a single bus. The MCU is programmed internally with a temperature limit set. If the detected temperature exceeds the upper limit of the lamps, the digital PWM control signals are sent to control the output current's size of the constant-current driving circuit. After the controlling chip in the constant-current driving circuit receives PWM signals sent by MCU controller, it will control the time the power switch tube in the switch driving circuit opens and adjust the output current of the driving circuit so as to achieve the regulation brightness of the LED lamps. For example, when the controller finds the brightness of the lamps is too high, or the lamp temperature exceeds the normal operating temperature, then it will send a small PWM signal, control the driving circuit, and reduce the output current, so as to achieve energy conservation and protect the lamps (Figure 6.124).

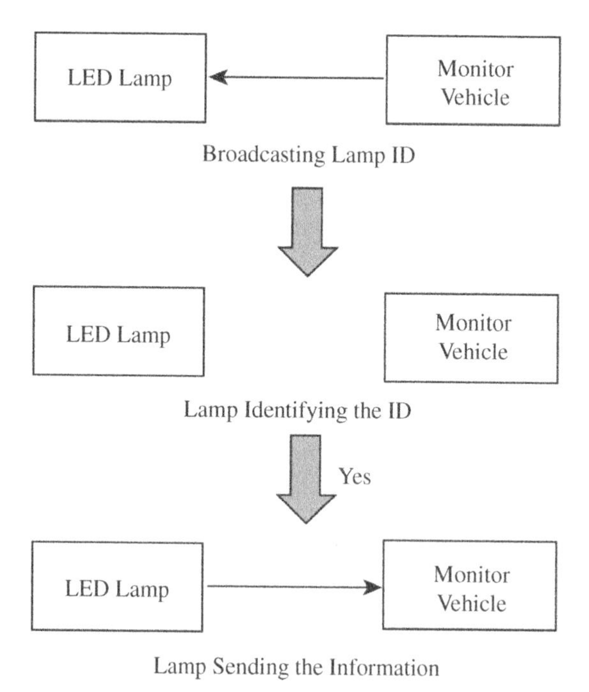

Figure 6.124 Wireless health management monitoring for LED light systems.

The system also has the function of malfunction detection. It is equipped with memory. When the MCU central controller detects the operating parameters of the lamps, such as the LED chip temperature, lighting brightness, LED chip voltage, and current, all these will be saved to the memory. The system contains a wireless communication module which can send out the lamp operating parameters in the memory. A malfunction detection vehicle with wireless communication module can be sent regularly to detect the LED lamps in turn. After the wireless communication modules in the lamps receive the detecting signals, the operating parameters of the lamps will be sent to the malfunction detection vehicle. The malfunction detection vehicle can analyze the received data and judge the working state of the LED lighting; if the lamp is judged as working incorrectly, the appropriate maintenance can be made. Therefore, the problems can be found as soon as possible to avoid further damage to lighting, to avoid the entire lamp being scrapped caused by abnormal situations like overheating of the lamps. In addition, prompt maintenance can be made after identifying the problems so as to avoid complete failure of the lamps and ensure the quality of both the lighting and road traffic safety.

The system can also achieve remote wireless lighting administration. The wireless communication module which is equipped with the lamps can receive the wireless control signal from the malfunction detection vehicles and control center; then the central controller can control the driving circuit in accordance with the control signals, regulate the size of the output current of the driving circuit, and adjust the lighting brightness of the LED in order to achieve remote administration of the LED lighting System. There are some disadvantages in the traditional control and administration system of road lamps and tunnel lights. It uses the communication cables and controls the entire lighting system with some sort of protocol; the construction and maintenance costs are relatively high. This invention of wireless remote administration system has some advantages such as low cost, simple maintenance and management.

6.4 Summary

In this chapter, the design of LED packaging applications was introduced, with the focus on light control. Initially, various means of light control were presented such as reflectors, lenses, diffusers, and so on. Color design and control were first described, along with many examples provided in applications. Thermal management in the system level of the application was presented, with types of heat dissipation to the environment. A one fin type heat sink was designed and optimized and was applied to several cases. An intelligent light control system was proposed, integrated with the wireless control and drive circuit.

References

[1] Evans, D.L. (1997), "High-luminance LEDs replace incandescent lamps in new applications," *Proc. SPIE, Light-Emitting Diodes: Research, Manufacturing, and Applications*, San Jose, CA, USA, 142–153.

[2] Craford, M.G. (2005). LEDs for solid state lighting and other emerging applications: status, trends, and challenges. *5th International Conference on Solid State Lighting*. I.T. e. a. Ferguson. San Diego, CA, USA: 594101.1–594101.10.

[3] Chen, G., Craven, A., Kim, A.M., Watanabe, S., Camras, M., Götz, W. and Steranka, F. (2008), "Performance of High-power III-nitride Light Emitting Diodes," *Physica Status Solidi (a)* **205**(5): 1086–1092.

[4] Jiang, J.B. and To, S. (2008). LED secondary optics design. *5th China International Exhibition & Forum on Solid State Lighting*. Shenzhen, China.
[5] Wang, K., Liu, S. and Luo, X.B. (2007a). A sensor integrated ultra-long span LED street lamp system. *8th International Conference on Electronics Packaging Technology*. Shanghai, China.
[6] Liu, S., Luo, X.B., Gan, Z.Y., Chen, M.X., Wang, K. and Liu, Z.Y. (2008). Ultra-long span high power LED street lamp. *China Patent*. **ZL200720073271.5**.
[7] Schlöder, U. (2007), "New optical concepts for headlamps with LED arrays," *SAE Papers* 2007-01-0869.
[8] Dross, O., Mohedano, R., Benítez, P., Miñano, J.C., Chaves, J., Blen, J., Hernández, M. and Muñoz, F. (2004), "Review of SMS design methods and real world applications," *Proceedings of SPIE*, Bellingham, WA, USA, 35–47.
[9] Wang, K., Liu, S., Luo, X.B., Liu, Z.Y. and Chen, F. (2008). Optical analysis of a 3-W LED MR 16 lamp. *IEEE 9th International Conference on Electronics Packaging Technology & High Density Packaging*. Shanghai, China: 456–460.
[10] Sun, C.C., Lee, T.X., Ma, S.H., Lee, Y.L. and Huang, S.M. (2006), "Precise optical modeling for LED lighting verified by cross correlation in the midfield region," *Optics Letters* **31**(14): 2193–2195.
[11] Wang, K., Luo, X.B., Liu, Z.Y., Zhou, B., Gan, Z.Y. and Liu, S. (2008), "Optical analysis of an 80W light-emitting diode street lamp," *Optical Engineering* **47**(1): 013002.
[12] Benítez, P., Miñano, J.C., Blen, J., Mohedano, R., Chaves, J., Dross, O., Hernández, M. and Falicoff, W. (2004), "Simultaneous multiple surface optical design method in three dimensions," *Optical Engineering* **43**(7): 1489–1502.
[13] Winston, R., Miñano, J.C. and Benítez, P. (2005), Nonimaging Optics, *San Diego, USA, Elsevier Academic Press*.
[14] Wang, K., Liu, S., Chen, F., Qin, Z., Liu, Z.Y. and Luo, X.B. (2009), "Freeform LED lens for rectangularly prescribed illumination," *Journal of Optics A: Pure and Applied Optics* Accepted.
[15] Ries, H. and Muschaweck, J. (2002), "Tailored freeform optical surfaces," *Journal of the Optical Society of America A* **19**(3): 590–595.
[16] Wang, L., Qian, K.Y. and Luo, Y. (2007), "Discontinuous free-form lens design for prescribed irradiance," *Applied Optics* **46**(18): 3716–3723.
[17] Ding, Y., Liu, X., Zheng, Z.R. and Gu, P.F. (2008), "Freeform LED lens for uniform illumination," *Optics Express* **16**(17): 12958–12966.
[18] Ries, H. and Rabl, A. (1994), "Edge-ray principle of nonimaging optics," *Journal of the Optical Society of America A* **11**(10): 2627–2632.
[19] Piegl, L. and Tiller, W. (1997), *The NURBS Book 2nd ed*, Berlin, Springer.
[20] Moreno, I. and Sun, C.C. (2008), "Modeling the radiation pattern of LEDs," *Optics Express* **16**(3): 1808–1819.
[21] Wang, K., Liu, S., Chen, F., Liu, Z.Y. and Luo, X.B. (2009), "Effect of manufacturing defects on optical performance of discontinuous freeform lenses," *Optics Express* **17**(7): 5457–5465.
[22] West, R.S., Kobijn, H., Sillevis-Smitt, W., Kuppens, S., Pfeffer, N., Martynov, Y., Yagi, T., Eberle, S., Harbers, G., Tan, T.W. and Chan, C.E. (2003), "High brightness direct LED backlight for LCD-TV," *SID International Symposium Digest of Technical Papers* **43**(4): 1262–1265.
[23] Wang, K., Liu, S., Chen, F., Liu, Z.Y. and Luo, X.B. (2009). Novel application-specific LED packaging with compact freeform lens. *IEEE 59th Electronic Components & Technology Conference*. San Diego, CA, USA: 2125–2130.
[24] Chang, S.I. and Yoon, J.B. (2006), "Microlens array diffuser for a light-emitting diode backlight system," *Optics Letters* **31**(20): 3016–3018.
[25] Hu, Y.J., Lee, J.C., Wang, Y.P., Wu, Y.F. and Sheu, L.G. (2007), "Diffuser array for a light-emitting diode backlight system," *Nonimaging Optics and Efficient Illumination Systems IV* San Diego, CA, USA.
[26] Wang, Y.N. (2008). Application Study on Illumination Quality Improvement of Tri-chromatic White Light Emitting Diodes. College of Electrical Engineering. Chongqing, Chongqing University. *Master of Engineering*: 78. (In Chinese).
[27] Wang, S.M. (2009). LED applications in Wuxi Lingshan Buddhist temple. *3rd China International Forum on Novel Light & Energy Sources*. Shanghai, China.
[28] Luo, X.B., Cheng, T., Xiong, W., Gan, Z.Y. and Liu, S. (2007), "Thermal analysis of an 80W light-emitting diode street lamp," *IET Optoelectronics* **1**(5): 191–196.
[29] Song, S., Lee, S. and Au, V. (1994). Closed-form equation for thermal constriction/spreading resistances with variable resistance boundary condition. *Proceedings of the 1994 International Electronics Packaging Conference*. Atlanta, Georgia, USA.

[30] Lee, S., Song, S., Au, V. and Moran, K.P. (1995). Constriction/spreading resistance model for electronics packaging. *ASME/JSME Thermal Engineering Conference*. Maui, Hawaii, USA.

[31] Luo, X.B., Xiong, W., Cheng, T. and Liu, S. (2009). Design and optimization of horizontally-located plate fin heat sink for high power LED street lamp. *59th Electronic Components and Technology Conference*. San Diego, CA, USA.

[32] Hollands, K.G.T. and Uuny, S.E. (1976), "Free convective heat transfer across inclined air layers," *Journal of Heat Transfer*, **98**: 189–190.

[33] Segiel, R. and Norris, R.H. (1957), "Tests of free convection in a partially enclosed space between two heated vertical plates," *Trans ASME of Heat and Mass Transfer* **79**: 663–674.

[34] Luo, X.B., Xiong, W. and Liu, S. (2008). A simplified Thermal resistance network model for high power LED street lamp. *IEEE 9th International Conference on Electronics Packaging Technology & High Density Packaging*. Shanghai, China.

[35] Liu, S., Lin, T., Luo, X.B., Chen, M.X. and Jiang, X.P. (2006). A microjet array cooling system for thermal management of active radars and high-brightness LEDs. *56th Electronic Components & Technology Conference*. San Diego, CA, USA: 1634–1638.

[36] Luo, X.B. and Liu, S. (2006). A closed micro jet cooling system for high power LEDs. *IEEE 7th International Conference on Electronics Packaging Technology*. Shanghai, China: 592–598.

[37] Luo, X.B. and Liu, S. (2007), "A microjet array cooling system for thermal management of high-brightness LEDs," *IEEE Transactions on Advanced Packaging* **30**(3): 475–484.

[38] Luo, X.B., Chen, W., Sun, R.X. and Liu, S. (2008), "Experimental and numerical investigation of a microjet based cooling system for high power LEDs," *Heat Transfer Engineering* **29**(9): 774–781.

[39] Holman, J.P. (1997), Heat Transfer, *8th Edition International Edition, Boston, McGraw-Hill Book Company.*

[40] Bar-Cohen, A. and Rohsenow, W. M. (1984), "Thermally optimum spacing of vertical, natural convection cooled, parallel plates," *Journal of Heat Transfer* **106**: 116–123.

[41] Churchill, S. W. and Usagi, R. (1972), "A general expression for the correlation of rates of transfer and other phenomena," *AIChE Journal* **18**(6): 1121–1128.

7

LED Measurement and Standards

7.1 Review of Measurement for LED Light Source

Light emitting diodes (LEDs) have been widely used in full-color displays, traffic signals, and backlighting for cell phones and also have considerable potential in general lighting applications [1]. It is highly desirable for LED manufacturers to have LED's characteristic parameters, including photometric parameters such as luminous intensity (Iv) and total luminous flux (Φ_V), colorimetric parameters such as chromaticity coordinates (x, y), peak wavelength (λ_p) and dominant wavelength (λ_d), and electrical parameters such as forward voltage (V_F), and reverse current (I_R), because of the necessity for quality management [2]. Measurement of LEDs differs greatly from conventional light sources because of the distinctive properties of LEDs such as asymmetrical light-emission profile, narrow spectral bandwidth, and temperature sensitivity [3–5]. The spectral power distributions and colorimetric parameters of LEDs are very important quantities for the characterization of LED light sources. The CIE (the Commission Internationale de l'Eclairage) has published a technical report, CIE 127, as a guide to the photometric and colorimetric measurement of LEDs. In CIE 127, colorimetric parameters are derived from spectral power distributions (SPD) measurement results. Accordingly, high-accuracy spectral measurement is especially important. The SPD measurements can be obtained by using the imaging spectroscopic technique [6]. Spectrometers in the colorimetric measurement of the light source include two types: mechanical scanning spectrometer and array spectrometer [7]. Compared to mechanical scanning spectrometers, the array spectrometers based on the charge-coupled device (CCD) can capture the whole spectral image at once. There are relevant research efforts on LED's photometric parameters testing and colorimetric parameters testing [8,9].

With the improvement of the high power LED's performance and its luminous efficiency, LEDs have begun to gradually replace traditional light sources in the field of functional lighting with a rapid and strong momentum, especially in road lighting. Presently, a great number of light fixture manufacturers in the world have pioneered the trial use of LED road lights in the market and have launched demonstration projects in many places. However, the relevant measurement standards of LED road lights have not been issued at home or abroad

LED Packaging for Lighting Applications: Design, Manufacturing and Testing, First Edition. Sheng Liu and Xiaobing Luo.
© 2011 Chemical Industry Press. All rights reserved. Published 2011 by John Wiley & Sons (Asia) Pte Ltd.

because of various reasons, resulting in great differences in the performance of LED road lights and a negative influence on solid state lighting because of some light fixtures and projects of a low quality. Therefore, a measurement standard for LED road light is urgently needed within this industry to regulate products. Presently, some national and regional organizations have proposed tentative drafts of specifications or standards for public comments, such as CSA (China Solid-state-lighting Alliance) approved recommendation—*Measurement Method for Integral LED Road Lights*, Taiwan regional standard—*Fixtures of Roadway Lighting with Light Emitting Diode Lights* and Guangdong regional standard—*LED Road Lights*. The CSA approved recommendation *Measurement Method for Integral LED Road Lights* is attached as an appendix. The following is a preliminary comparison of these three standards.

Measurement Method for Integral LED Road Lights has presented all major specifications of parameters for road lights and details which are used to evaluate the performance of LED road lights; it has also explained the measurement methods for every parameter, including the requirements of the measuring environment and the measuring instruments, but it has not pointed out the desirable level that every parameter should reach. *Fixtures of Roadway Lighting with Light Emitting Diode Lights* has briefly presented the major parameters of LED road lights and the requirements that these parameters should meet; it has also prescribed the measurement methods for the parameters. The Guangdong regional standard—*LED Road Lights*, has prescribed the measurement methods for every performance parameter; however, compared with Taiwan regional standard—*Fixtures of Roadway Lighting with Light Emitting Diode lights*, this standard covers fewer measuring parameters.

Based on the description above, we feel obligated to present the measurement methods of LED devices and LED road lights.

7.2 Luminous Flux and Radiant Flux

The radiant flux characterizes the physical properties of electromagnetic radiation by radiometric units. However, the radiometric units are irrelevant when it comes to the light perception of humans. Luminous flux characterizes the optical properties of LEDs by photometric units. The luminous flux represents the light power of a light source as perceived by the human eye. The unit of luminous flux is the lumen (lm). The luminous flux can be expressed as:

$$\Phi_V = 683(lm/W)\int_{380}^{780} S(\lambda)V(\lambda)\mathrm{d}\lambda \tag{7.1}$$

where $S(\lambda)$ is the spectral power distributions (SPD), that is, the light power per unit wavelength emitted in all directions, and the factor 683 lm/W is a normalization factor; and $V(\lambda)$ is spectral luminous efficiency function of the human eye.

In order to accurately measure the luminous flux of LEDs, the light emitted from LEDs need to be collected by a human photopic vision detector. Then the luminous intensity is converted into photocurrent via the human photopic vision detector. The spectral sensitivity of the detector should match the spectral luminous efficiency function $V(\lambda)$ of the human eye.

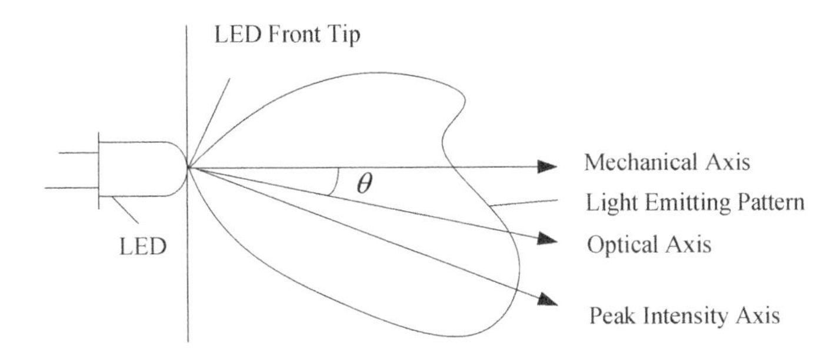

Figure 7.1 Definition of mechanical axis, optical axis, and peak intensity axis. (Reprinted with permission from M. Bürmen, F. Pernuš and B. Likar, "LED light sources: a survey of quality-affecting factors and methods for their assessment," *Measurement Science and Technology*, **9**, 12, 122002, 2008. © 2008 IOP Publishing.)

7.3 Measurement for Luminous Intensity

The luminous intensity, which is a photometric quantity, represents the luminous intensity of a light source, as perceived by the human eye. The luminous intensity is measured in units of candela (cd). As the luminous intensity of LEDs varies with the angle, the luminous intensity of the measured LEDs can depend on the detector. The luminous intensity of LEDs can be given by:

$$I_V = \mathrm{d}\Phi_v / \mathrm{d}\Omega \tag{7.2}$$

where $\mathrm{d}\Phi_V$ is the luminous flux captured by the detector, and $\mathrm{d}\Omega$ is the solid angle corresponding to the input aperture of the detector.

The measured luminous intensity is correlated with the detector size, shape, and distance and direction from the LED front tip to the detector [10]. In order to solve this problem, three reference axes through the LED front tip were recommended for luminous intensity measured by the CIE Technical Committee (TC2–46). Figure 7.1 presents the recommended reference axes.

As shown in the figure, the optical axis lies in the direction of the centroid of the optical radiation pattern. The mechanical axis lies in the direction of the axis of symmetry of the emitter body. The peak intensity axis lies in the direction of the maximum intensity. Any of these three axes could be used as the reference axis for luminous intensity measurement.

The CIE recommends two standard conditions, including condition A and condition B, for the measurement of the average LED luminous intensity under near-field conditions. The standard conditions recommended by CIE for measuring average luminous intensity are shown in Table 7.1.

Table 7.1 Standard conditions recommended by CIE for measuring average luminous intensity

CIE Standard Condition	Distance from LED Front Tip to Detector (mm)	Solid Angle (s_r)	Planar Angle (°)
Condition A	316	0.001	2
Condition B	100	0.01	6.5

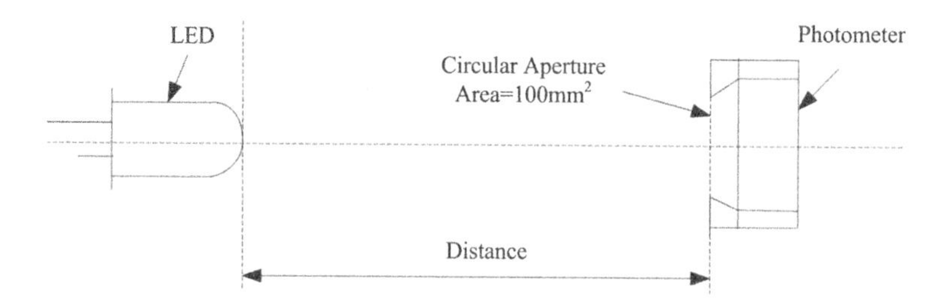

Figure 7.2 Schematic of CIE standard conditions for the measurement of average LED intensity.

Figure 7.2 is the schematic of CIE standard conditions for the measurement of average LED intensity. The measured LEDs are adjusted to make the mechanical axis pass through the center of the detector aperture. The spectral sensitivity of the detector should be calibrated to match the spectral luminous efficiency function $V(\lambda)$ of the human eye. The distance d should be set according to the standard conditions A and B recommended by the CIE. Under these two conditions, the detector used is required to have a circular input aperture with an area of $100\,mm^2$ with a corresponding diameter of 11.3 mm. Moreover, with regard to the distance from the LED front tip to the detector, the solid angle and planar angle are different.

7.4 LED Chromaticity Coordinates

According to the tri-stimulus theory of color perception, color can be represented by three parameters, which can be derived from spectral power distributions of a light source. If the spectral power distributions function $S(\lambda)$ is calculated, the chromaticity coordinates of LEDs can be calculated. It is given by:

$$\begin{cases} X = k\int_{380}^{780} S(\lambda)\bar{x}(\lambda)\mathrm{d}\lambda \\ Y = k\int_{380}^{780} S(\lambda)\bar{y}(\lambda)\mathrm{d}\lambda \\ Z = k\int_{380}^{780} S(\lambda)\bar{z}(\lambda)\mathrm{d}\lambda \end{cases} \tag{7.3}$$

Generally, the above equation is expressed as:

$$\begin{cases} X = k\sum_{\lambda=380}^{780} S(\lambda)\bar{x}(\lambda)\Delta\lambda \\ Y = k\sum_{\lambda=380}^{780} S(\lambda)\bar{y}(\lambda)\Delta\lambda \\ Z = k\sum_{\lambda=380}^{780} S(\lambda)\bar{z}(\lambda)\Delta\lambda \end{cases} \tag{7.4}$$

where X, Y, and Z are the tri-stimulus values; k is a normalizing constant; $S(\lambda)$ is the SPD function of measured LEDs; $\bar{x}(\lambda)$, $\bar{y}(\lambda)$ and $\bar{z}(\lambda)$ are the CIE 1931 color-matching functions; and $\Delta\lambda$ is the wavelength interval. For a light resource, Y is luminance of the light resource, and k can be calculated by adjusting the Y value of the light source to 100, that is:

$$k = \frac{100}{\sum S(\lambda)\bar{y}(\lambda)\Delta\lambda} \tag{7.5}$$

The chromaticity coordinates of LED are described as:

$$\begin{cases} x = \dfrac{X}{X+Y+Z} \\ y = \dfrac{Y}{X+Y+Z} \\ z = \dfrac{Z}{X+Y+Z} \\ x+y+z = 1 \end{cases} \tag{7.6}$$

Apparently, the z chromaticity coordinate value can be obtained from x and y. Therefore, the z coordinate is redundant, and does not need to be used.

7.5 Dominant Wavelength Determination Algorithm

The dominant wavelength of the LED source is defined as the wavelength located on the perimeter of the chromaticity diagram which appears to be the closest to the color of the test LED source. It can be determined by drawing a straight line from the equal-energy point (W_E) with coordinates of 0.333314, 0.333288 to the chromaticity coordinates (x, y) of the measured LED source, and then extending the straight line to the perimeter of the chromaticity diagram. The intersection point is the dominant wavelength of the LED source. The procedure is schematically shown in Figure 7.3.

7.5.1 Curve Fitting Method

In this method, the cubic spline function is used to fit the perimeter of the chromaticity diagram [11]. When the perimeter of the chromaticity diagram ranges from 380 nm to 507 nm, we can obtain the function as follows:

$$y = 0.64 - 9.94x + 57.92x^2 - 125.54x^3 \tag{7.7}$$

When the perimeter of chromaticity diagram ranges from 508 nm to 520 nm, the function is expressed as Equation 7.8:

$$y = 0.67 + 7.58x - 127.10x^2 + 767.33x^3 \tag{7.8}$$

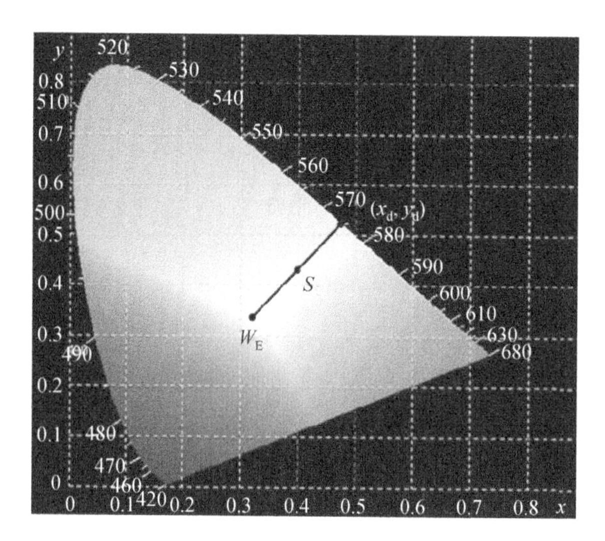

Figure 7.3 1931 CIE-XYZ chromaticity diagram showing the determination of the dominant wavelength of the LED light source with chromaticity coordinates (x, y) using the equal-energy point with coordinates of 0.333314, 0.333288 as the white-light reference.

When the perimeter of chromaticity diagram ranges from 520 nm to 540 nm, the function is described as Equation 7.9:

$$y = 0.81 + 0.89x - 7.55x^2 + 11.87x^3 \tag{7.9}$$

When the perimeter of chromaticity diagram ranges from 540 nm to 700 nm, we employ a linear function to match the edge, and the function is given as Equation 7.10:

$$y = -0.9956x + 0.9968 \tag{7.10}$$

After the chromaticity coordinates of the LED sample are measured, we can obtain a linear equation through the chromaticity coordinates (x, y) and equal-energy point (W_E). By combining the linear equation and the fitting function, the chromaticity coordinates corresponding to the dominant wavelength of the LED can be calculated and thus the dominant wavelength of the LED can be quickly determined. The measurement results for the dominant wavelength of measured LED samples are shown in Table 7.2.

7.6 LED Color Purity

After measuring the chromaticity coordinates of LEDs and the dominant wavelength, the color purity of LEDs can be calculated according to 1931 CIE-XYZ chromaticity diagram.

The equal-energy point (W_E) with chromaticity coordinates of 0.333314, 0.333288, as shown in Figure 7.3, is chosen as the white-light reference. The ratio of the distance from the

Table 7.2 Measurement results for dominant wavelength of measured LED samples

		Dominant Wavelength (nm)		
Part No.	Color	Standard Value	Measurement Value	Error
1	Blue	480.7	481.16	0.46
2	Blue	480.2	480.72	0.52
3	Blue	481.4	481.38	– 0.02
4	Green	514.4	514.85	0.45
5	Green	513.6	513.74	0.14
6	Green	513.3	513.82	0.52
7	Red	626.0	626.03	0.03
8	Red	626.6	627.48	0.88
9	Red	625.2	625.95	0.75

equal-energy point W_E (x_0, y_0) to chromaticity coordinates $S(x, y)$ of the measured LED and the distance from equal-energy point W_E (x_0, y_0) to λ_d (x_d, y_d) is used to represent purity of the measured LED. It is given by:

$$P_e = \frac{\sqrt{(x - x_0)^2 + (y - y_0)^2}}{\sqrt{(x_d - x_0)^2 + (y_d - y_0)^2}} \tag{7.11}$$

where (x, y) is the calculated chromaticity coordinates of the measured LED; (x_d, y_d) is the chromaticity coordinates corresponding to dominant wavelength of LEDs; and (x_0, y_0) are the chromaticity coordinates of the equal-energy point.

7.7 Color Temperature and Correlated Color Temperature of Light Source

The color of the radiation of the light source at a temperature T is the same as the color of blackbody radiation at the temperature T_c, and T_c is defined as the color temperature of this light source. The relative spectral power distributions of the blackbody are defined by Planck's law:

$$P(\lambda, T) = c_1 \lambda^{-5} (e^{\frac{c_2}{\lambda T}} - 1)^{-1} \tag{7.12}$$

In the formula, T is the absolute temperature of the blackbody; λ is the wavelength; c_1 is the first radiation constant; c_2 is the second radiation constant; and $P(\lambda, T)$ is the radiant flux emitted per unit area per unit wavelength interval.

For non-radiation light sources, such as an LED, fluorescent light, and high voltage sodium light, their spectral power distributions are much different from blackbody radiation and their chromaticity coordinates are not always on the locus of the blackbody radiation, but are close to that line. The color which is the closest to the blackbody radiation at a certain temperature is

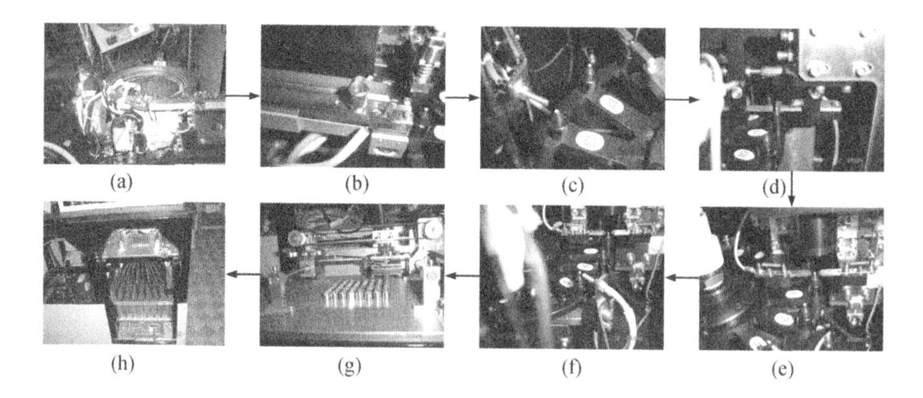

Figure 7.4 Process flow of a sorting system.

defined as the correlated color temperature (CCT). In a uniform chromaticity diagram, if a light source is provided with the closest color to the blackbody at a certain temperature, the absolute temperature value of the blackbody is the correlated color temperature of the light source.

The color temperature or correlated color temperature of LEDs can be determined provided that the chromaticity coordinates and spectral power distributions of LEDs are calculated. When the chromaticity coordinates of the light source are on the coordinate's locus of the blackbody, the calculated result is the value of the color temperature of the light source. On the contrary, when the chromaticity coordinates of the light source are off the coordinate's locus of the blackbody, the calculated result is the correlated color temperature.

7.8 Automatic Sorting for LEDs

The LED automatic sorter mainly consists of basic transportation equipment (disk vibration conveyer, linear vibration conveyer), interval feeding equipment, testing holder, and sorting equipment. A testing holder is set up in order to measure characteristic parameters of the LED for production line. When LEDs are transported to the testing holder, testing probes begin to clamp terminals of the LEDs. Once the LED samples are firmly clamped by the probe, the light emitted from the measured LED, which is illuminated under a constant pulse current operation condition, is captured by a spectrometer and detector. The process flow of the LED automatic sorting is shown in Figure 7.4, and a brief explanation of each step is given below.

(a) Vibration feeding: uses a high-frequency vibration system which is composed of a circular vibrating conveyer and a linear vibrating conveyer connected with a static electricity eliminator;
(b) LED pick up: uses a servo motor-driven eccentric slider mechanism, which can suck LEDs at the exit of the feeding mechanism and place them onto the rotating plate;
(c) Inspecting of LEDs: uses sensors to inspect whether LEDs are on the rotating plate;
(d) Locating of LEDs: correct the position of the picked LEDs to the testing position; and
(e) Testing holder: A rotating plate controlled by a servo motor is used for the testing holder. Sixteen vaccum suction nozzles are distributed on the perimeter of the rotating

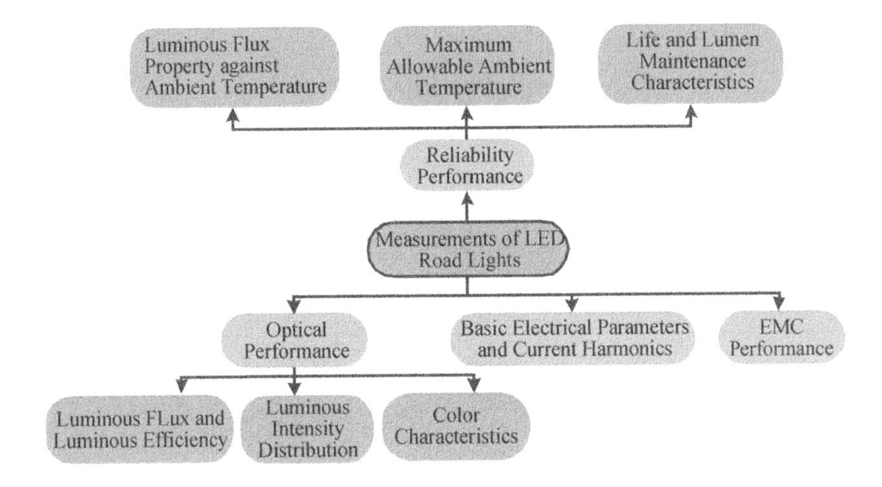

Figure 7.5 Diagram of measurement items of LED road light fixture.

plate, and LED samples are located on the perimeter of the rotating plate. The optical element from the testing instrumentation setup, including the spectrometer and the human photopic vision detector, are fixed on the top of the rotating plate. The electrical connection with the LEDs is achieved by two testing probes. When the testing probe clamps the terminals of the LEDs, The SOT (Start of Trigger) signal is sent to the testing instrument. After the end of the testing process, the EOT (End of Trigger) signal and sorting code signal generated from the testing instrument setup are sent to the sorter;

(f) Inspection of LEDs: use sensors to inspect whether the measured LEDs have been detached from the rotating plate;
(g) Sorting of LEDs: move the servo motor-driven platform to drive receiving catheter to the corresponding receiving bin according to the sorting signal code; and
(h) Receiving of LEDs: uses compressed air to blow the measured LEDs into the receiving bin.

7.9 Measurement for LED Road Lights

The measurement items of LED road lights are shown in Figure 7.5.

The CSA approved recommendation, *Measurement Methods for Integral LED Road Light,* attached as an appendix, which was drafted by a group of experts led by Professor Jiangen Pan *et al.*, can be a reference for the measurement methods of the items above. Referring to the Taiwan regional standard—*Fixtures of Roadway Lighting with Light Emitting Diode Lights*, measurement parameters are as follows.

7.9.1 Electrical Characteristics

The power factor of an LED road light must be over 0.95, harmonic distortion of input current should be less than the values in Table 7.3, the total harmonic distortion of current should not be more than 33%.

Table 7.3 Allowable value of harmonic distortion

Harmonic Order (n)	Maximum Allowable Value of Harmonic Distortion (Marked by the Percentage of the Basic Wave of Input Current)
2	2
3	$30 \times \eta$ (η, Power Factor)
5	10
7	7
9	5
$11 \leq n \leq 39$	3

7.9.2 Color Characteristics

The color temperature marked on the light should be within the range of the color temperature grades shown in Table 7.4.

The range of the chromaticity coordinates corresponding to different color temperatures is the area surrounded by four points in the CIE1931 chromaticity diagram (shown in Figure 7.6); the concrete values are shown in Table 7.5.

7.9.3 Light Distribution Characteristics

According to the types of light distribution, LED road lights can be divided into full cut-off luminaries, half cut-off luminaries (A and B), and non-cut-off luminaries. Their light distribution should meet the requirements in Table 7.6.

7.9.4 Dynamic Characteristics

Driven by the alternating current from 90% to 110% of the rated input voltage, the central luminous intensity shift of an LED road light should be within the range of $\pm 5\%$.

The angles in the table above are defined in Figure 7.7:

Table 7.4 Correlated color temperature grades

Color Temperature Grade	Range of Color Temperature
2700 K	2725 ± 145
3000 K	3045 ± 175
3500 K	3465 ± 245
4000 K	3985 ± 275
4500 K	4503 ± 243
5000 K	5028 ± 283
5700 K	5665 ± 355
6500 K	6530 ± 510

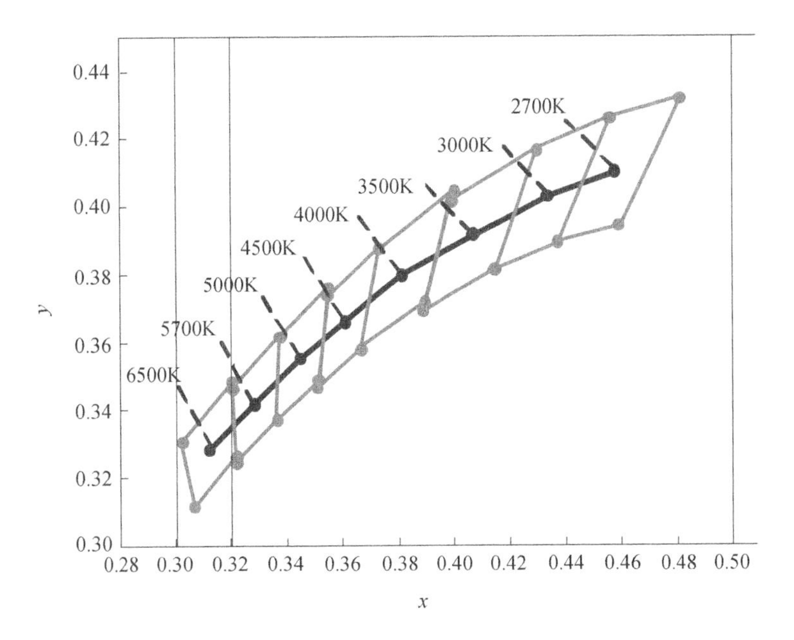

Figure 7.6 Ranges of the chromaticity coordinates corresponding to different color temperatures in CIE1931 chromaticity diagram. (Reproduced from CNS: Fixtures of roadway lighting with light emitting diodes lamps.)

Table 7.5 Chromaticity coordinates corresponding to different color temperatures and the ranges of error

Color Temperature	2700 K		3000 K		3500 K		4000 K	
Coordinate Axis	*x*	*y*	*x*	*y*	*x*	*y*	*x*	*y*
Chromaticity Coordinates of the Central Point	0.4578	0.4101	0.4338	0.4030	0.4073	0.3917	0.3818	0.3797
Allowable Range of Chromaticity Coordinates	0.4813	0.4319	0.4562	0.4260	0.4299	0.4165	0.4006	0.4044
	0.4562	0.4260	0.4299	0.4165	0.3996	0.4015	0.3736	0.3874
	0.4373	0.3893	0.4147	0.3814	0.3889	0.3690	0.3670	0.3578
	0.4593	0.3944	0.4373	0.3893	0.4147	0.3814	0.3898	0.3716
Color Temperature	4500 K		5000 K		5700 K		6500 K	
Coordinate Axis	*x*	*y*	*x*	*y*	*x*	*y*	*x*	*y*
Chromaticity Coordinates of the Central Point	0.3611	0.3658	0.3447	0.3553	0.3287	0.3417	0.3123	0.3282
Allowable Range of Chromaticity Coordinates	0.3736	0.3874	0.3551	0.3760	0.3376	0.3616	0.3205	0.3481
	0.3548	0.3736	0.3376	0.3616	0.3207	0.3462	0.3028	0.3304
	0.3512	0.3465	0.3366	0.3369	0.3222	0.3243	0.3068	0.3113
	0.3670	0.3578	0.3515	0.3487	0.3366	0.3369	0.3221	0.3261

Table 7.6 Requirements of light distribution of LED road light (unit: c_d/Klm)

Type of Lights	Vertical Angle 90° Horizontal Angle 90°	Vertical Angle 80° Horizontal Angle 90°	Vertical Angle 70° Horizontal Angle 65°–95°	Vertical Angle 65° Horizontal Angle 65°–95°	Vertical Angle 60° Horizontal Angle 65°–95°
Full Cut-off	<10	<30	-	-	<180
Half Cut-off A	<30	<120	-	>90	-
Half Cut-off B	<60	<150	-	>150	-
Non Cut-off	<100	-	>150	-	-

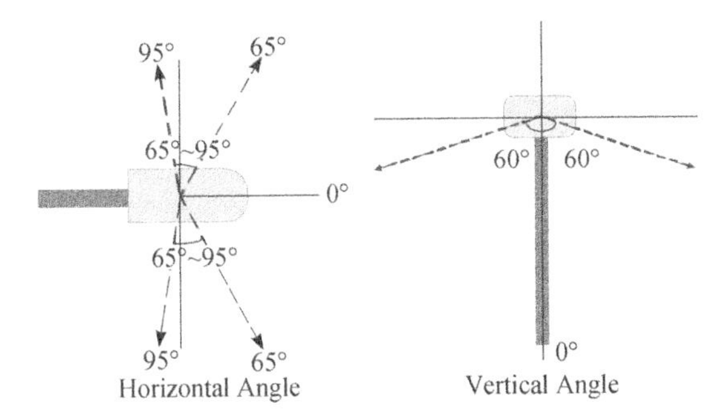

Figure 7.7 Spacial angles' schematic of the light distribution of LED road light. (Reproduced from CNS: Fixtures of roadway lighting with light emitting diodes lamps.)

7.9.5 Test of Reliability

The test of reliability mainly includes a maximum allowable ambient temperature test, lighting maintenance test, light waterproof test, and anti-vibration test.

7.10 Summary

In this chapter, LED measurement methods and standards were presented, with the objective of stimulating more discussion, as it is believed that measurement methods and standards are essential before any new technology and product could be used in volume, reliably and with high quality. Those optical and color concepts were discussed again in terms of the integrity of chapter organization. Measurement for LED road lights was discussed in terms of electrical characteristics, color characteristics, light distribution characteristics, dynamic testing, and test of reliability.

References

[1] Fujii, T., Gao, Y., Shama, R., Hu, E.L., DenBaars, S.P., and Nakamura, S. (2004), "Increase in the extraction efficiency of GaN-based light-emitting diodes via surface roughening," *Applied Physics Letters* **84**(6): 855–857.

[2] Park, S.C., Lee, D.H., Kim, Y.W., and Park, S.N. (2007), "Uncertainty evaluation for the spectroradiometric measurement of the averaged light-emitting diode intensity," *Applied Optics* **46**(15): 2851–2858.

[3] Miller, C.C., Zong, Y.Q., and Ohno, Y. (2004), "LED photometric calibrations at the National Institute of Standards and technology and future measurement needs of LEDs," *Fourth International Conference on Solid State Lighting*, Proceedings of SPIE 5530: 69–79.

[4] Young, R. (2006), "Measuring light emission from LEDs," *Advanced LEDs for Solid State Lighting*, Proceedings of SPIE 6355 (63550H): 1–11.

[5] Liu, M.Q., Zhou, X.L., Li, W.Y., Chen, Y.Y., and Zhang, W.L. (2008), "Study on methodology of LED's luminous flux measurement with integrating sphere," *Journal of Physics D: Applied Physics* **41**: 1–5.

[6] Johansson, T. and Pettersson, A. (1997), "Imaging spectrometer for ultraviolet-near-infrared microscopy," *Review of Scientific Instruments* **68**(5): 1962–1971.

[7] Shen, H.P., Pan, J.G., Feng, H.J., and Liu, M.Q. (2009), "Stray light errors in spectral color measurement and two rejection method," *Metrologia* **46**: 129–135.

[8] Ho, C.H. (2001), "A practical and inexpensive design for measuring the radiation patterns and luminescent spectra of optoelectronic devices," *Review of Scientific Instruments* **72**(7): 3103–3107.

[9] Manninen, P., Hovila, J., Karha, P., and Ikonen, E. (2007), "Method for analysing luminous intensity of light-emitting diodes," *Measurement Science and Technology* **18**: 223–229.

[10] Bürmen, M., Pernuš, F., and Likar, B. (2008), "LED light sources: a survey of quality-affecting factors and methods for their assessment" *Measurement Science and Technology* **19**: 1–15.

[11] Zhou, S.J., and Liu, S. (2009) "Transient measurement of LED characteristic parameters for production lines," *Review of Scientific Instruments* **80**(095102): 1–7.

Appendix

Measurement Method for Integral LED Road Lights Approved by China Solid State Lighting Alliance

The China Solid State Lighting Alliance (CSA) has recently approved recommendation *Measurement Method for LED Road Lights*, which was drafted by a group of experts led by Professor Jiangen Pan in 2008, and revised in 2009.

1 Scope

This recommendation regulates the measurement method for basic characteristics of Integral LED Road Lights.

This recommendation applies to integral LED road lights used for roadway lighting, which operate stably when driven by an internal controller (self-ballasted) or an external controller and supplied by alternating current with 50 Hz/220 V.

This recommendation is of reference to those LED road lights which are above the scope of this recommendation or other similar products.

2 Normative References

The items, cited by this recommendation in the following documents, become its items. This recommendation does not apply to all modified documents (excluding the corrected contents) or revised editions, if the citations are dated. However, all the parties involving in reaching agreement on this recommendation are encouraged to conduct research on whether they can

LED Packaging for Lighting Applications: Design, Manufacturing and Testing, First Edition. Sheng Liu and Xiaobing Luo.
© 2011 Chemical Industry Press. All rights reserved. Published 2011 by John Wiley & Sons (Asia) Pte Ltd.

make use of the latest editions of these documents. As to the undated citations, the latest editions are suitable for this recommendation. Relevant documens are as follows:

JJG 211-2005. Luminance meter calibrating regulations;
JJG 245-2005. Illuminance meter calibrating regulations;
GB/T 5702-2003. Evaluation methods of color rendering properties of light sources;
GB/T 7922-2003. Measuring methods of the color of light sources;
GB 17625.1-2003. Electromagnetic compatibility (EMC), limits value, measurement method of harmonic current emission (equipment input current ≤ 16A) (IEC 61000-3-2, IDT);
GB 17743-1999. Limits and measuring methods of radio disturbance characteristics of electrical lighting appliance and similar equipment (idt CISPR 15);
GB/T 18595-2001. Requirements of electromagnetic compatibility and immunity of general lighting equipment (idt IEC 61547);
CIE 15-2004. Colorimetry;
CIE 70-1987. The measurement of absolute luminous intensity distribution;
CIE 84-1989. The measurement of luminous flux;
CIE 102-1993. Recommended file format for electronic transfer of luminaries photometric data; and
CIE121-1996. Photometry and goniophotometry of luminaries.

3 Definitions

This recommendation adopts the following definitions.

3.1 Integral LED Road Light

The integral LED road light is an integrated light source device which is used for road lighting. An LED which is used as the light source, is composed of components such as optics, mechanical, electric, and electronic components. The LED is integrated with the light and it is a non-removable and irreplaceable part of the light; for clarification, it is called the LED road light. It is driven by an internal controller (self-ballasted) or an external controller.

3.2 Self-ballasted Integral LED Road Light

It is the integral LED road light which is accompanied by an internal controller and can be connected to the power supply directly; called a self-ballasted LED road light for short.

3.3 Externally Controlled Integral LED Road Light

It is the integral LED road light which is connected to the power supply through external controller; called an externally controlled LED road light for short.

3.4 LED Reference Controller

The external controller which provides reference working conditions for an externally controlled LED road light.

3.5 Standard LED Road Light

A stable and reproducible LED road light, which is calibrated with basic photoelectric performance parameters. It can calibrate the measuring equipment with the substitution method. A standard LED road light should be provided with a temperature monitoring point and the reference working temperature of this point.

3.6 Initial Values

The photometric, colorimetric, and electric quantities of an LED road light, providing stable lighting under specified conditions.

3.7 Reference Axis

The reference axis refers to the axis which goes through the center of the luminous surface and is vertical to the luminous surface.

3.8 Photometry Center

The photometry center refers to the center of the luminous surface of the LED road light. In the goniophotometric measurement, the photometry center of the tested LED road light should be at the rotation center of the goniophotometer.

3.9 Measuring Half-plane (C Plane)

Measuring half-plane (C plane) refers to the half-plane which goes through the reference axis of the LED road light with the reference axis as its initial rotating line. It is called the measuring plane, or C plane for short, to avoid confusion. For road lights, the planes parallel to the longitude axis of the road are defined as C0° and C180° planes, and a C90° plane is defined as the half plane vertical to the longitude axis at the roadside, and a C270° plane is the half plane vertical to the longitude axis at the houseside, as shown in Figure A1.

3.10 Auxiliary Axis

The auxiliary axis goes through the photometry center of the LED road light and is vertical to the reference axis; in practial applications it is always parallel to the direction of the road. The auxiliary axis and the reference axis define the C0°/C180° plane.

3.11 Third Axis

The third axis goes through the photometry center and is vertical to reference axis; in practical applications it is always vertical to the direction of the road. The third axis and reference axis define the C90°/C270° plane.

3.12 Standard Measurement Attitude

The luminous surface of the LED road light is horizontal and spreads light downward.

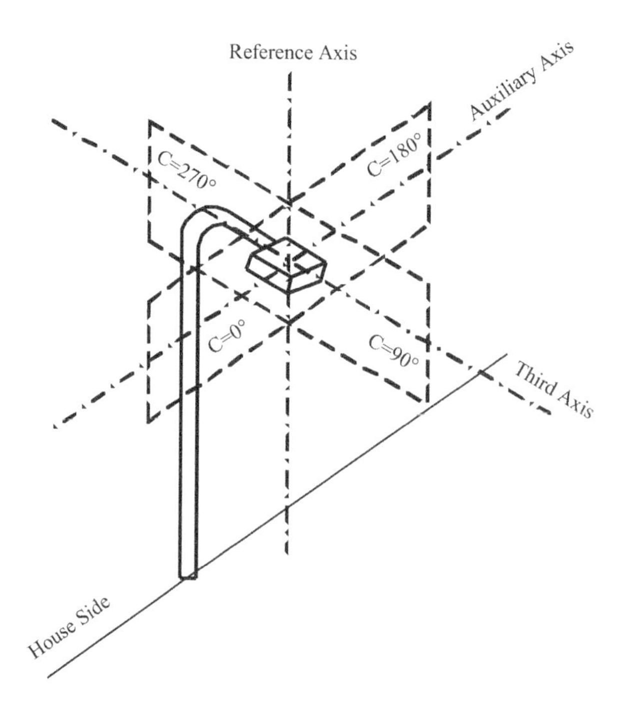

Figure A1 Coordinate system of the light distribution of LED road light.

3.13 Right Downward Point

It refers to the point which is just below the LED road light and usually on the reference axis.

3.14 Measurement Distance

It refers to the effective optical distance from the rotating center of the axes to the receiving surface center of the goniophotometer or the goniospectroradiometer.

3.15 Half-peak Side Angle

Taking the light center as the origin, the angle between the reference axis and the direction in which the LED road light has a luminous intensity of 50% of the maximum value. (If two or more directions are of 50% maximum luminous intensity, take the largest angle.)

3.16 Total Luminous Flux

The total amount of luminous flux of all directions generated by the LED road light; it is called luminous flux to avoid confusion.

3.17 Upward (Downward) Flux Fraction

It refers to the luminous flux of the LED road light in the directions above or below the horizontal plane of the photometry center.

3.18 Luminous Efficacy

It refers to the quotient of the total luminous flux by the power consumed of an LED road light, unit: 1 m/W.

3.19 Flashed Area

It refers to the projected area of the luminous part observed in the direction of $C = 0°$ or $180°$, $\gamma = 76°$.

3.20 Average Color Nonuniformity

It refers to the color difference between the average color after all light emitted from an LED road light is mixed, and the color of the light in the direction of the reference axis.

3.21 Maximum Color Nonuniformity

It refers to the maximum color difference between the light emitted from any direction in the half peak beam angle, and the light from the reference axis direction.

3.22 Lumen Maintenance

It refers to the ratio of the luminous flux at a given time in its life time and the initial luminous flux when the LED road light is operated under specified conditions, expressed as a percentage value. For simplicity of measurement, the illuminance at a right downward point can replace the luminous flux to calculate the lumen maintenance.

3.23 Color Shift

It refers to the difference between the colorimetric quantities at a given time in its life time and the initial quantities when the LED road light is operated under specified conditions. It can be represented by the coordinate difference in the CIE 1976 uniform color space of the average color or the right downward point at a certain distance of the LED road light.

3.24 Allowable Ambient Temperature Range

It refers to the range from the lowest temperature at which LED road light can be lit up normally, to the highest temperature at which LED can operate normally.

4 Main Measurement Items

4.1 Basic Electric Characteristics

4.1.1 Voltage

4.1.2 Current

4.1.3 Power

4.1.4 Power Factor

4.1.5 Frequency

4.2 Electromagnetic Compatibility (EMC)

Electromagnetic compatibility of a self-ballasted LED road light mainly includes radio disturbance characteristics, harmonic performance of input current, and immunity characteristics. Electromagnetic compatibility of an LED road light with an external controller includes radio disturbance characteristics and immunity characteristics of the input port.

4.3 Optical Performance

4.3.1 Luminous Flux and Efficacy

4.3.2 Luminous Intensity Distribution Characteristics

4.3.3 Flashed Area and Average Luminance Characteristics

4.3.4 Colorimetric Characteristics

4.3.4.1 Average Chromaticity

4.3.4.2 Average Color Rendering Index

4.3.4.3 Spatial Colorimetric Distribution Characteristics

4.3.4.4 Spatial Color Non-uniformity

(a) Average Color Non-uniformity
(b) Maximum Color Non-uniformity

4.4 Temperature Characteristics and Luminous Maintenance

4.4.1 Luminous Flux versus Temperature Curves

4.4.2 Highest Allowable Ambient Temperature

4.4.3 Endurance of High-Low Temperature Cycling

4.4.4 Luminous Maintenance Characteristics

4.4.4.1 Lumen Maintenance

4.4.4.2 Color Shift

4.4.5 On/Off Characteristics

5 General Requirements and Equipment Requirements for Measurment

5.1 Working Conditions

Unless otherwise specified, tests or measurements shall be conducted under the working conditions detailed below.

5.1.1 Laboratory Environmental Conditions

The Electric, photometric, and colorimetric measurement should be conducted with the LED road light in an environment maintained at an ambient temperature of 25°C ± 1°C, relative humidity of ≤65% and should be draught-free. There should be no air movement in the vicinity of the test LED road light.

For luminous maintenance measurement and on/off test, the ambient temperature shall be in the range of 25°C ± 5°C, relative humidity of ≤65% and should be draught-free.

The measurement point of ambient temperature should be at the same level of the photometry center of the LED road light, 0.5 meters away from the tested LED road light; thermo-probe should not be projected by the LED road light.

Air pressure: 86 kPa–106 kPa.

5.1.2 Power Supply Requirements

Self-ballasted controlled LED road lights should be tested or measured at specified voltage (if the specified voltage has a range, choose the medium value) and frequency. During stabilization, the voltage should be constant within ±0.5%; during the measurement, the power voltage should be constant within ±0.2%, and the fundamental frequency error shall not exceed 0.1% and the total harmonic content shall not exceed 3% of the fundamental; as for the aging and life test, it should be constant within ±2%.

An externally controlled LED road light shall operate under the drive of a reference control gear or equivalent driven condition. The output voltage/current/power of the special facility shall be maintained constant within ±0.2%, and the total harmonic content and the frequency error shall be specified if the facility is AC output, generally, the frequency error shall not exceed 0.1%, and the harmonic content shall not exceed 3% of the fundamental.

5.1.3 Operation State Requirements of the Measured LED Road Light

The optical characteristics of an LED road light are sometimes restricted by the work attitude due to heat dissipation issue, so when it is tested or measured with no special requirements, the LED light should be placed in a free space in the specified standard measurement attitude. During the sampling in the measurement, LED road lights should be kept static.

LED road lights should operate in a state of thermal equilibrium. The temperature of an LED road light itself should be monitored at the same time while monitoring the ambient temperature to keep a good reproducibility. If it is possible to monitor the junction voltage of LED lights, it should be monitored. Otherwise, the temperature of the designated position of the shell should be monitored.

5.1.4 Stable Working Condition Requirements of the Measured LED Road Light

The photometric, colorimetric, and electric quantities shall not be measured until the LED road light attains stable conditions. The condition of determining the stable working of an LED light is as follows: the variation of luminous flux or luminous intensity is less than 0.5% within two continuous periods of 15 minutes.

5.2 Requirements of Measurement Equipments

5.2.1 Power Supply

DC power supply: The stability shall be within 0.1% and the ripple coefficient should not exceed 0.5%. An AC power supply should have a very low impedance so that the voltage drop produced by the power supply should not exceed 0.1% of the specified voltage value when connected to an LED road light load. The power supply should comply with the requirements of IEC 61000-3-2 when measuring harmonics and the power factor. Generally speaking, only a pure sine-wave power supply with variable frequency can meet all the above requirements.

5.2.2 Electrical Measuring Instruments

The accuracy of DC electrical measuring instrument shall be better than 0.1%.

The voltage sampling input impedance of an AC electrical measuring instrument should be at least 1 MΩ. The current sampling impedance should be small enough to make sure that the voltage drop produced less than 0.1 V in the current sampling resistor. The precision of the electrical measuring instrument should meet the requirement that the practical measurement uncertainties of the measured voltage, current, and power is less than 0.5%. In general, a Class Index 0.5 meter cannot meet this requirement. An instrument of 0.2% or of higher accuracy is recommended.

When harmonics and power factor measurements are conducted, the digital power meter should meet the requirements of IEC 61000-3-2.

5.2.3 Electromagnetic Compatibility Testing Instrument

Test instruments for RFI emissions should comply with the requirements of CISPR 15.

Test instruments for input harmonics and power factors shall comply with IEC 61000-3-2.

Test instruments for electromagnetic immunity shall comply with requirements of IEC 61547.

5.2.4 Thermometer

The Grade A temperature detector is recommended; The thermometer should have at least three digits display. The accuracy shall be better than ±0.3°C and resolution shall be better than 0.1°C. A semiconductor temperature probe has a certain degree of photosensitivity, thus it must be used with caution.

5.2.5 Photometer Detector

A photometer detector should meet the requirement of standard class of JJG245-2005 with its $V(\lambda)$ mismatch index f_1' less than 3.5%. A photometer detector should have cosine correction performance, but for the photometer detector measuring luminous intensity at long distance, the cosine correction is not required, so as to obtain higher sensitivity.

5.2.6 Photometer

A photometer shall be equipped with a photometer detector specified in 5.2.5, it should have readings of at least four significant digits, user calibration function, and post-calibration lock protection function; Except for indication error, the other performance parameters should meet the requirement of standard class of JJG245-2005.

Note: If the system comprises a standard light with high enough accuracy, the calibration and correction of the photometer can be completed by users.

5.2.7 Spectroradiometer (Spectrometer)

A spectroradiometer is the equipment measuring the radiation power of light at each wavelength. It is an essential instrument for measuring the spectral power distributions, chromaticity, color rendering index, and other related photometric quantities of the LED road light. Spectroradiometers can be classified as being a mechanical scanning type and an array type. The former can have high accuracy but long measurement time, it is not suitable for applications which require high speed. The latter has the advantage of a short measurement time. Depending on the difference in manufacturing, technical, and the accuracy of the adopted key components, for example, gratings and detectors, an array type spectroradiometer can also be classified as being a basic type and a high accuracy type, with the former having low signal-noise-ratio, low sensitivity, and narrow linear dynamic range, thus not being able to satisfy the requirements for high accuracy LED measurement; while the latter (high accuracy array type spectroradiometer) usually adopts the high end concave grating and scientific grade TE-cooled array detector, and it has a high signal-noise-ratio, high sensitivity, and wide linear dynamic range.

The spectroradiometer shall be calibrated by spectral radiant intensity or irradiance standard lamps. After calibration, the measurement accuracy of chromaticity coordinates shall be better than 0.003, the chromaticity resolution and reproducibility of stable standard light source shall be better than 0.0002. The functions of light source color measurement and color rendering indices analysis of the spectroradiometer shall meet the requirements of standards GB/T 7922-2003 and GB/T 5702-2003 which are based on CIE published documents. Besides the general calibration function, the spectroradiometer is also supposed to have calibration functions against standard LED road lights.

5.2.8 Goniophotometer

A goniophotometer measures the photometric quantities (such as luminous intensity and illuminance, and so on) in different spatial angles, it usually includes a mechanical structure for the support and positioning of a measured light source, a photometer detector, and other necessary sensors and signal processing system. The basic performance and measurement conditions of a goniophotometer should meet requirements of those technical documents CIE 70-1987, CIE 84-1989, and CIE 121-1996. For the accurate photometric measurement of LED road lights, a goniophotometer should meet the following requirements:

1. During the measurement, the measured LED road light has always been measured in a standard attitude, and the measured LED road lights remain static or only rotate around the

reference axis to switch the measuring planes. When measured in a certain plane, LED road lights should be static. If the goniophotometer keeps the LED road light moving, or not in the specified attitude during the measurement, corrections shall be made, details refer to Appendix A.2 (omitted due to the space limitation);

2. The angle precision of the goniophotometer should not be less than 0.2°, and the smallest angle step shall not exceed 0.2°;

3. The goniophotometer should be calibrated by standard luminous flux lamp or standard luminous intensity lamp;

4. The reference goniophotometer measuring the total luminous flux should meet the requirements of CIE 84-1989, with its photometer detector rotating around the LED road light, directly receiving the light beam of measured LED road light and the detector having good cosine correction. Details refer to Appendix A.1 (also omitted due to the space limitation);

5. The goniophotometer measuring the LED road lights intensity distribution should be able to achieve the required distance measurement;

6. The reflectance of the mirror in mirror type goniophotometer non-spectral-sensitivity, or V (λ) matching of spectral responsivity of the photometric detector should take the mirror's spectral reflectance into account; and

7. The goniophotometer should have comprehensive software, at least the following data and curves shall be provided: the total luminous flux, regional luminous flux, upward (downward) flux fraction, efficacy, luminous intensity distribution (curve), isocandela curve, isolux curve, analysis of road illumination uniformity, luminance, and glare analysis. The data output format shall comply with the CIE 102-1993.

5.2.9 Integrating Sphere (Integrating Photometer, Integrating Spectroradiometer)

An integrating sphere is an instrument to rapidly measure luminous flux, spectral distribution, color, and color rendering index of LED road lights by the substitution method. The combination of the integrating sphere and photometer is called the integrating photometer. The combination of integrating sphere and spectroradiometer is called an integrating spectroradiometer. The integrating sphere should be large enough; the reflectance of internal surface coating should have good uniformity and spectral neutrality, and be insensitive to changes of temperature and humidity. The light blocking objects within the integrating sphere should be reduced to a minimum.

An integrating sphere should meet the requirements of CIE 84-1989. For an integrating photometer, the reflectance of the interior coating is recommended to be around 80% so as to keep good spectral flatness and reflectance stability. For an integrating spectroradiometer, the requirement on spectral characteristics of the interior coating is not so strict, and the reflectance more than 80% is allowed. The integrating sphere with high reflectance coating helps to improve the measurement accuracy of the system if the sphere size is small or the sensitivity of the instrument is low. However, if the reflectivity is too high, the long term stability of the output light and the effective transmittance of blue light will be adversely affected.

It is recommended to open a sampling port at the side of the integrating sphere for convenient operation (Figure A2(a) and Figure A2(b)). Although one can use an integrating sphere with the sampling port on the top, which is convenient to realize the standard attitude of the LED road light, dust can easily accumulate and it is inconvenient to operate. Because LED road lights

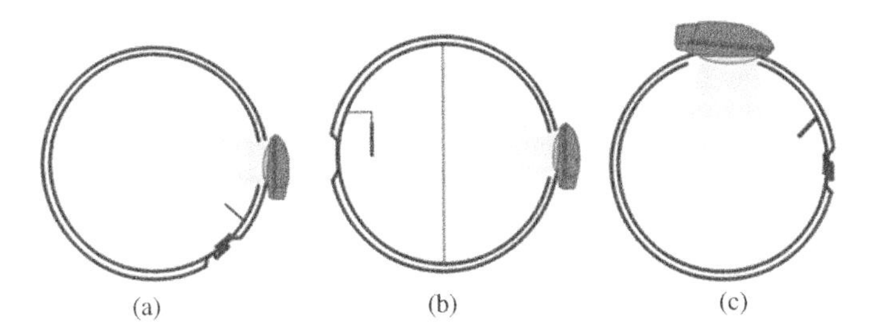

Figure A2 Schematic of LED road light measurement by integrating sphere.

themselves often have a great light blocking object (shell), measuring LED road lights in the integrating sphere is not recommend. The sampling port at the side will bring errors due to the fact that the LED road light is not in the standard attitude. Corrections shall be made according to Appendix A.2 (omitted due to the space limitation).

5.2.10 Goniospectroradiometer

A goniospectroradiometer replaces the photometric detector with a high accuracy array spectroradiometer, with details in Appendix A.4 (omitted due to the space limitation).

The goniospectroradiometer should meet the requirements of the spectroradiometer in 5.2.7 and goniophotometer in 5.2.8 (requirements of photometric detector not included). It is recommended that the spectroradiometer directly receives the beams from the LED road lights, and that the spectroradiometer is required to have a measurement speed fast enough, and has the synchronous sampling function to achieve the synchronization of spectral radiation intensity measurement with the angle rotation of the goniospectroradiometer.

5.2.11 Imaging Luminance Meter and Near-Field Goniophotometer

An imaging luminance meter uses a two-dimensional photoelectrical device as the detector (for example, CCD). The luminous of every point in a measurement area can be measure through one time sampling.

An imaging luminance meter should have a high linearity. There shall be optical components in front of the sensitivity area of the photoelectrical device to make the spectral response of all pixels in the photoelectrical device match the $V(\lambda)$ function. The $V(\lambda)$ mismatch factor shall meet the requirement of class 1 or above specified by JJG211-2005.

Replace the photometer detector in the goniophotometer by an imaging luminance meter to constitute a near-field goniophotometer which can comprehensively and accurately measure the luminance characteristics, illuminance, and luminous intensity distribution characteristics of LED road lights(details can be found in Appendix A.3 of original document, omitted due to the space limitation).

The near-field goniophotometer should meet the requirements of the goniophotometer in 5.2.8 (the requirements of a photometric detector are not included) and the above requirements on the imaging luminance meter. The imaging luminance meter in the near-field goniophotometer is recommended to directly receive the beams from the LED and should have the synchronous sampling function.

5.2.12 Temperature Controllable Thermostatic Chamber and Measurement Equipment for Luminous Flux against Temperature Characteristics

The temperature controllable thermostatic chamber can make the measured LED road lights operate in the chamber in required attitude. The controlled temperature is from −30 °C to 100 °C; temperature control accuracy is ±3 °C; temperature measurement accuracy is ±1 °C. The space in the chamber shall be large enough with an even temperature field. The temperature measurement point inside the chamber should be on the same horizontal level of the photometry center, and 0.5 meters away from the LED road lights. The temperature measurement probe should not be directly illuminated.

The measurement equipment of luminous flux against temperature characteristics is composed of a temperature controllable thermostatic chamber and measurement equipment of relative changes of luminous flux in the test chamber and it has the recording function for the relative values of luminous flux in different chamber temperature.

6 Measurement Methods

6.1 Measurement of the Basic Electrical Properties and Harmonic Current

Measure DC supplied externally controlled LED road lights by voltmeter and ammeter.

Voltage, current, power, power factor, frequency, and input harmonic current of LED road lights (externally controlled LED road light being generally attached by a reference external controller) are measured by a digital power meter with the function of measuring voltage, current, power, power factor, frequency, and input current harmonic, and so on.

Because the voltage sampling connected in parallel has a certain by-pass current, and the current sampling connected in series has a certain voltage drop, the application of the ammeter internal connecting method or external connecting method is determined according to the practical voltage and current of the measured LED road light. When the current is relatively large, or the lead is relatively long, a quadric-pole method can be used for voltage sampling.

6.2 Test of Electromagnetic Compatibility

6.2.1 An RFI Emissions Test for Self-Ballasted LED Road Lights and Externally Controlled LED Road Lights (Including Designated External Controller) is Carried Out According to CISPR 15

6.2.2 An EMC Immunity Test for LED Road Lights is Carried Out According to the Requirements of IEC 61547

6.3 Measurement of Luminous Flux and Luminous Efficacy

The measurement of luminous flux include the illuminance integration method, the luminous intensity integration method, and the substitution method using an integrating sphere, in which the illuminance integration method can be used as reference measurement of the total luminous flux for its high accuracy. When there is doubt about the total flux measured by the luminous intensity integration method or the integrating sphere method, the measurement result of illumination integration should be preferred.

6.3.1 Reference Measurement Method of Total Luminous Flux

In the photometric dark room, measure the total luminous flux of an LED road light by reference goniophotometer specified in 5.2.8.

Mount the LED road light within the reference goniophotometer for flux in the specified burning attitude, and its photometric center should be as close to the rotation centre of the goniophotometer as possible.

Measure the illuminance of the points on the surface of the imaginary sphere around the LED road light in enough measuring planes with an angle step small enough. The angle interval is generally 5° between planes and the angle steps in a plane is generally 1°. When the size of the measured LED road light is relatively large or the beam angle is relatively narrow, smaller plane intervals and angle steps are required to ensure complete sampling of illuminance distribution.

The calculation equation of total luminous flux Φ_{tot} is:

$$\Phi_{tot} = \int_{(S_{tot})} E dS = \int_0^{4\pi} r^2 E(\varepsilon, \eta) d\Omega = \int_0^{2\pi} \int_0^{\pi} r^2 E(\varepsilon, \eta) \sin \varepsilon d\varepsilon d\eta \quad (A1)$$

where r is the radius of the imaginary sphere, S_{tot} is the total area of the imaginary sphere, and (ε, η) is the space angle, as shown in Figure A3.

6.3.2 Measurement of Luminous Flux by Luminous Intensity Integration Method

In the photometric dark room, measure the spatial luminous intensity distribution of an LED road light by a goniophotometer specified in 5.2.8, and calculate the total luminous flux, regional luminous flux, and the upward (downward) luminous flux fraction of the LED road light by the numerical integration method.

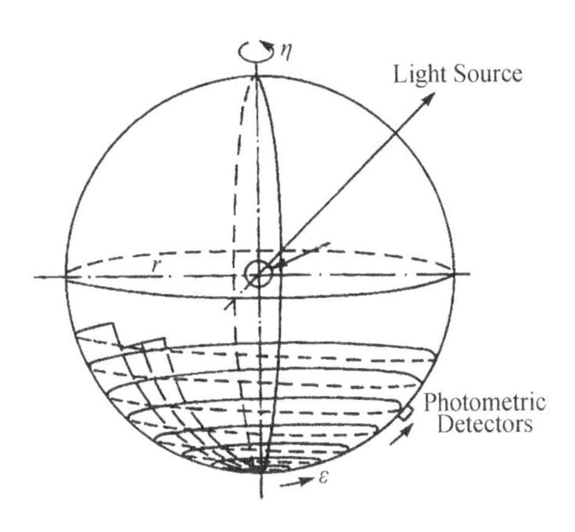

Figure A3 Schematic diagram of the calculation of luminous flux.

Mount the LED road light within the goniophotometer in the specified burning attitude, and its photometry centre shall be at the rotation centre of the goniophotometer. Measure the luminous intensity of the LED road light in all spatial directions, with a small enough angle step (generally 5° or less) in enough measuring planes (with the plane interval of 10°or less). The luminous flux is calculated by the Equation A2:

$$\text{Total luminous flux:} \quad \Phi_{\text{tot}} = \int_0^{2\pi} \int_0^{\pi} I(\varepsilon, \eta) \sin \varepsilon \, d\varepsilon \, d\eta \tag{A2}$$

$$\text{Upward luminous flux:} \quad \Phi_{\text{up}} = \int_0^{2\pi} \int_{\pi/2}^{\pi} I(\varepsilon, \eta) \sin \varepsilon \, d\varepsilon \, d\eta \tag{A3}$$

$$\text{Downward luminous flux:} \quad \Phi_{\text{down}} = \int_0^{2\pi} \int_0^{\pi/2} I(\varepsilon, \eta) \sin \varepsilon \, d\varepsilon \, d\eta \tag{A4}$$

$$\text{Regional luminous flux:} \quad \Phi_{\text{zone}} = \int_{\eta_1}^{\eta_2} \int_{\varepsilon_1}^{\varepsilon_2} I(\varepsilon, \eta) \sin \varepsilon \, d\varepsilon \, d\eta \tag{A5}$$

where Φ_{tot}, Φ_{up}, Φ_{down}, and Φ_{zone} are total luminous flux, upward luminous flux, downward luminous flux, and regional luminous flux respectively, and (ε, η) are spatial angles.

6.3.3 Measurement of Luminous Flux by Integrating-sphere Method

Measure the luminous flux of an LED road light quickly by the substitution method using the integrating-sphere specified in 5.2.9. The luminous flux measurement with an integrating sphere includes the following three methods:

(1) Integrating method: a photometer detector is arranged at the detection window ofintegrating sphere to measure the luminous flux;

(2) Spectrophotometry method: a sampling device of a spectroradiometer is connected to the detection window of the integrating sphere to measure the luminous flux with a spectroradiometer; and

(3) Spectrum-photometer combined method: a photometer detector and a sampling device of a spectroradiometer are arranged in the detection window of an integrating sphere. A spectral correction factor is calculated according to the measurement result to correct the value obtained by the photometer. The spectral correction factor is calculated as:

$$\text{K1} = \frac{\int P(\lambda)_t V(\lambda) d\lambda}{\int P(\lambda)_t s(\lambda)_{\text{rel}} d\lambda} \times \frac{\int P(\lambda)_s s(\lambda)_{\text{rel}} d\lambda}{\int P(\lambda)_s V(\lambda) d\lambda} \tag{A6}$$

where $V(\lambda)$ is the known CIE standard spectral luminous efficiency function; $s(\lambda)_{\text{rel}}$ is the product of the known relative spectral response and integrating-sphere equivalent transmittance; $P(\lambda)_s$ is the known relative spectral power distributions of standard lamp; and $P(\lambda)_t$ is the relative spectral power distributions of the LED road light measured by the spectroradiometer.

A high accuracy array spectroradiometer with wide linear dynamic range and high measurement speed is recommended to measure the total luminous flux.

If there are large differences between the measured LED road light and the standard lamp in shape and size, an auxiliary lamp is needed for the self absorption correction. And a stable LED road light with good reproducibility, which has similar luminous intensity and spectrum distribution with the measured one, are recommended to be the standard lamp for calibrating the measuring instrument.

6.4 Measurement of Luminous Intensity Distribution and Beam Angle

In a photometric dark room, measure the luminous intensity distribution and beam angle of an LED road light by a goniophotometer specified in 5.2.8.

The LED road light shall be mounted within the goniophotometer in the specified burning attitude. Align the test LED road light with laser, or a more effective method, to make its photometric centre exactly at the rotation centre of the goniophotometer. Take readings in the specified measuring planes with an angle step less than 1/20 of the half peak side angle.

6.5 Measurement of Flashed Area and Average Luminance of an LED Road Light

6.5.1 Simple Method for Flashed Area and Average Luminance

The flashed area can be measured by the following method: place a closed rectangle frame on front of the test LED as close as possible; the four sides of the frame are movable by sliding and the frame size is bigger than the lighting area of the LED road light. Read the luminous intensity of the LED road light in the direction of $C=0°$, $\gamma=76°$, when there is no blockage in front of the LED light. Then move each side of the rectangular box slowly towards the lighting area, until the new reading after every slide is 0.98 of its latest. The left area of the rectangular frame is the so called "flashed area".

Calculate the average luminance in the 76° direction in the C0° and C180° plane according to the measurement result of luminous intensity in 6.4 of this appendix.

$$L(0°/180°,76°)_{avg}=\frac{I(0°/180°,76°)}{A_{flash}(0°/180°,76°)} \tag{A7}$$

where $L(0°/180°,76°)_{avg}$ is the average luminance in the flashed area; $I(0°/180°,76°)$ is the luminous intensity in the direction of flashed area; $A_{flash}(0°/180°,76°)$ is the luminous area of LED road light.

6.5.2 Measurement of Flashed Area and Luminance Characters by Imaging Luminance Meter

Measure the luminance distribution of an LED road light in the 76° direction in C0° and C180° planes by a near-field goniophotometer or an image luminance photometer. The photosensitive surface of the imaging luminance photometer should directly facethe LED road light with its optical axis passes through the photometry center of the LED road light. Determine the flashed area with the same principle as described in the simple method. Calculate the average luminance and maximum luminance of the LED road light in the flashed area.

6.6 Measurement of Colorimetric Performance

6.6.1 Measurement of Colorimetric Performance by Goniospectroradiometer

In the photometric dark room, measure the colorimetric quantities of an LED road light by a goniospectroradiometer, as specified in 5.2.10.

The LED road light shall be mounted within a goniospectroradiometer in the specified burning attitude, and its light centre should be mounted at the rotation centre of the goniophotometer.

Measure the relative spectral power distributions of the LED road light in all directions on enough luminous planes with small enough angle steps. The angle interval between measuring planes is generally 10° and the angle step i is generally 5°. When the size of the measured LED road light is relatively large or the beam angle is relatively narrow, smaller plane intervals and angle steps should be applied. Calculate the colorimetric quantities in all spatial directions according to CIE 15-2004; the quantities of the LED road light include: chromaticity coordinates, correlated color temperature, color rendering index, color tolerance, and so on. The average colorimetric quantities of an LED road light are calculated by the numerical method.

6.6.2 Measurement of Average Color by Integrating Sphere

It is convenient to measure the average colorimetric quantities of an LED road light by the substitution method with an integrating spectroradiometer. The measurement method is similar to measuring the luminous flux by the integrating sphere method.

Calibrate the integrating spectroadiometer by a standard spectral radiant flux lamp.

6.6.3 Measurement of Color Non-uniformity

The colorimetric quantities are measured in various spatial directions of the LED road light according to 6.6.1. Calculated the average color non-uniformity and the maximum color non-uniformity according to their definition.

6.7 Measurement of Luminous Flux Property Against Temperature

The relative measurement method can be used to measure the luminous flux property against temperature. Operate the measured LED road light in a temperature controllable thermostatic chamber at its standard measurement attitude, under the specified voltage/current/power or the maximum value in the range of the specified voltage/current/power. Control the temperature in the chamber to make it increase from the minimum temperature of allowable ambient temperature range. The change of relative luminous flux is measured at interval of five when the LED road light has been stable at least for 15 minutes at the temperature. Take the luminous flux value at 25°C as 100%, and record the relative changes of luminous flux at each temperatures. If there is no specific requirement, the maximum temperature in the measurement is the maximum ambient temperature in the range of allowable ambient temperature for LED road light.

6.8 Test of Maximum Allowable Ambient Temperature

Operate the LED road light in a temperature controllable thermostatic chamber at its standard attitude and at the maximum allowable ambient temperature for 100 hours. There should be no mechanical damage of any component, no abnormality of any LED, and the luminous flux of the measured LED road light should be better than 95% of rated value when returned back to 25°C ambient temperature.

6.9 Test of Endurance of High-Low Temperature Cycling Behavior

Operate an LED road light in a temperature controllable thermostatic chamber at its standard measurement attitude, and under the specified voltage or the maximum value in the range of the specified voltage. Control the temperature of the chamber to make it rise from the room temperature to 50 °C and keep it at this temperature for 16 hours. Then drop the temperature to −5 °C and keep it at this temperature for 16 hours, consequently, rise the temperature to the room temperature again. The rate of the temperature changing is 0.5 °C/min to 1 °C/min. After two such cycles, there should be no mechanical damage to any component of the LED road light, no abnormality should occur for any LED, and the luminous flux of the measured LED road light should be higher than 95% of rated value.

6.10 Measurement of Light Maintenance

The light maintenance property measurement includes the measurement of lumen maintenance and color shift.

LED road light is subjected to aging under specified conditions, until at least 6000 hours, with the both the total luminous flux and the chromaticity parameters recorded at least every 1000 hours. The quantities measured after 1000 hours shall be compared with the value obtain at 1000 hours. To simplify the measurement, the illuminance at a given right downward point can be used to replace the total luminous flux of the LED road light for the calculation of lumen maintenance, and the color sift can be expressed by the coordinate difference of (u', v') in CIE 1976 according to CIE 15-2004.

It is suggested that LED road light continues operating to eliminate the possible influence cause by on/off modulation. During the aging testing, care should be exercised to examine or automatically monitor if the LED road light fails. If failure occurs, an examination should be performed to check if the LED road light really fails, or if the failure is caused by other auxiliary or fixtures.

6.11 Test of On/Off Characteristics

Under the specified operation conditions, a switching cycle is defined to be 30 seconds open and 30 seconds off. Repeat the switching cycle until the specified number of cycles is reached. The LED road light is expected to operate normally after the cycles.

Index

LED Packaging for Lighting Applications: Design, Manufacturing and Testing, First Edition. Sheng Liu and Xiaobing Luo.
© 2011 Chemical Industry Press. All rights reserved. Published 2011 by John Wiley & Sons (Asia) Pte Ltd.

Printed and bound by CPI Group (UK) Ltd, Croydon, CR0 4YY